THE TWENTY YEAR WAR

Ballast Books, LLC
Washington, DC
www.ballastbooks.com

ISBN 978-1-7334280-9-5

Library of Congress Control Number has been applied for

Printed in Canada

Published by Ballast Books
www.ballastbooks.com

For more information, bulk orders, appearances or speaking requests, please email info@ballastbooks.com

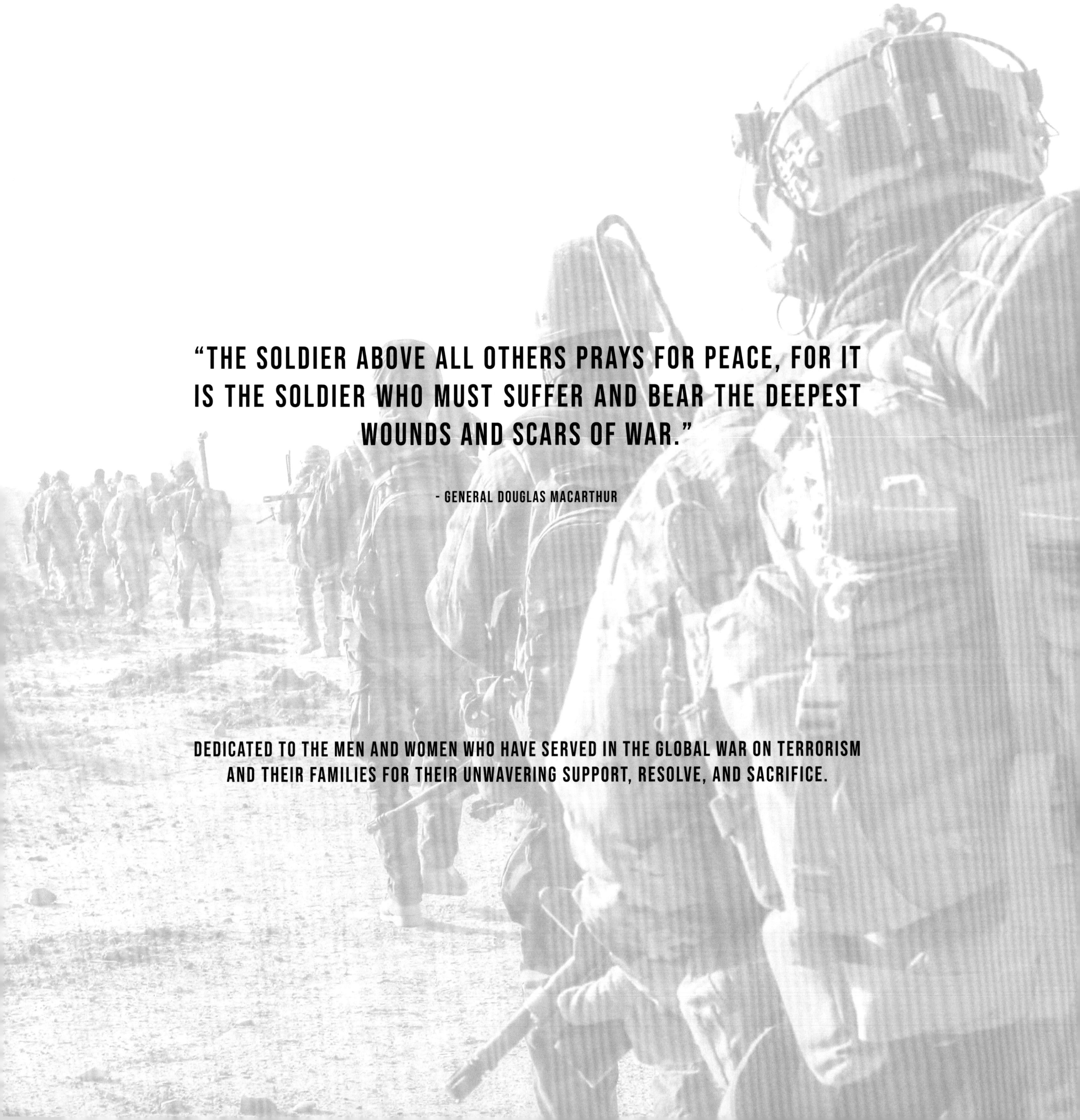

“THE SOLDIER ABOVE ALL OTHERS PRAYS FOR PEACE, FOR IT IS THE SOLDIER WHO MUST SUFFER AND BEAR THE DEEPEST WOUNDS AND SCARS OF WAR.”

- GENERAL DOUGLAS MACARTHUR

DEDICATED TO THE MEN AND WOMEN WHO HAVE SERVED IN THE GLOBAL WAR ON TERRORISM AND THEIR FAMILIES FOR THEIR UNWAVERING SUPPORT, RESOLVE, AND SACRIFICE.

FOREWORD

JOSEPH L. VOTEL

GENERAL, US ARMY (RETIRED) 1980 - 2019

On September 11th, 2001, I finished doing morning "PT" with one of my officers and was puttering around in my office before taking a shower and getting on with the day's schedule. Around 8:45 a.m., my Secretary, Linda Davis, knocked on the door and told me to turn on the TV – a plane had crashed into one of the World Trade Center Towers in New York City. Like many across our country, I could not take my eyes away from the images we were all seeing in real-time over the next several hours. As the recently installed Commander of the US Army's 75th Ranger Regiment, I had a strong suspicion that this would prompt a military response – but at the time, I did not fully appreciate how profoundly this would change my life and the lives of so many others. Within a month – we were deployed, and my unit began a continuous combat deployment that endures to this day. I am confident I was not alone in the initial assessment I made that morning. In the 20 years since that fateful day, our Nation, our fellow Americans, and indeed the lives of millions worldwide changed in ways that defied expectation and were uniquely different for each person. None have been affected more than those who served in military uniform during this period.

The Twenty-Year War by Dan Blakeley, Beau Simmons, and Tom Amenta tells the story of dozens of those who served their Nation in the wake of that fateful day. It is a small but powerfully representative subset of the hundreds of thousands who answered the call and were participants in the wars since 9/11. Through interviews and photographs, they capture the individually unique impacts from the all-to-common experience of deploying to war. In doing so, they reveal the often-misunderstood realization that there is no typical response to combat. Everyone handles this experience differently. Some thrive, and some struggle. Some are motivated, and others inhibited. Some are changed and some not. Some suffer in silence, and others are outwardly expressive about their experience. But in all these stories, we can find inspiration and life lessons.

Our authors are uniquely qualified to help tell these stories. Dan and Beau are lifelong best friends – meeting for the first time at the age of five. One became an Army Ranger, and the other pursued a passion for photography. Both joined or attempted to join the service before reaching the age of majority – requiring their parents' approval – begrudged in one case and for a combat military specialty outside of the "agreement" with his family. Tom came along later, capitalizing on a service-related connection and added another perspective to the project.

Beau is an accomplished photographer who has worked with celebrities, fashion models, and commercial brands in Los Angeles and New York City. He became a gallery-represented fine artist creating custom artwork for clients around the world. At the top of his game – he sold it all, packed a car, and drove to Raleigh to be closer to his best friend. It is said that a picture tells a story of a thousand words. We have all seen images in our lives that require no narrative or explanation because the photographer's keen eye and composure do the work for us. This is Beau's singularly unique skill.

Dan joined the North Carolina National Guard after his active-duty service and eventually completed an undergraduate and two graduate degrees from Appalachian State University. He found a solar car racing team – providing a convenient way to "forget" about his military experience. Dan struggled mightily but, with the help of an understanding wife – avoided the fate that entraps many of our Veterans and which often leads to suicide. He realized that his experience as a small unit leader in combat could benefit others with lessons and observations that transcend the military. Anyone who has spent time with a Ranger Squad Leader appreciates that they possess unique skills in dissecting, diagnosing, and understanding different personalities. This is Dan's particular skill.

Tom enlisted in the US Army at the age of 18 against his family's wishes. Serving for five years as a member of the 75th Ranger Regiment, he left active service, continued in the National Guard, and earned a degree in political science from the University of Illinois. Tom embraced his veteran status by becoming the COO of a successful veteran lifestyle apparel company. Along the way, he struggled with his view of the veteran community and, after engaging with Dan and Beau, saw a new purpose in this project. I knew Tom on active duty and in his role as our rear-detachment NCO – he befriended my wife, serving not only my family well but indeed the entire organization. He is a trusted interlocutor to those who were navigating the post-9/11 world. This skill served him well and brought him to Dan when they served together in the National Guard. Connections are Tom's unique skill.

Between the three of them are multiple tours in combat, successful careers, difficult transitions, unique skills – but more importantly, a realization that the Twentieth Anniversary of 9/11 offered an opportunity to document the after-stories of those who served. This is not a book about war stories. Those stories are well documented. But in many ways – it is a volume about war. The war that many veterans have fought after their military service. Our authors capture the intricate interaction between teamwork, comradeship, mutual reliance, exuberance, and tragedy of service – juxtaposed with the loneliness, loss, fear, isolation, redemption, and rebirth that often comes in the aftermath. In doing this, they help all of us understand the veterans that live among us. This book is a worthy undertaking and one I am sure will leave you in awe. But it will also make you cringe, cry, laugh, and most importantly, think hard.

Continued on next page

In the end, the reader is left with the conclusion that military service is not an end unto itself - but instead a platform through which we can define ourselves going forward. These are not just veterans – but veterans, and you fill in the blank. We all have our perceptions of veterans.

Most of us grew up having interactions with veterans from World War II, Korea, or Vietnam. Some are open about their service, and some are not. Some entirely define themselves by their status as veterans, and others take a different path. All were affected in different ways. We have all heard the stereotypes of broken veterans who seemingly cannot cope and now understand this notion is misguided. This book tells the stories of a new generation of veterans. A generation that came to service after a horrific attack on our country. They joined for many reasons – but as you will see in these stories, it was primarily because of an overriding recognition that they must serve. The Twenty-Year War reminds us of the vital role that veterans play in our society and Democratic experiment. It also reminds us that sacrifice and suffering do not end when you leave uniformed service. We have made a lot of progress in the Veterans Affairs arena, but there is much more to do, as these stories will attest.

As a veteran, what has always been clear to me is how the experience of uniformed service can fundamentally change people. Thirty-nine years in service ingrained habits and perceptions that I believe serve me well but can be annoying, frustrating, or confusing to others. But my story is, in fact, just that: my story, unique to me as the seventy-one stories in this book are to each of my fellow veterans. As you peruse this book, I hope you will read the stories and study the pictures' faces. I believe they will inspire you, promote understanding, and most of all, I hope you will find a story that speaks to your path.

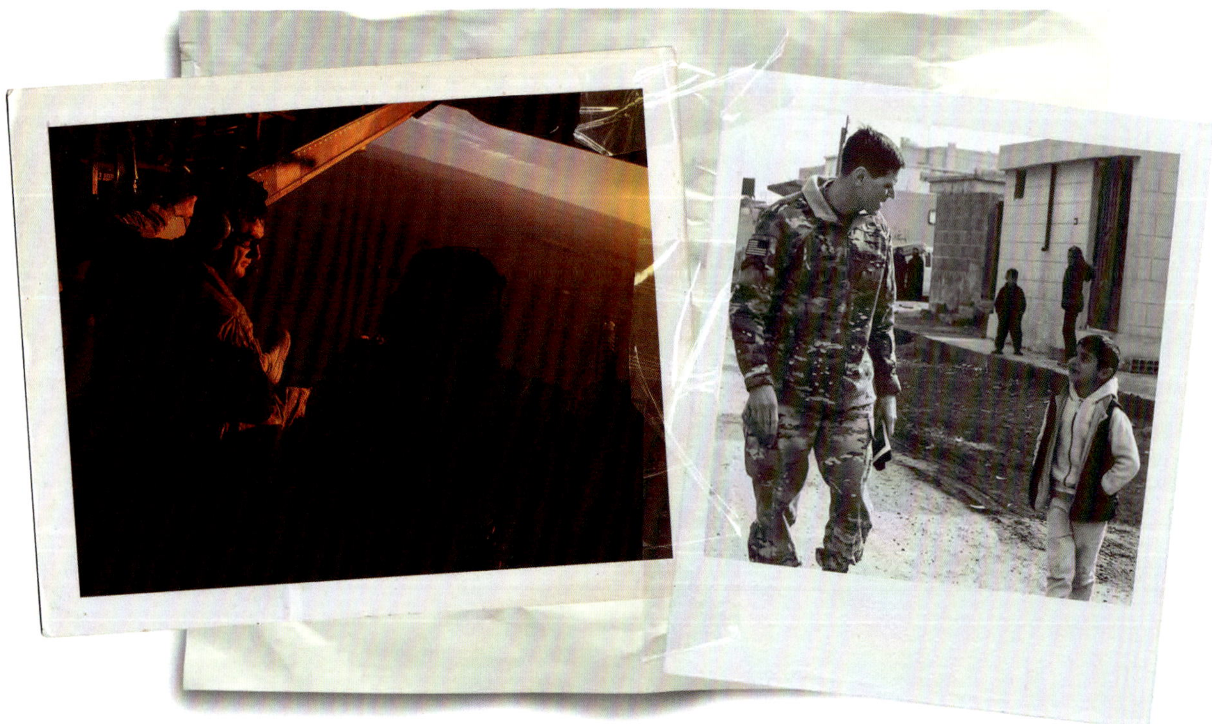

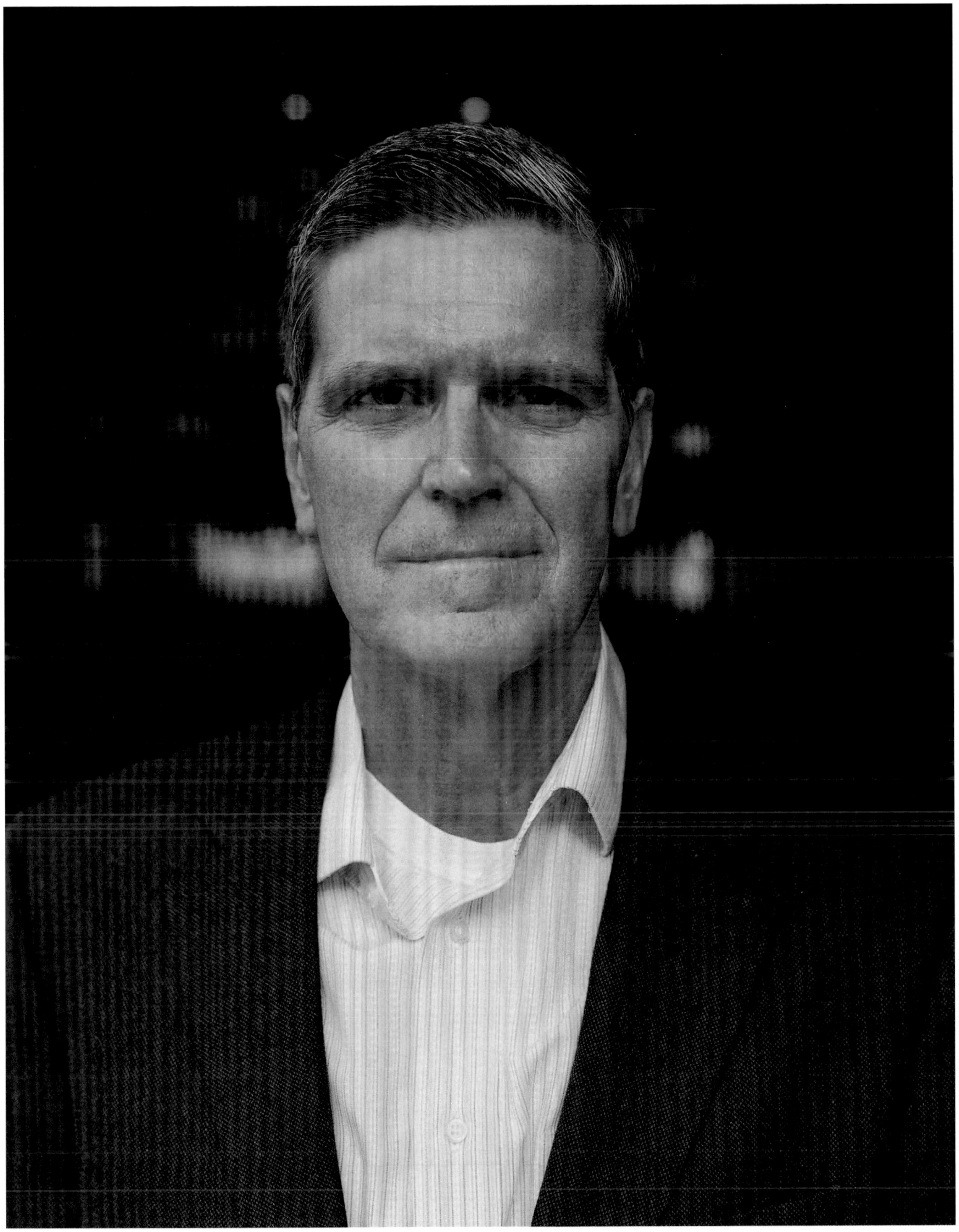

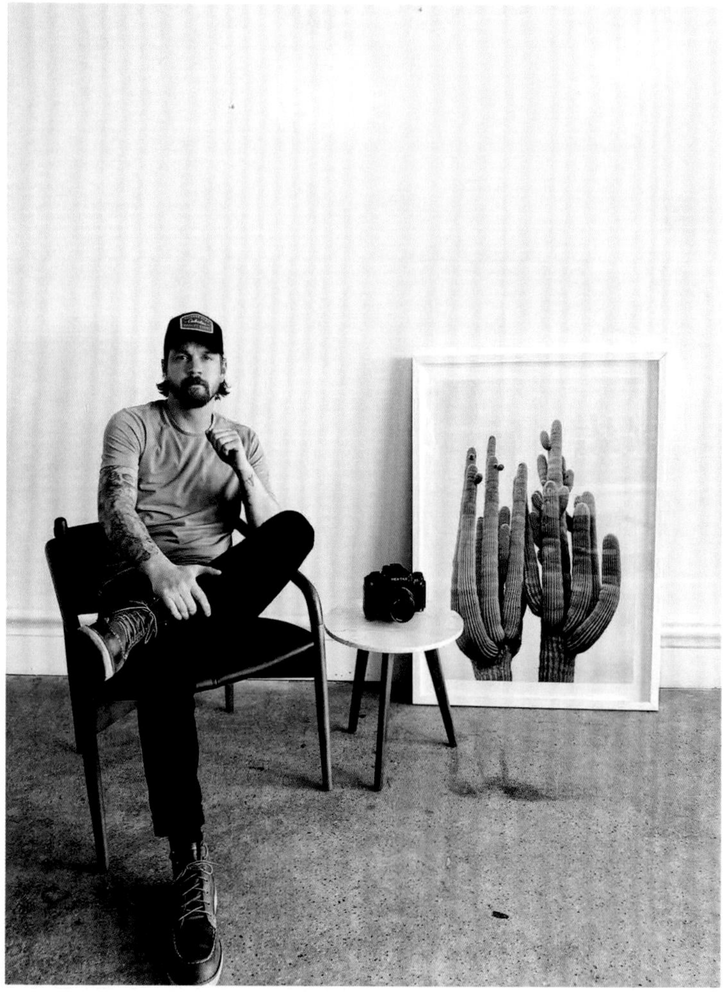

BEAU SIMMONS, FINE ART PHOTOGRAPHER

42 states.

61 days.

16,000 miles of travel.

71 unique veteran stories.

This book has been far and away the most rewarding and amazing photography project of my life. My journey into photography started when I was sixteen, on a family vacation to Hawaii. By twenty-one, it was my full-time career, shooting in the fashion and celebrity industries. I was bouncing back and forth between Los Angeles and New York City working with some of the top modeling agencies in the world. For just over ten years taking pictures of pretty girls and fashion was my life, but it wasn't fulfilling. I needed something much deeper. Something real and with a purpose.

I cut ties with the industry, literally overnight. I spent the next month capturing landscape and Americana-inspired photos. Soon after, I started producing large-format framed artworks to sell in a fine art gallery in Los Angeles. Something was still missing.

This feeling just wouldn't go away. So, I decided to move from California to North Carolina, where my childhood friend Dan lives with his wife and family. Dan is a former Army Ranger with multiple combat tours, and I somehow wanted to find a way to connect my photography to a veteran cause. Knowing that 2021 was the twentieth anniversary of 9/11, we thought about doing something for the anniversary. The title, as we kicked the idea around, just jumped out at us: The Twenty-Year War. My calling and purpose landed right at my feet. This book idea felt like it was fate.

Fate stepped in again when Dan and I connected with Tom Amenta. Tom and Dan had served together, and he instantly believed in this project, just as we did. With his expertise in writing about and working in the veteran world, he was the perfect fit. There was a ton of work to do, and it was already January of 2021. So, once again, it was back into the car. Once again, a drive across the United States. All the way back to California, only to turn around and head back to North Carolina. This time, there was purpose. This time, there was a clear vision for how to use my skills and talents.

Through freezing temperatures, snowstorms, and giant mountain passes. Navigating swampy marshes, open prairies, congested cities, and hot desert regions. Sleeping only a few hours in cheap roadside motels, or in a tent under the stars, to interview multiple veterans a day. This was just the experience I needed. No matter how exhausted I became, or how much ground I covered each day, I was so alive. To hear each of these veteran's stories and to photograph the raw portraits of these men and women — this was my purpose.

Each portrait in this book has been photographed using medium format film. Each veteran helped me choose a place or subject they would like to have their portrait taken in, an effort to truly reflect their life and interests individually. Some are photographed using black and white film. Others are photographed in color. There was no rhyme or reason to it, just what we thought would work best for each situation. There is a big difference between that and taking a polished photograph of a fashion model who has had their makeup, hair, and wardrobe styled so perfectly. There is none of that in these portraits. Each veteran photograph is raw, with no lights or a styling team. These stories have left a lifelong impression on me. There were times I was full of joy, happiness, excitement, and love. Other days I was filled with tears, anger, and confusion while hearing these stories. It has taught me just how much veterans are stereotyped. That the struggles they have faced trying to find their place among our society back home can be just as harrowing as their experiences on deployment. Each veteran in this book has faced or is still facing their own struggles on a daily basis.

We hope these stories encourage other veterans to continue finding connection and purpose, that we can have a tiny, positive, effect on bridging the gap between the veteran and civilian communities. Roughly 1 percent of Americans have served overseas in support of the Global War on Terrorism over the past twenty years. This book is a tiny glimpse into the lives of these men and women. I am truly honored to have captured these portraits and to now call many of these veterans my friend.

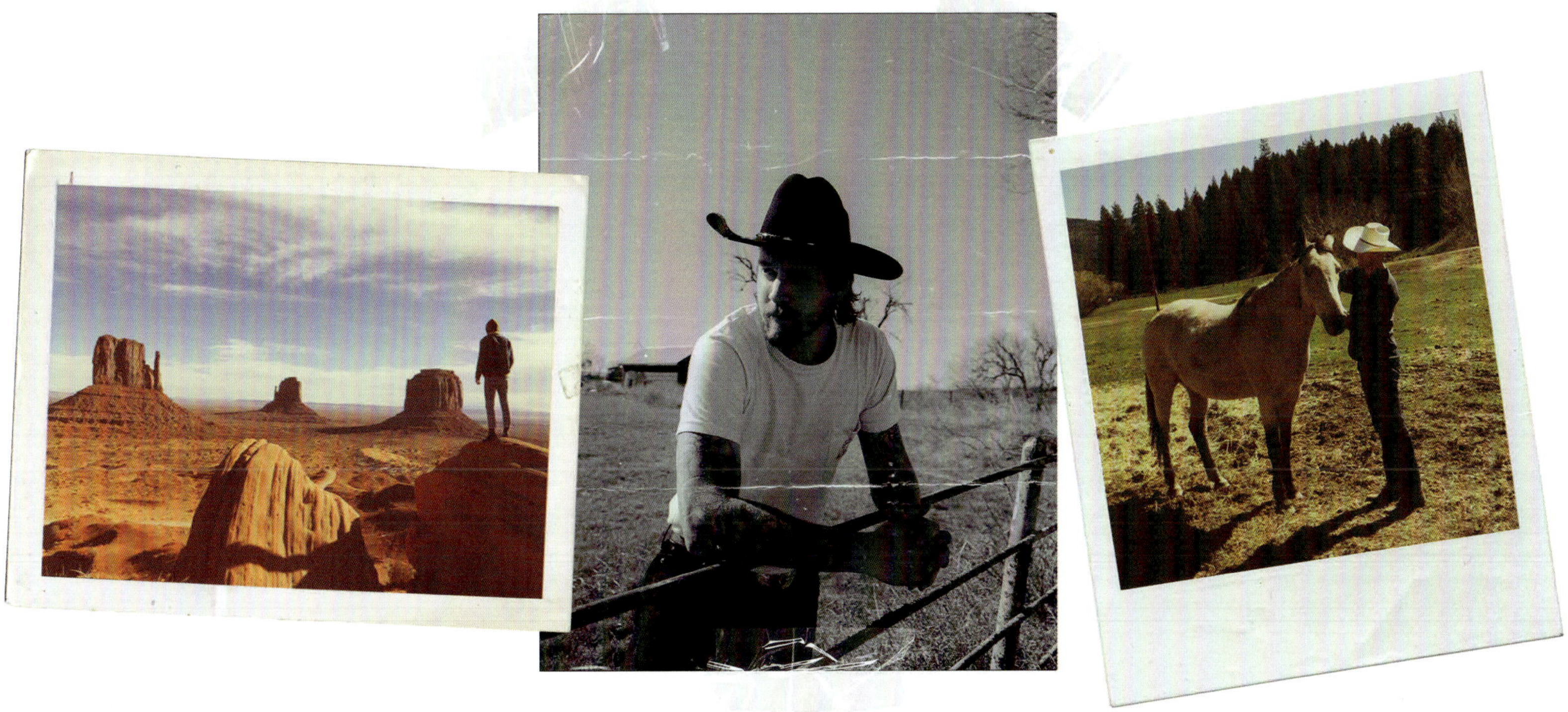

In the spring of 2011, I was returning from my fifth combat rotation as a newly minted Staff Sergeant in the 2nd Battalion, 75th Ranger Regiment. I had every intention of reenlisting. In the middle of the Army's Advanced Leaders Course, it happened. I fell out of a run. We were one mile in when I started to feel a sharp pain in my chest. I tried to shake it off, but it only intensified. The cadre circled back to me and threw me in the back of the fall-out van. Then, I was on my way to the ER. A few months later I found out I had a small hole in my heart. Nothing could be done. The doctors said it was "nothing to worry about ... unless something else happens." After serving six years and deploying to Iraq and Afghanistan six times, my wife and I made the hard decision that I would not reenlist. It wasn't because of a lack of desire to serve.
I loved being a Ranger.

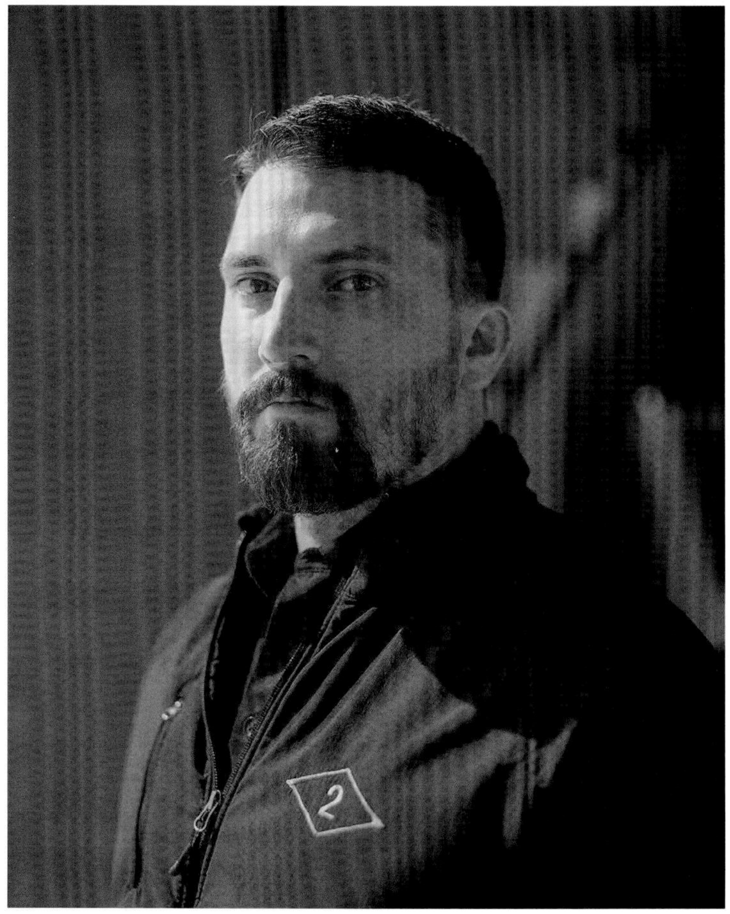

DAN BLAKELEY, 2/75 RANGER

On paper, I had an excellent transition. I went to college, earned a few degrees, and landed a great job. In reality, my transition created a void in me. To fill it, I told myself that success was measured in titles and the number of zeros in my salary.

That was a mistake. I handicapped myself and those around me by not leaning into my military experience. I limited myself and those around me by closing that part of myself off. Looking back, I realize there were three phases of transition.

Phase One: Out of the fight.

To me, no longer being able to deploy was like taking an all-star player off the field right before the playoffs. I had trained for it. I was great at it. But I could no longer lead and fight with my squad. That realization sucked.

Phase Two: No uniform, now what?

Once I turned in all my gear and fully separated from the Army, I had a plan for the next mission. I knew what I was doing next: getting a college degree. However, I had a naive belief that no matter who I encountered, fellow veteran or not, they would fail to understand what I had experienced and what I left behind.

So, I pushed away my military experiences. I didn't wear military memorabilia and I stopped saying things like "chow hall" and "roger." Hell, I didn't even reach out to my brothers in arms I had served with. I tried to create a new identity. That identity was a weak attempt to fill that void. I tried to fill my time with more academic and extracurricular activities, which I hoped would distract me from the sense that I was missing something. I was attempting to forget my past and create a new future. Thing is, the past never forgets you.

Phase Three: Self-reflection.

In 2020, I had a lot of time to think. I'm sure many people did. The fast-paced world around me slowed down. I realized what the void in my soul was: service. I had been missing the piece that had inspired me to join the military in the first place. I started to realize I was no longer serving a greater cause. It was that realization and the desire to start giving back again, the relentless draw toward leadership, the missing mission, which inspired The Twenty-Year War. I knew I couldn't serve and fight as I once had, but that didn't mean I lost all the knowledge and drive I gained while serving in the military. Losing my physical abilities didn't mean I had lost my mental aptitude for leadership. I was redefining myself again, but with a new vision for my future. Don't always believe what you see in the mirror.

The transition from service is a fate every veteran goes through, planned or not. Most stories in this book are about transition, and I've learned something about myself and my transition from their stories. That's the point: we all learn from each other. It doesn't matter what "side" you're on. Those lines of division fade once you put on a military uniform. It's one of the experiences that separates veterans from a majority of the population. But that experience is not exclusive to the veteran population. In fact, veterans and civilians alike can grow collectively. So, as you read, find a piece of yourself in each of these stories, feel the connection, and learn from each other.

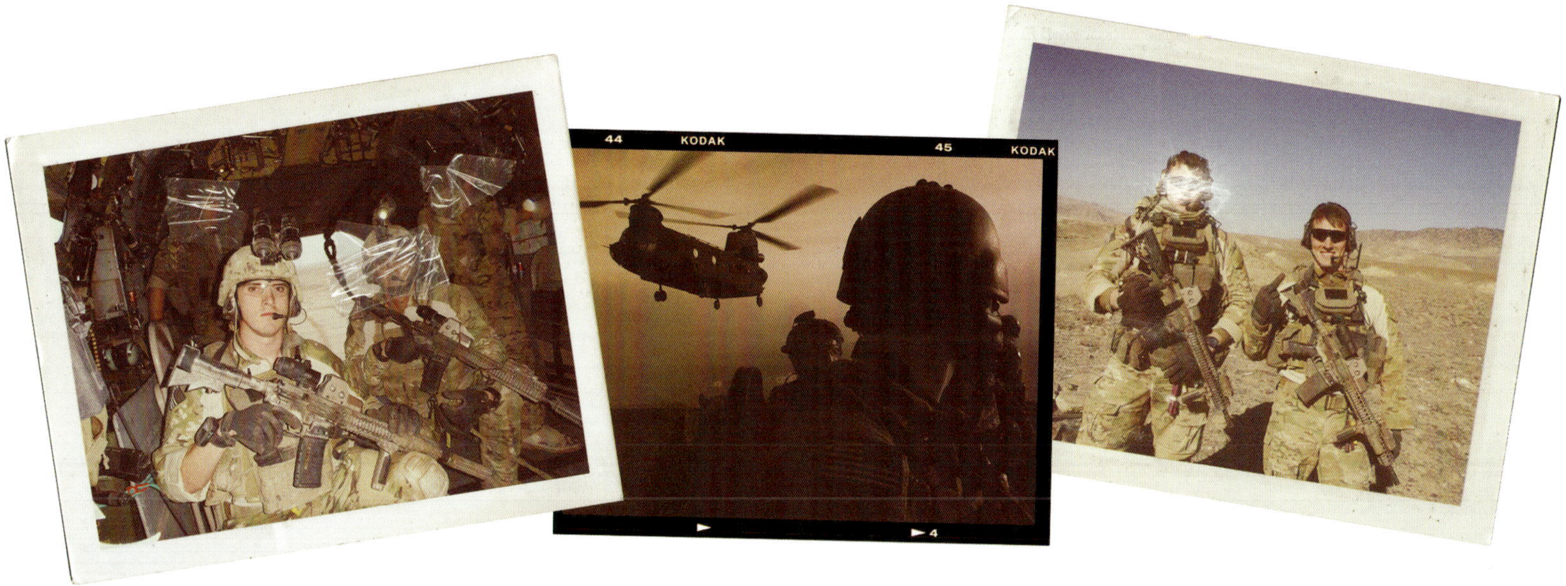

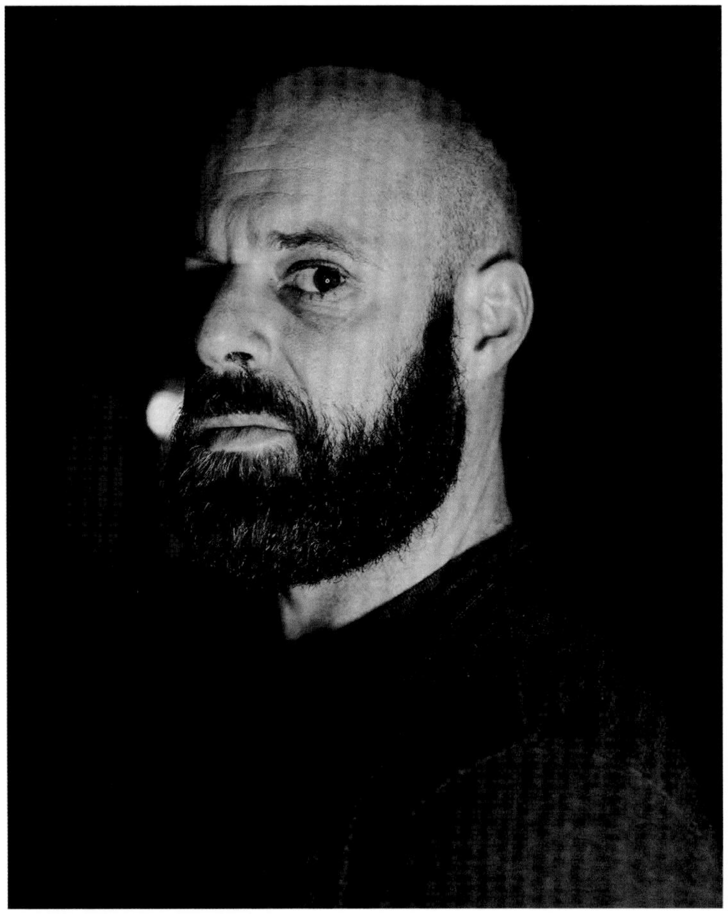

TOM AMENTA, 2/75 RANGER

I was in Germany on 9/11 as a member of Bravo Company, 2nd Ranger Battalion. B-Co got stuck on a weeklong training exercise supporting the 173rd Airborne Brigade and we were super pissed about it. The rest of the guys in the battalion were going to get three of those days off to see Europe, do some light training during the day, and heavy drinking with the Belgians in the evening.

As we were getting ready to go out into the training area, the news about planes hitting the World Trade Center started spreading. The base immediately locked down. From the back of a Blackhawk helicopter, I was listening to the air crew share updates, hearing how bad it was, how another plane had hit the Pentagon. We were so close to starting the training exercise, we didn't even get to see the CNN feed until we got done, a full week later.

I was dirty, tired, and hungry walking into a makeshift mess hall for breakfast and ... there it was. That plane smashing into the WTC over and over and over again as the reporters talked about it. America had seen it so many times, they said, it had become almost numb to it. For B-Co, standing in that chow line, it was the first time we'd seen it. A couple hundred Rangers, silent as church mice, staring at that screen, seething. Rumors were already going around that we were headed to Afghanistan. All operations to that part of the world had to go through Germany, and we were already there. Instead, we were told that we were headed back to the US. We didn't make it to Afghanistan until right after The Battle of Tora Bora in early 2002.

I separated from active duty in 2004 to attend the University of Illinois where I got my degree in Political Science. Freshly graduated from college, I decided the best idea in the middle of a global recession was to help start a lifestyle apparel company focused on veterans and patriotic Americans called Ranger Up. Just like when I told my family I was going to enlist, they thought I was absolutely nuts. In fairness, I might have been. We grew Ranger Up into the premiere lifestyle apparel company in the veteran space, and we created the first fully veteran-produced-and-starred-in Hollywood movie, Range 15. We were number one on Amazon's movie list three weeks in a row.

That might sound like a hell of a success story, but my transitions were really rough. In college, I barely made it, almost dropping out and, at one point, I nearly committed suicide. I was lost and I was scared. You might be thinking, "Wow, that's a lot!" but I don't see it that way. Eventually I found the path that led me to working on this book, and after writing these seventy-one veteran stories, I find mine completely unremarkable. Most of the veterans featured in this book feel the same way about their own experiences.

We don't think we're special for serving. It's something we all chose to do. A few, like me, made that choice pre-9/11. Most answered the call after being profoundly affected by the events of that day. We don't see ourselves as heroes; it was just the job. It was about those we served with once we got to our units.

If you're looking for answers as to why the United States has been at war for twenty years, this isn't the book for you. That is far too complicated and far too broad a topic for our scope. But if you want to hear who we veterans used to be, read our stories and listen to our voices. Then, look into our eyes. You'll see the age lines and maturity that Beau captured beautifully in our faces. And you'll see who we are now.

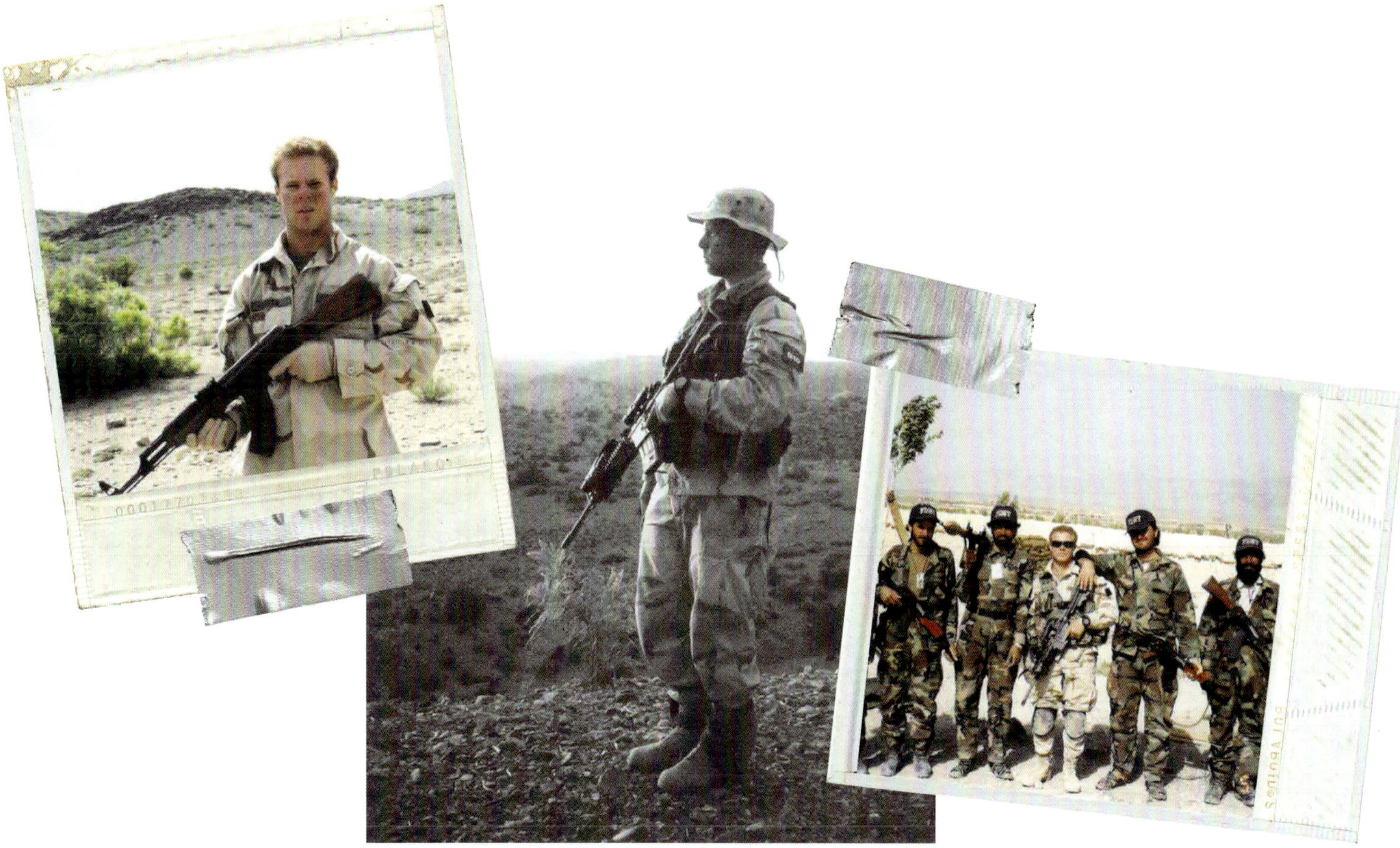

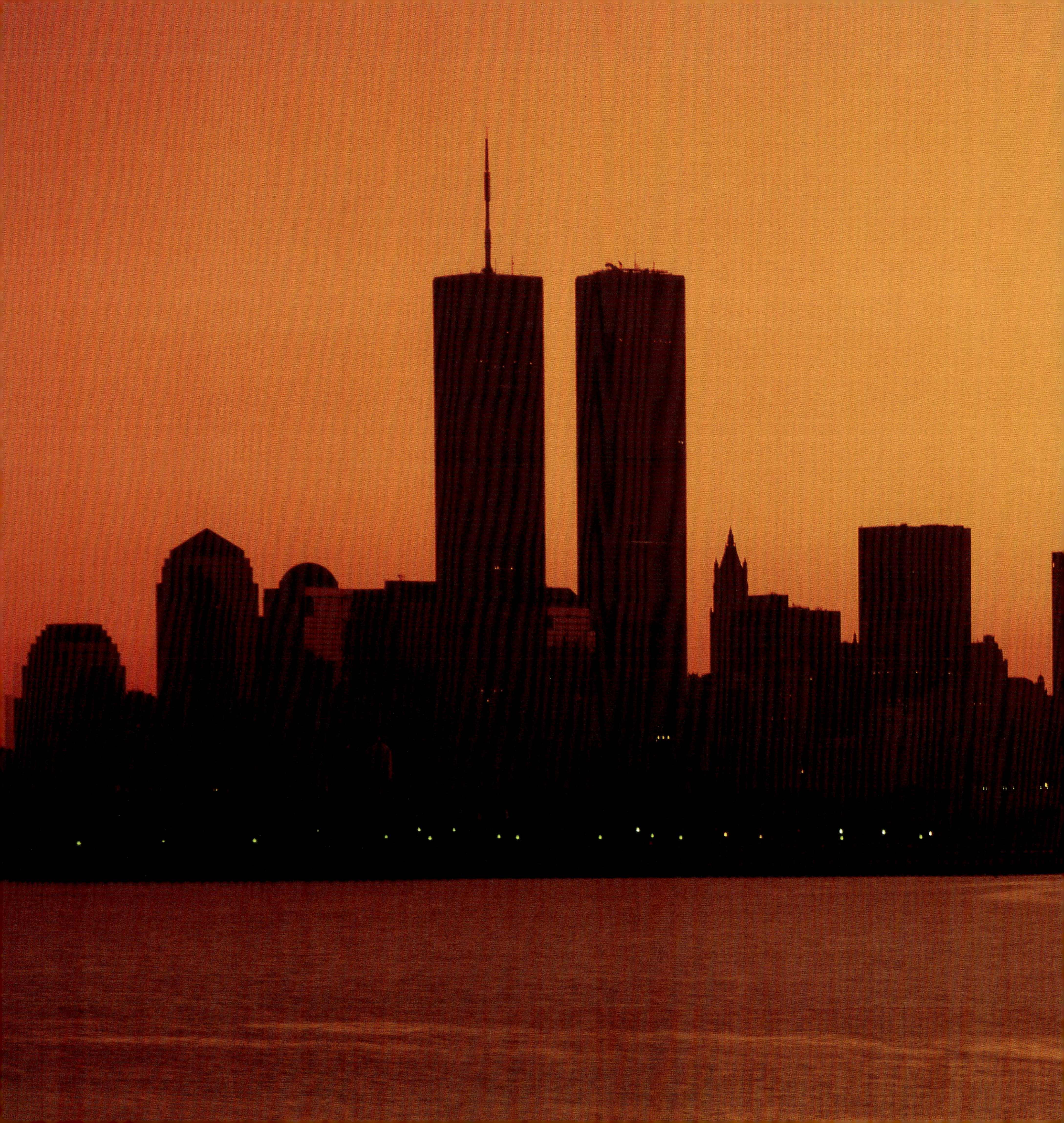

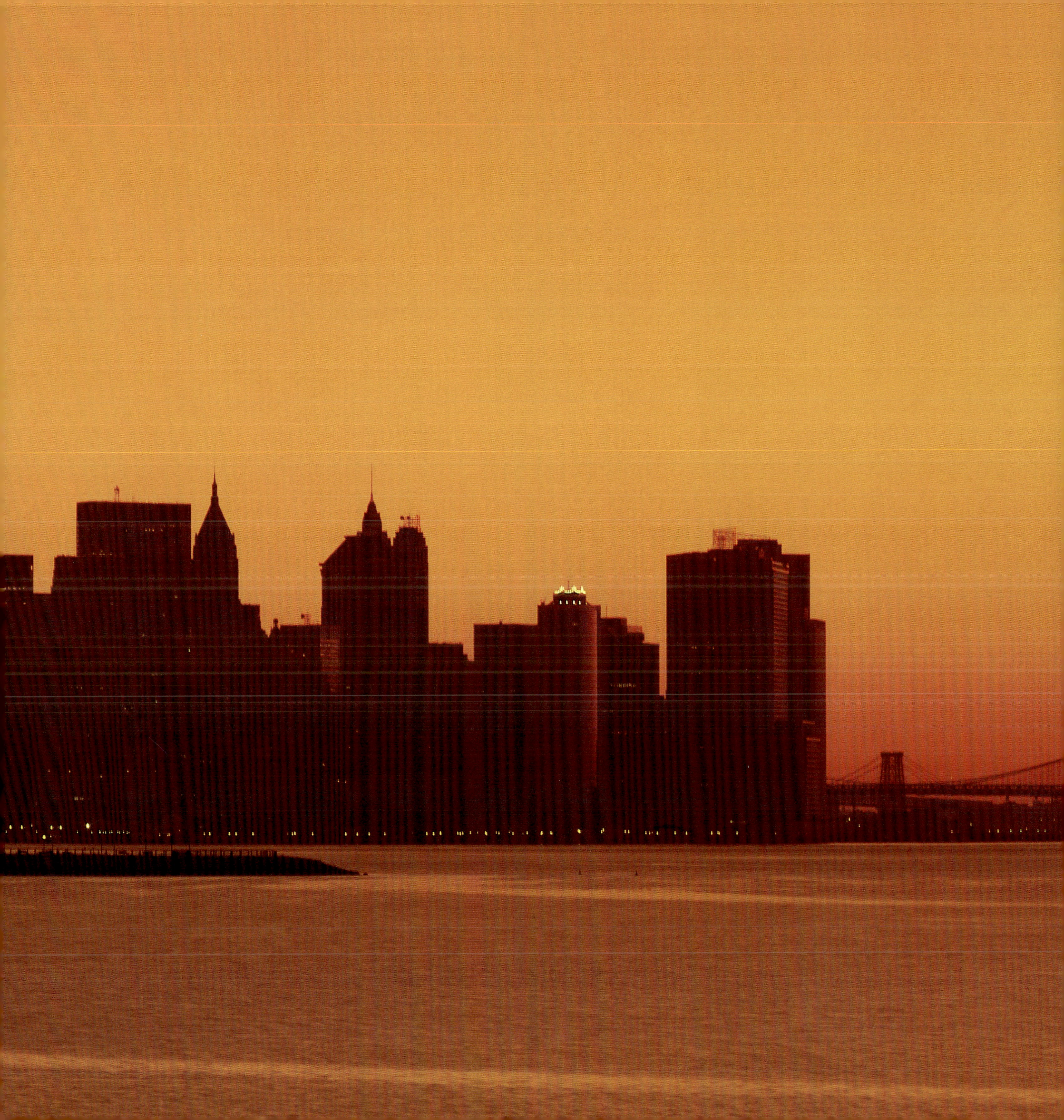

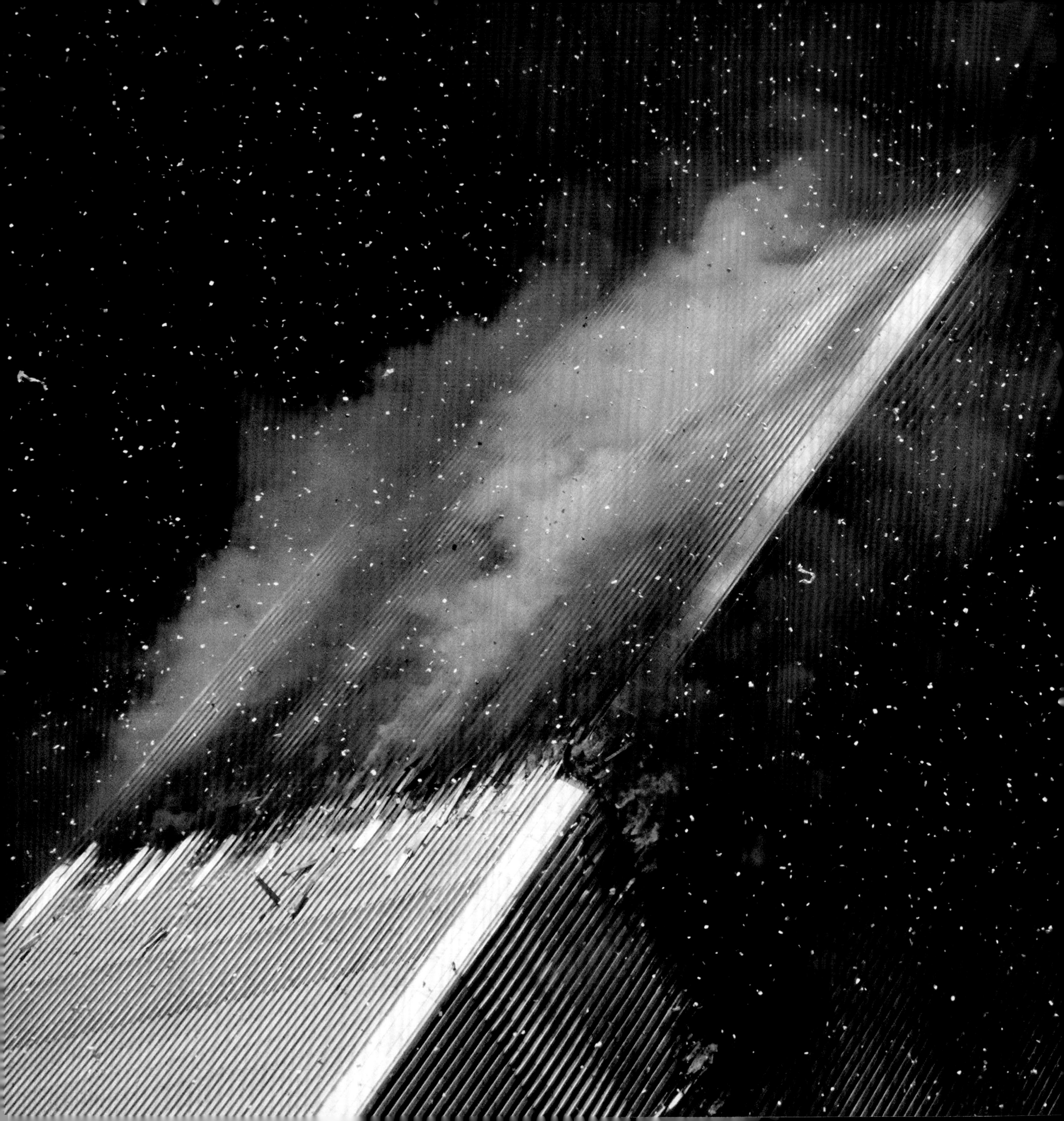

September 11th, 2001 was the day that started what has become known as the Global War on Terrorism. Little did we know that event would kick off the longest war in our country's history.

The men and women in this book are a representation of those who have served and fought in the Global War on Terrorism over the last twenty years.

These are their stories.

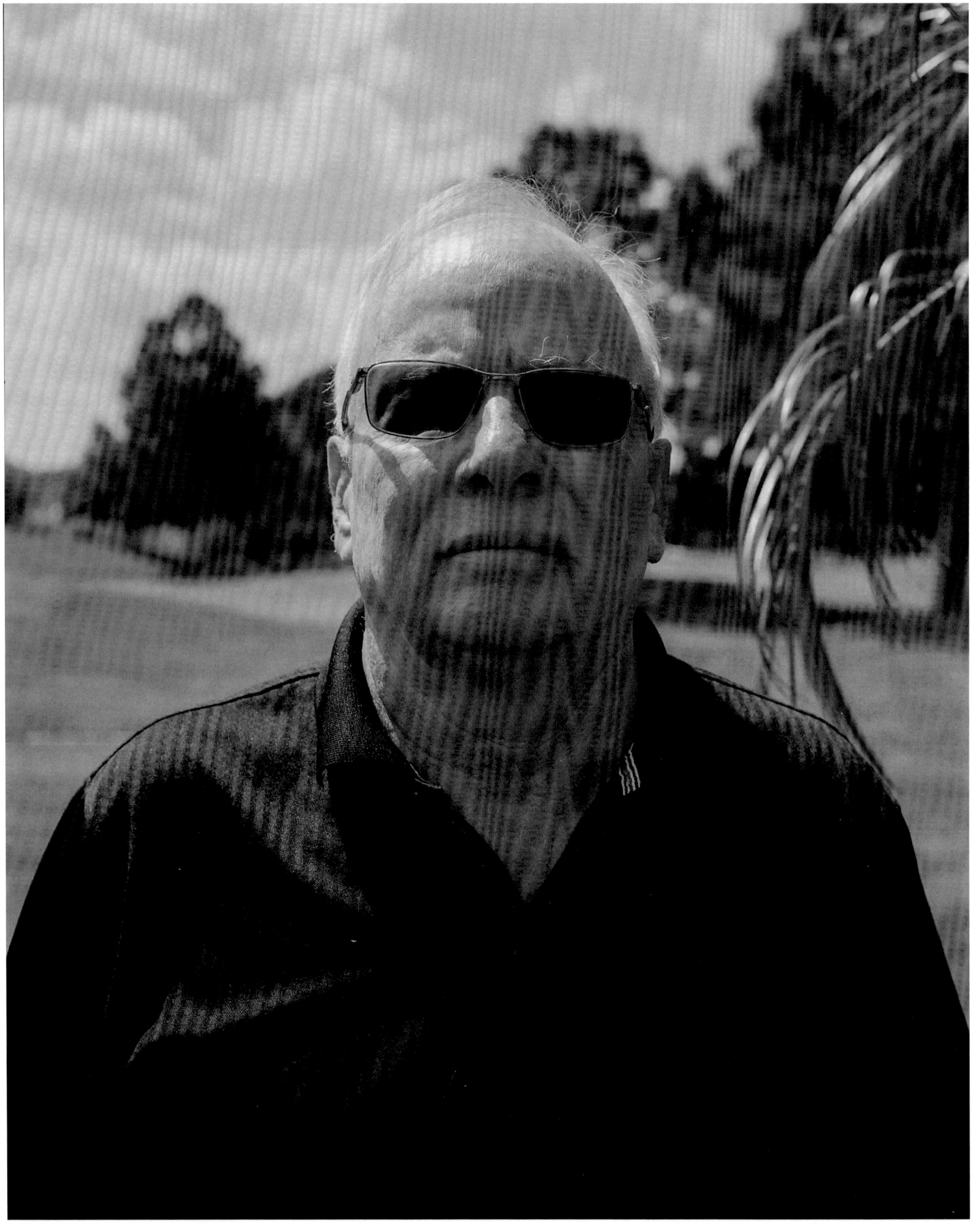

Jack Tilley, Sergeant Major
Army - 12th Sergeant Major of the Army
Years of service - 36 years
Deployments - 16 [Global War on Terror]

My name is Jack Tilley and I retired after thirty-six years of service as the twelfth Sergeant Major of the Army. I was born into impoverished conditions in Vancouver, Washington. Childhood made me a pugnacious kid who endured difficult times: parents' divorce, bouncing in and out of trouble, spending ten months in reform school, adopting a loser's mentality of avoiding life's responsibility by doing the minimum ... I was unsure of myself and what path to pursue. My upbringing wasn't particularly patriotic or traditional military, even though my father had been drafted and spent a few years in the service. But the journey to becoming Sergeant Major of the Army really began fresh out of high school. I was oblivious to the Vietnam War; my main considerations in joining were to explore the world and become a soldier, but the 1968 Tet Offensive taught us all what we were made of. Believe me: my achievements have been hard-earned, but if having a less auspicious career would mean that my childhood buddy had survived Vietnam, I would take that trade in a heartbeat.

On 9/11 I was in my office, getting ready to go out and talk with some military lawyers at Fort Myers. As I was getting ready to leave, one of my NCOs said, "Turn on the TV," and we saw the second plane hit the WTC. I remember telling my staff, "This doesn't look right, I'm heading out, but keep me in the loop." On my drive back to the Pentagon, we saw the plane hit it. There were guards trying to keep us from getting into the building and I remember running around them and up to my office to check on my staff. They were gone and safe, thankfully. Then we went around to where the plane had struck the building and it was complete chaos. Just tons of confusion and being told there was another plane coming. We all moved to a safer area and things stopped for a moment. Then people just started showing up: fire and rescue and LEOs and first responders ... the next ten or twelve hours were a blur, just trying to help where I could. At one point I tried to go into the building to see if we could help anyone. I barely made it a hundred feet. There was just too much fire and smoke. Once they put the fires out, they had to get the remains of the victims. Initially, they wanted the soldiers of the Old Guard to do it. I instantly said, "No, that needs to be mortuary affairs. That's their job. The Old Guard can move the body bags." It was tough asking those things of soldiers. There were 125 people killed in the Pentagon and sixty-four Americans on the plane.

As people were moving bodies and clearing rubble over the subsequent days, I mostly let people talk as they needed to. The job of the Sergeant Major of the Army is to listen to people and their troubles. Any coordination or calm that I could offer for our people when things got the most stressful — that was my mission. During periods of reflection, other traumatic military events flood my consciousness. As a young soldier in Vietnam, my position was overrun, and I recall stepping over bodies of men who weren't much older than me. You never get tough enough to accept some of the things you see in a warzone. Not all battle wounds are visible.

Continued on next page

Exiting the military was extremely tough. Fortunately, my wife Gloria and my children were there to make the transition a lot smoother. They had been there during deployments and TDYs, and in retrospect, it saddens me to have missed so much time as my children grew up. I owe Gloria Tilley a great amount of gratitude for allowing me to stay laser-focused on the military and the Army, because the last fifteen years of my career consisted of multiple positions with increased levels of responsibility: Brigade, Sergeant Major, USCENTCOM CSM, and ultimately the SMA. For three and a half years as the SMA I traveled over 800,000 miles. No, I did not think the war would go this long.

American Freedom Foundation is an organization I helped start to assist transitioning veterans with scholarships and their military separations. We just started the Your Next Mission podcast to increase our ways to help. It's a weekly show, and it's the first step in trying to give veterans more support. The podcast is the crawl phase. In the walk phase, we'll do Facebook Live weekly. In the run phase, we'll have live video and you'll be able to call in and ask questions. I had initially worked for someone else after leaving the military, but I quickly decided I wanted to work for myself. So, I started J. Tilley, Inc and Pinnacle Five. We do consulting, contracting, and leadership development. At the time my undiagnosed PTSD was addressed with anger and frustration. As I've grown, I've learned that talking and communicating helps. So does golf: I've had three holes in one, and I'm totally hooked.

The military is such a learning experience in what America is: constantly in pursuit of a "more perfect union." It's okay to disagree, but we are losing the way to do it respectfully. More and more people live in this little circle and they never get out. In the military, you're forced to get out of that circle, just like in life. If you want to be successful, you need to step out and keep stepping out every day. In the end it comes down to the Golden Rule. Treat people the way you want to be treated: with dignity and respect.

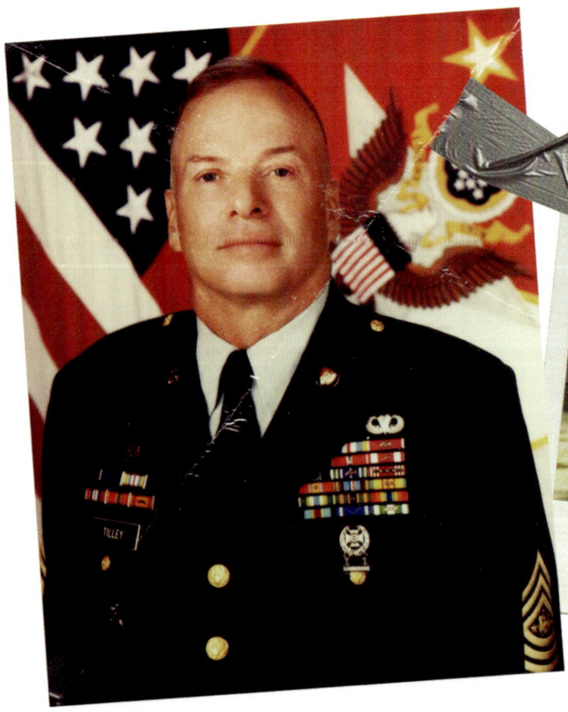

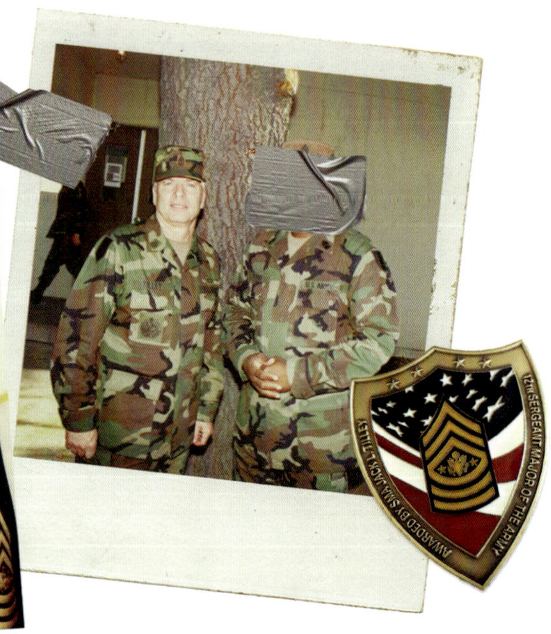

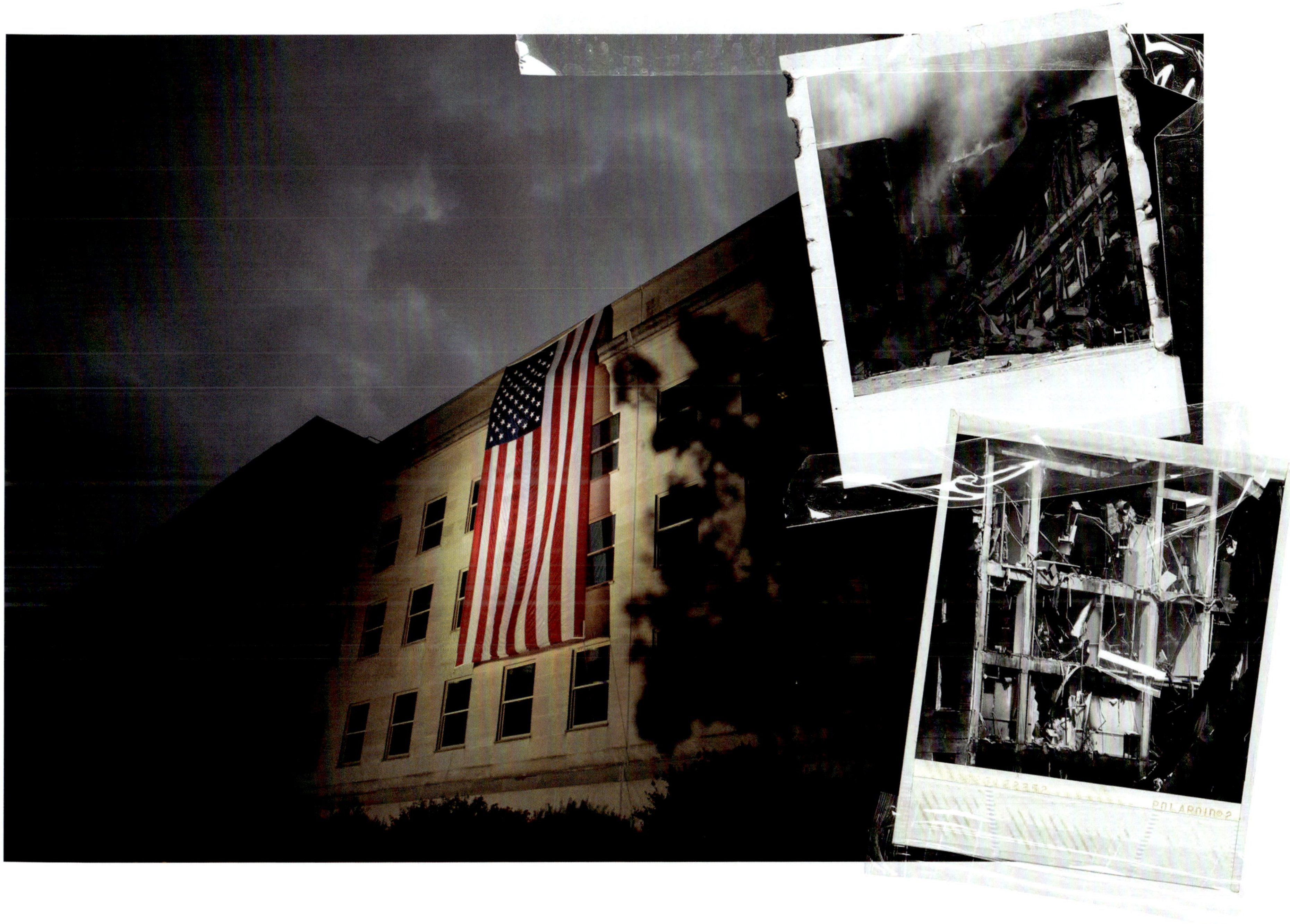

My name is John Rodriguez, and I was an infantryman in the United States Army. I joined because of 9/11. The military had been on the radar, my brother had wanted me to join, and initially I said no. But as a New Yorker, when the World Trade Center was hit, I joined right away. I deployed in support of the Global War on Terror in 2003 to the Horn of Africa. I separated from service in 2005. Then, a few months before my term was up, I was recalled to the Inactive Ready Reserve (IRR) and deployed with the Texas National Guard in 2010 in support of Operation Iraqi Freedom. I got blown out of my socks by indirect fire. The rocket hit about fifty feet from me. Sent me face-first into the lava rock we put down to lessen the dust when the helicopters landed.

Separating was hard. It was hard at home, but I was committed to remaining a good husband and father. I also found help that worked for me. Finding help is like a fingerprint, it's a niche. So even if you haven't found your place yet, keep looking. Help is out there. For me, that was Merging Vets and Players (MVP) and the Vet Center. The counselors at the Vet Center were great to me post-service. They're a part of the VA, but it's treated like a separate entity. They don't share files, and that is important. I didn't want a stigma or to worry about losing my job. I've been an NYPD officer since 2006. I'm a detective and I teach tactics at the academy. When I got back, I was going through a rough time, but I found out about the Vet Center and it fit me. The world is gonna keep on turning, so you might as well keep walking. The average person is into their own stuff. Keep walking, keep pushing.

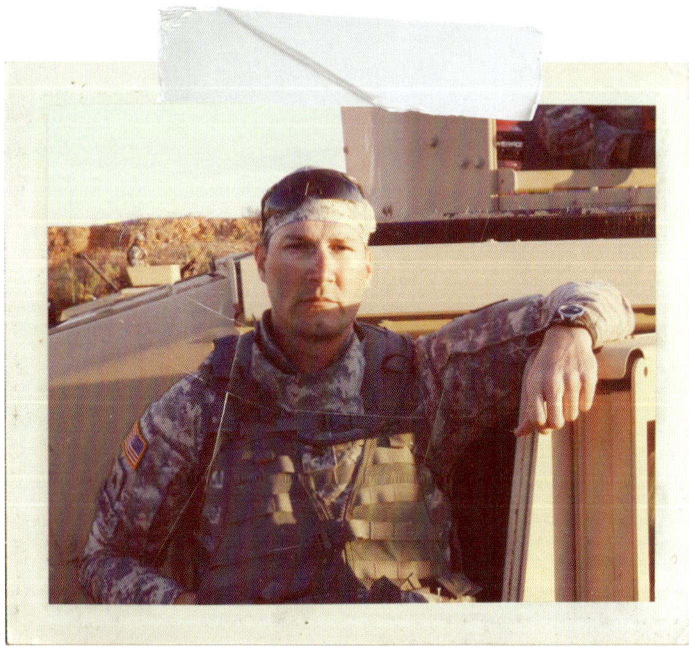

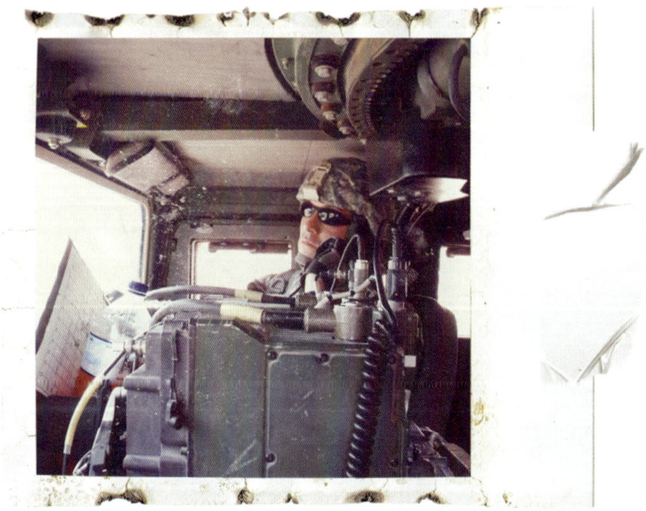

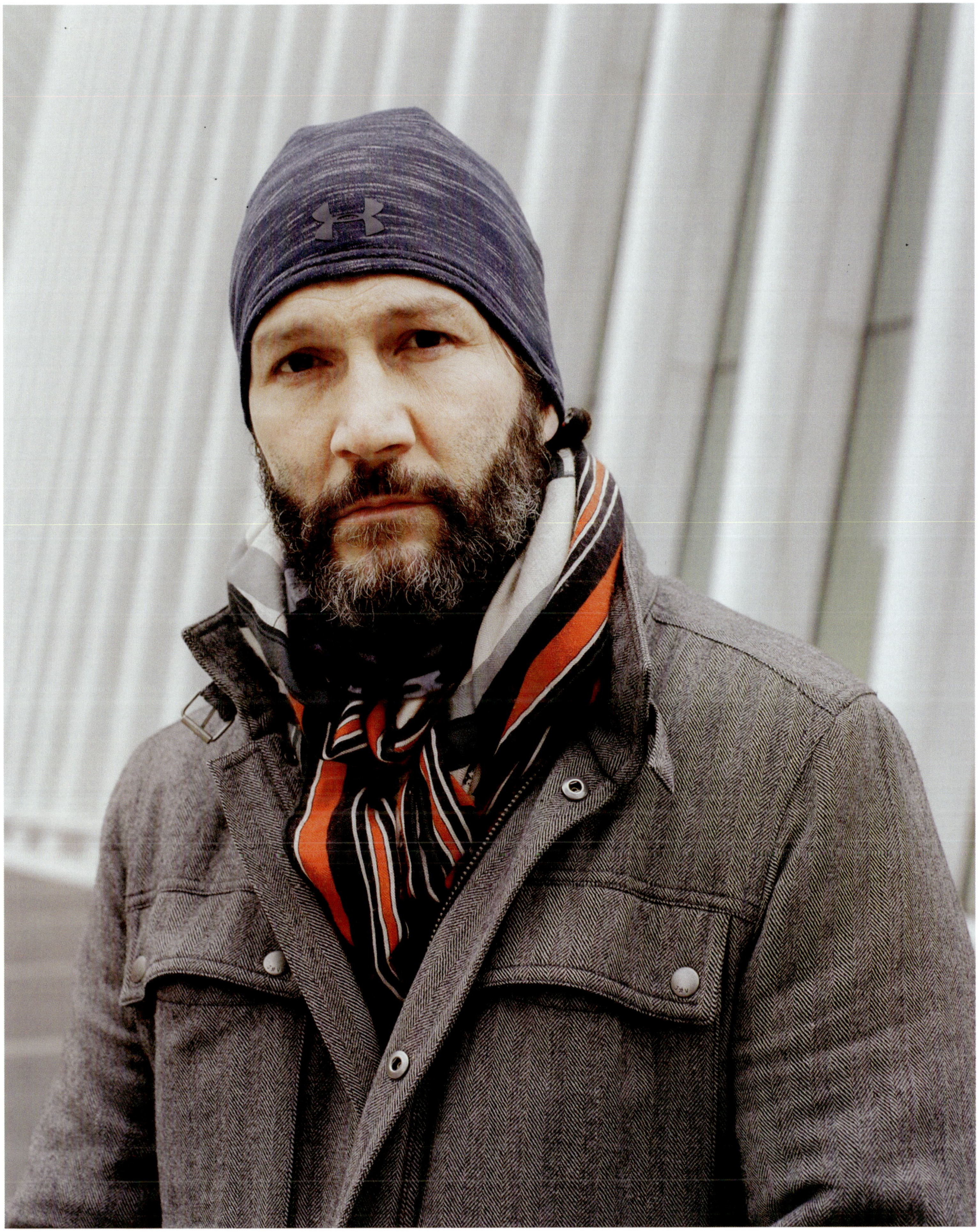

John “J-Rod” Rodriguez, Staff Sergeant
Army - 11B / Infantryman
Years of service - 5 years
Deployments - 2 [Operation Iraqi Freedom]

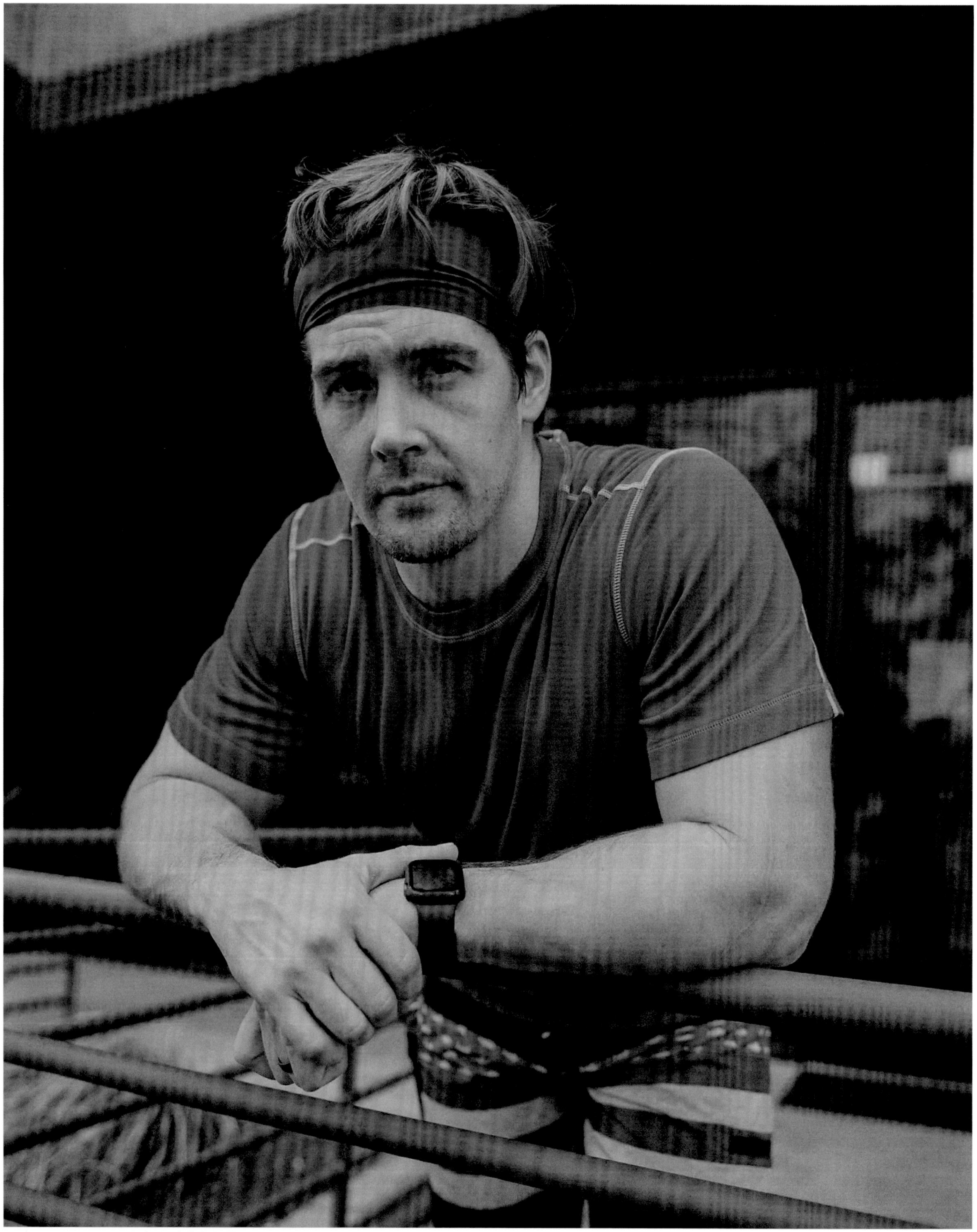

Justin Lakin, Staff Sergeant
Army - 18B / Special Forces
Years of service - 8 years
Deployments - 3 [Operation Enduring Freedom]

My name is Justin Lakin, and I served as an 18B Special Forces Weapons Sergeant in the United States Army. I was a part of multiple missions in support of Operation Enduring Freedom and the Global War on Terrorism, including Afghanistan and South America. I also went back overseas as a government contractor multiple times. I always felt a calling to serve, and I still do. My grandfather was a Marine, and my father also served in the Air Force. 9/11 inspired me to follow in their footsteps even more. Initially, I listened to my family's recommendations to at least try college first. A year of college didn't change my mind, so I enlisted under the 18X contract. I had no idea how challenging my military career would be both during and after my time in service. I have some very dark parts in my personal history. I grew up around drugs and gangs on the wrong side of town. The first time I was introduced to marijuana I was nine years old, and then at the ripe age of twelve I was smoking crystal meth. While this may seem extreme to most, it was just part of my family dynamic. Seeing the effects firsthand of what a life of addiction looked like, I knew I had to succeed, in college or the military. Going down the same road as many of my family and friends wasn't an option. By the grace of God, I've been successful, both in the military and private sector. While my family is doing better, the aftermath has left a lot of destruction in its wake. My family now frequently turns to me as a sort of anchor. I consider this an honor. It's hard, though.

Transitioning into defense contracting post-service was really easy, so many of the same people or types of people that I served with did it. But going to the more traditional civilian job after doing all that I did, and was a part of, was very hard. Having someone who had never been anywhere or done anything trying to boss me around or talk down to me was infuriating. Also, learning how to have a new relationship with my wife was difficult. When you're in or you're contracting, you are always leaving and coming back. There is this excitement, and on some levels, a freshness to the relationship. When you're home all the time, you almost have to relearn how to live with someone. There were a lot of challenges, trying to find a fit. I'm grateful I did find my fits. I do a few different things. I have a financial planning practice. When I was contracting, I got really into investing. So many guys in the military and first responder communities don't really know a lot about investing. With my background we could talk about shared experiences, and I can really break down what certain things can do for my clients. I'm also the executive director of TK Enterprises, working with Tim Kennedy.

There are so many stereotypes about veterans I'd like to change, but the biggest one is the idea that vets are just one-dimensional. People constantly discount what a veteran can bring to the table. The loyalty, depth of resilience, and capability that they can bring to a job or a relationship.

Continued on next page →

When I was studying psychology, I read Eric Greiten's book Resilience and really love his, "Identity, Action, Feeling" model. We identify who we truly are, then we act in accordance with who we are, and the feelings follow. Initially, I had a really hard time with that when I got out. Now, if I make a decision that goes against my integrity, I'm really hard on myself. If I'm working out with Tim, if I'm not sure how many repetitions I've done for sure, I'll discount my reps. My faith is my true north. I've only been saved by God's grace. I tell all my friends, even if you aren't religious or a Christian, you still have the Golden Rule. I live by that and try to be intentional and purposeful. In all the studies that have been conducted about dying and regrets, you generally hear the same few regrets. Not taking more risks is really common, but the one that always sticks with me is to invest more in relationships and be more intentional.

The top correlator with happiness and longevity is deep and meaningful relationships. I work really hard to try to mean it when I ask, "How are you doing?" I've had to work and learn to be transparent and vulnerable, especially with the way I grew up. But it's amazing the conversations and relationships you can have when you are. Contracting, I made a lot of money, but I was miserable. A lot of guys died on those jobs. Those dollar bills didn't go with them.

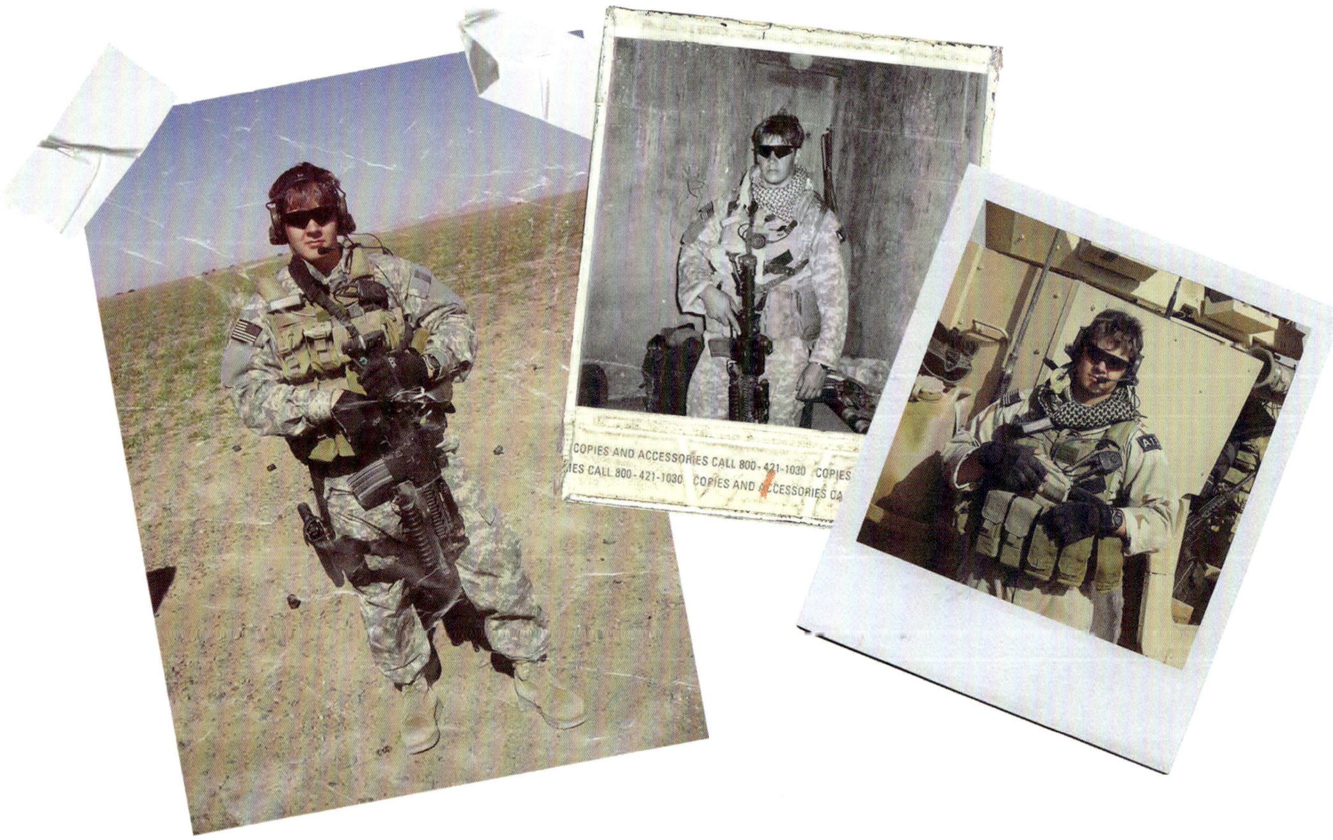

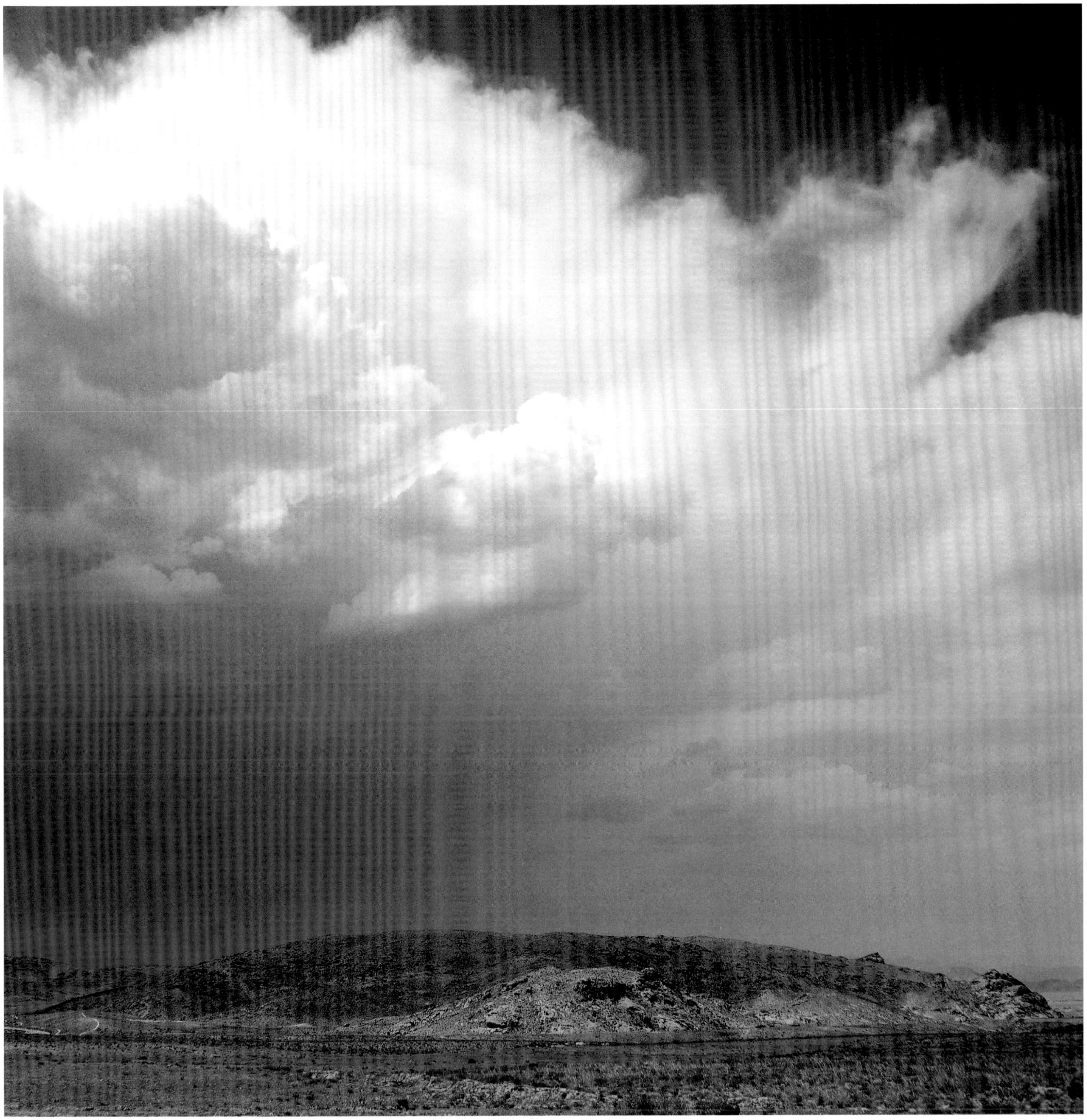

Justin: "Tim and I have been close for a long time. We met in college when I was working at a place called Cabo San Luis. It was a restaurant right across from the MMA gym Tim was training in at the time. I knew all the guys in there, and we'd trade some food for gym memberships. We ended up running in the same circles, and Tim and I went through the same recruiting station."

Tim: "I was just ahead of Justin in the pipeline. We ended up seeing each other going through the Qualification (Q) Course, and we both ended up in Seventh Special Forces Group. We went to different companies, so we weren't side by side, but we hung out and were around town together."

Justin: "I had been doing some bodybuilding and physique competitions in North Carolina while still a member of Seventh Group. Tim was looking to get off active duty to fight professionally full-time. He asked me if I'd do his strength and conditioning program, and we started getting even closer."

Tim: "Then the Texas National Guard offered me, basically, a part sponsorship, part pitch to be their recruiting poster boy."

Justin: "They offered me a little something, too. We both decided to do it. We got down there, having never lived in Texas and not really knowing the area. We lived in a house together and flash forward, now our kids are each other's nieces and nephews. Tim's wife is also the sister of my wife."

Tim: "Justin is so protective of me."

Justin: "I am."

Tim: "That might sound weird to some people who see me on TV or whatever, but I am a horribly trusting person at times."

Justin: "He has a huge network and loves to be around people so much and I've seen people take advantage of that. Tim's an interesting cat. He's a huge dreamer and much more extroverted than I am. I'm more introverted and we balance each other out."

Tim: "I fuck up often. I wear my heart on my sleeve and then Justin is an oak. He's grounding."

Justin: "And on the flip side, he forces me out of my comfort zone. I help him realize things and plot the course for all these dreams. To conceptualize how to do it and follow through on something. To be intentional in what we do and how we do it."

Tim: "There are a lot of people who are trying to work with me or ask things of me. Justin helps me sort that out, helps me ask questions like, 'Are they here out of friendship and loyalty, or something else?' And that's cool and powerful."

Justin: "Tim's truly hard to kill, like his show, but he has so much more depth than just some MMA guy or shooter. He's not just the guy people see on TV. It's not just Tim or me that are doing TK Enterprises, either; it's everyone who is a part of the business. Our employees and contractors who work with us, it's about them too."

Tim: "Starting in college and being silly kids in San Luis Obispo and then enlisting together, going to SF together, deploying together, moving to Texas together, there are a lot of layers there, a lot of trust, and it makes for a great partnership."

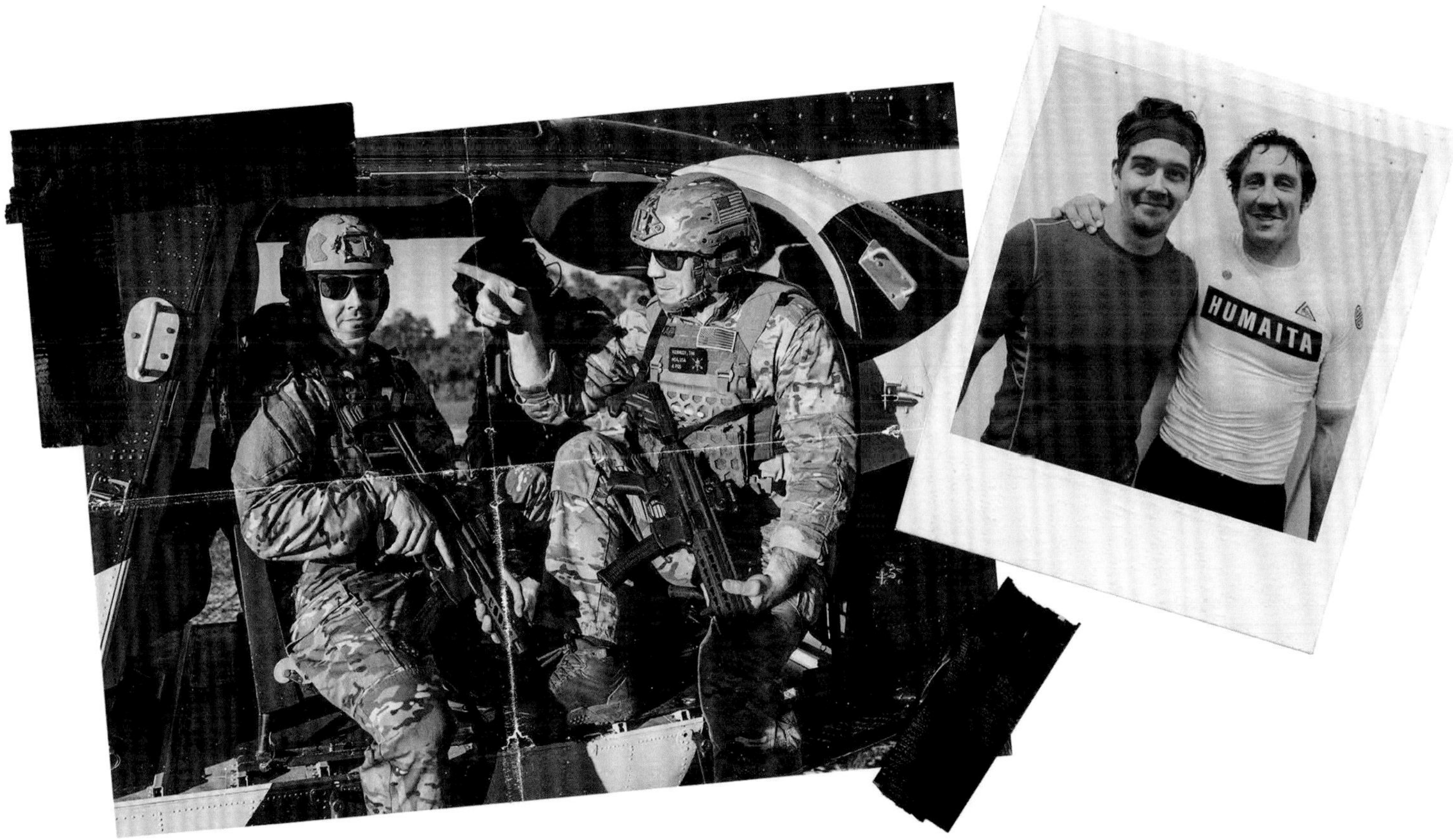

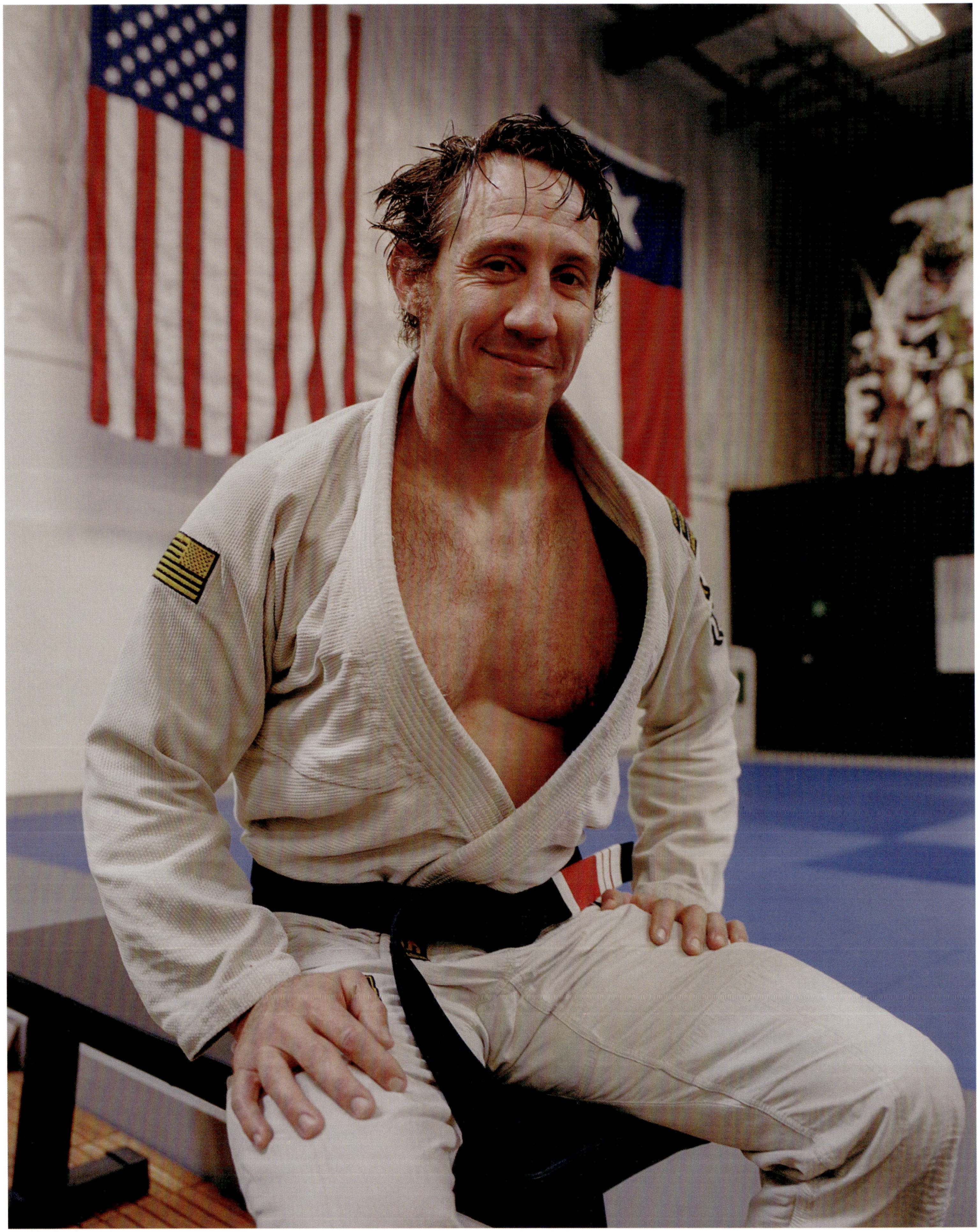

Tim Kennedy, Master Sergeant
Army - 18B, 18Z / Special Forces
Years in service - Active
Deployments - Multiple in support of GWOT

My name is Tim Kennedy, and I was an 18B Weapons Sergeant for almost ten years in the United States Army Special Forces. I am currently 18Z, Special Forces Team Sergeant (E-8) in the Texas National Guard. Throughout my career, I've deployed multiple times, including Operation Iraqi Freedom and Operation Enduring Freedom, in support of the Global War on Terror.

I was a shithead before 9/11. I was more worried about what pair of designer jeans I was going to wear after my next MMA fight. I was ranked top ten in the world at the time, while working at a dot.com in California. I had just gotten two women pregnant. I thought I had AIDS. I was totally out of control. Then I saw a bunch of Americans standing in a burning building in New York City looking out a window, looking back inside, and having to decide if they were going to burn to death or jump to their death. I was in the recruiter's office later that day. I was blown up in 2008. The truck right in front of us got hit by an IED that killed everyone. My gunner and the guys in the back of my vehicle got fucked up pretty bad, too. I was in the captain's seat of my vehicle, which is part of what saved me. I was shitting myself from overpressure sickness in the middle of a pretty gnarly gunfight for the next three days. I ran down, trying to save the driver from the first vehicle, and ended up taking shrapnel in my shoulder and back.

My transition was a big shift. Not just separating from active duty and going to fight but going from an elite Special Forces Commanders In-extremis Force (CIF) Company to the National Guard. I had a lot of ego when I first got onto active duty, and it took my team sergeant and the other leaders to really teach me how to be a good SF soldier. It's a gut shot now, thinking of how I was when I got to the Special Forces. I burnt so many bridges. Then I had to fight that same mental battle in a different way when I got to the Guard. There's a massive culture shock from having the best of everything ... from having a personal locker of ten elite, boutique, custom guns and having any optic, plate carrier, load system I wanted, night vision everything ... then, going into the Guard and they're like, "Well we can't really issue anything because we don't have anything, here's a musket. Good luck." There couldn't have been a better thing to help me in terms of humility and humbling instances. I'm not as cool as I think. I'm not as talented as I thought, and I don't deserve anything. The only things I will have is what I earn and take.

As I was adjusting to life in the National Guard, I was fighting professionally in the top tier of Mixed Martial Arts. I made two title runs in Strikeforce and came up short, so I'm the two-time almost world champion. I also had one run at the UFC, including dominating Michael Bisping in a main event. As my fight career progressed, I was able to get other opportunities and I transitioned to TV. I hosted three seasons of "Hunting Hitler" on The History Channel. I also had my own show on Discovery called "Hard to Kill," and have done some random episodes and things along the way.

Continued on next page →

I'm a serial entrepreneur, investing in companies or ideas I believe in. Currently, I have four companies and am working on the fifth. Sheepdog Response is my tactical shooting company. Ranger Up military lifestyle apparel, I was a founder of and continue to work with and support. I also work with Woobies Shoes, which is a surfer-style shoe company, but ours are better than most. And then, Apogee Cedar Park is a private school and we're getting ready to launch Apogee Strong, an online mentorship for young men on how to be leaders and prepare them for life. As a kid, I was kicked out of kindergarten and my teacher and the parents of other kids were always saying I needed to be medicated. I was belted, spanked, and literally duct taped to a chair to make me do homework. But once I was finally given freedom, I tested out of second and sixth grade, and graduated high school at fifteen and college at nineteen. I think the best way to learn is by doing. Allowing kids to have experiences and to grow. Guides, not teachers. Guides help steer the boat; it's the students' responsibility to drive it. If the boat is sinking, the guides will ask, "How do we fix this boat and keep sailing?" Everyone's experiences are so different and shame on us for making a standardized test across the board. We test an inner-city kid from Chicago the same as we test a ranch kid from Texas. Shame on us for trying to standardize their experiences and knowledge; they are wildly different.

I'm also committed to ending the stigma of "the unintelligent veteran." That vets and soldiers are just robots and unthinking. Right now, in the media, people are running to podcasts for nuance. We need to do a better job of that in the military community. In the military, you have the most brilliant people on the planet. It's so much harder to get into the military than to get into college. The public perception is still that these men and women are mindless idiots. We're not. We're creative and intelligent and philosophical. We're hard workers and passionate and committed, and we need to show that more.

For fun I still love training martial arts. I fought for almost twenty years and, to me, Brazilian jiu jitsu is like yoga, but I get to choke people. Seriously, when I leave jiu jitsu, I want to fight a velociraptor and fuck a tiger, I am so energized.

Individual responsibility is a real thing, and we all need to accept personal responsibility for our actions. I could go down the PTSD rabbit hole and drink too much. Or I can take the onus and responsibility to be a good husband, father, boss, and training partner in my life. All of those things complement each other and create a foundation to continue to grow.

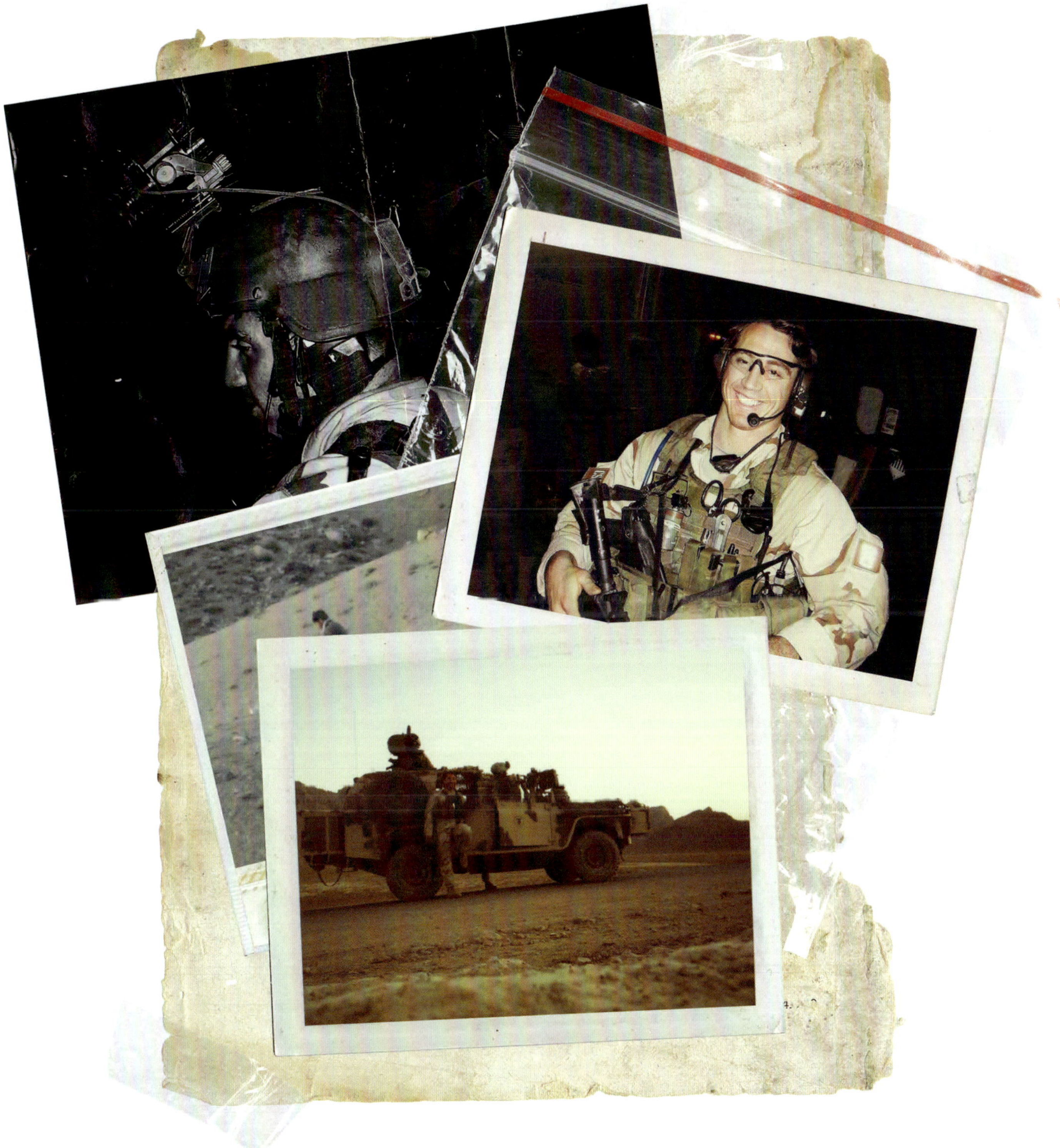

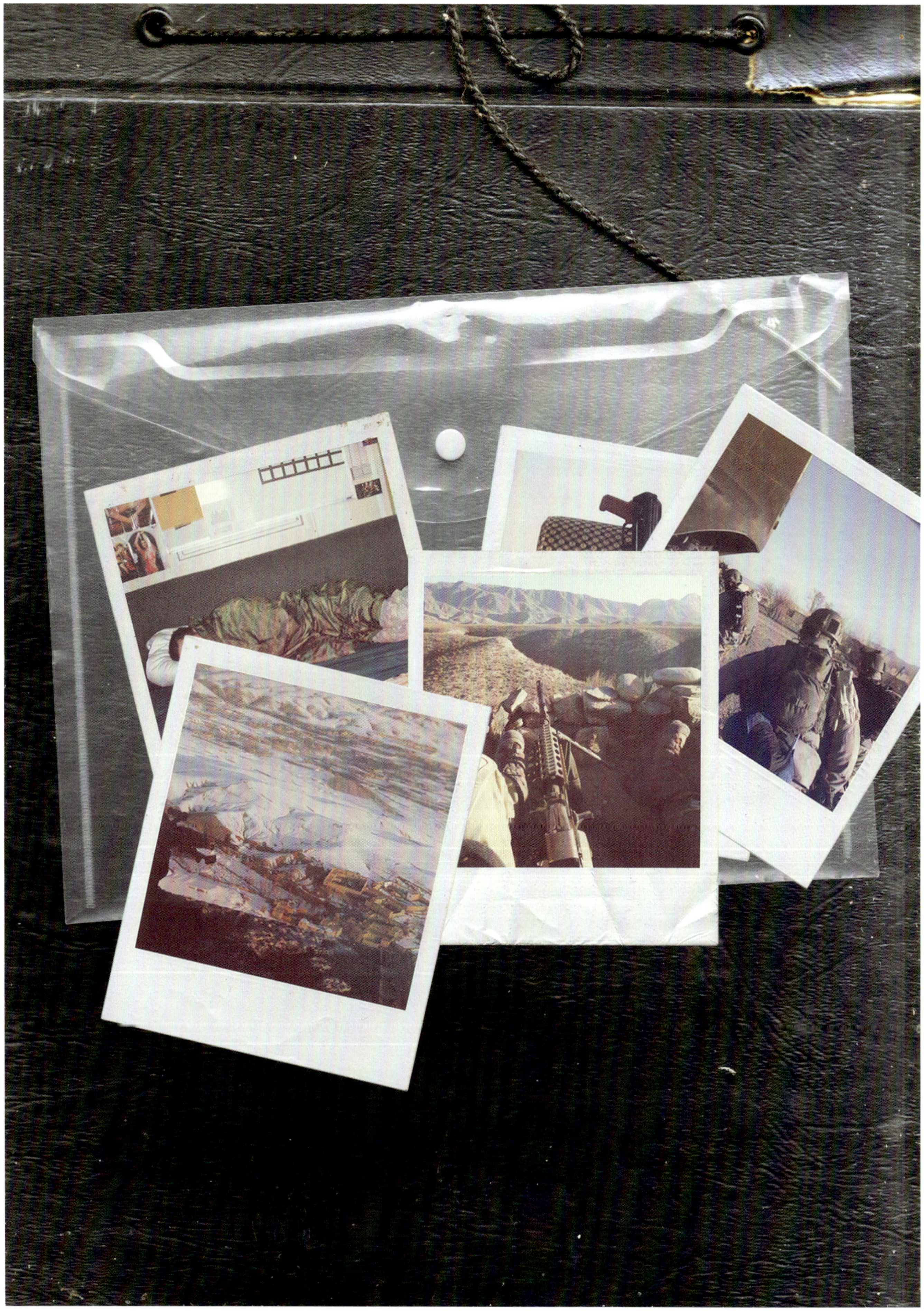

Oct 2011

Dear Family,

I know it has been a while since you've heard from me. Don't worry, I'm good. I can't tell you exactly where I am, or when I'll be home. But, I can tell you that I am as safe as I can be.

Honestly, we have it pretty good. I feel bad for others around the FOB. We get to spend most of our days when we're not on objective, sleeping, eating, getting stronger, training, watching movies, or playing video games. Don't get me wrong, it's not summer camp, but it also isn't like I'm always getting shot at or something.

When we go on a mission, we are looking for the worst of the worst. And we are making a difference, at least where I am. There have been fewer attacks on locals, their schools and businesses, and our base.

I find myself sometimes, thinking how beautiful of a place it is here. If I didn't know exactly where I was, I would think I was back home near Joshua Tree. The mountains and desert plains are simply stunning. When the sun sets, the colors you see from the clarity in the sky are something I'll always remember. Not to mention the night sky! Since there are no major cities around here, it seems I could see every star in the sky. I wish everyone had a chance to see it.

I hope everyone in the family is doing well, and I'm sure I'll see y'all during the holidays!

Love you all!
Dan

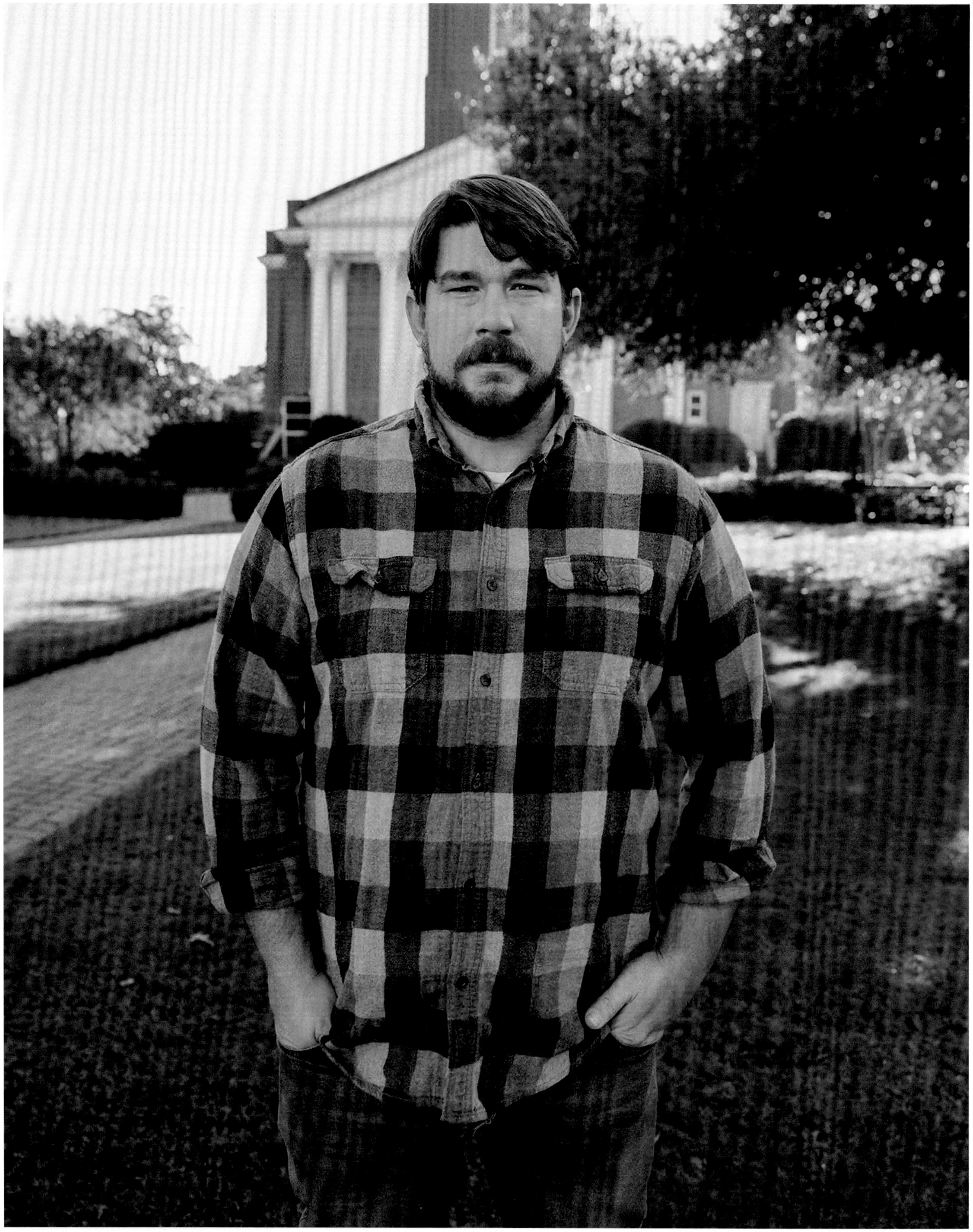

Kenneth Hagler, Lance Corporal
Marine Corps - 0311 / Infantryman
Years of service - 4 years
Deployments - 4 [Operation Iraqi Freedom & Operation Enduring Freedom]

My name is Kenneth Hagler. I served in the United States Marine Corps as an 0311 infantryman. I joined to defend people who couldn't defend themselves. Also, my grandfather served in World War II and I wanted to honor his sacrifice. I was in the delayed entry program pre-9/11, waiting to go to boot camp. I eventually got there in January of 2002. I deployed twice in support of Operation Iraqi Freedom and twice to the Horn of Africa in support of the Global War on Terror. I separated after four years of service as a Lance Corporal.

Readjustment and being away from my brothers was rough. I was always answering the door high-strung, a little paranoid and lonely. I struggled with alcoholism until 2015, which is when I got sober. I got out in 2006, lost a child in 2008, and really hit rock bottom. I was alone for years. I was so guarded, but even then, I knew there was something for me. It's easy to make yourself a victim, but somehow, I knew I was going to come out the other side. Eventually, I gave my life to Christ and now I have clarity in what I'm living for. Currently, I'm going to school to be a pastor. When someone's hurting or can't defend themselves, it's about stepping up to help; that's my greatest pleasure.

You gotta be willing to sacrifice for others. Get out of your comfort zone and be there for people. We judge so quickly; we rarely stop to ask what they might be going through, to have the humility to see it from their side. I know so many are suffering with alcoholism. I'm not judging; I've been there. It seems so innocent for vets to boast that we get together and drink. Then you don't have the group, but you still have the drinking and the loneliness. There were so many things that drinking ruined for me, and so much it cost me: jobs, family relationships, relationships in general.

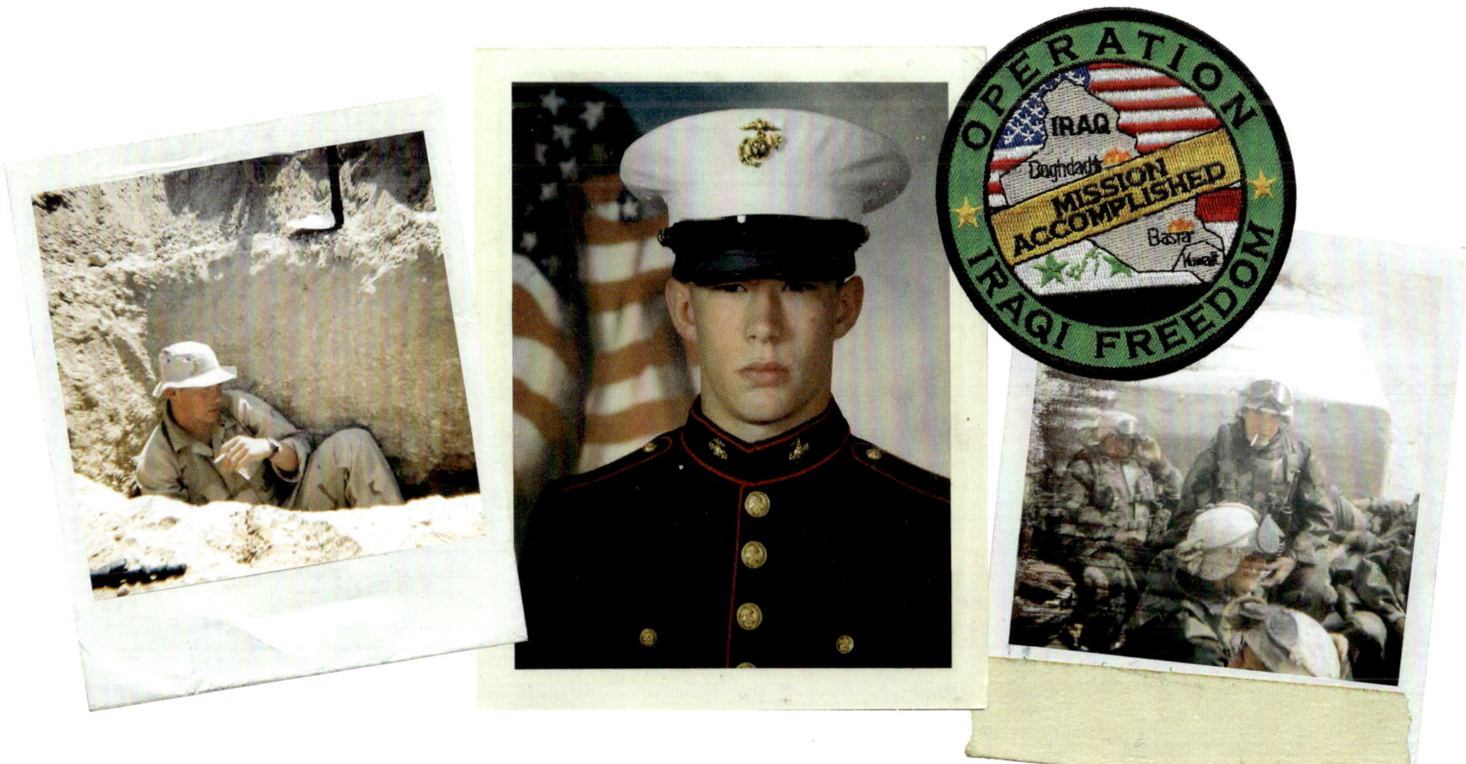

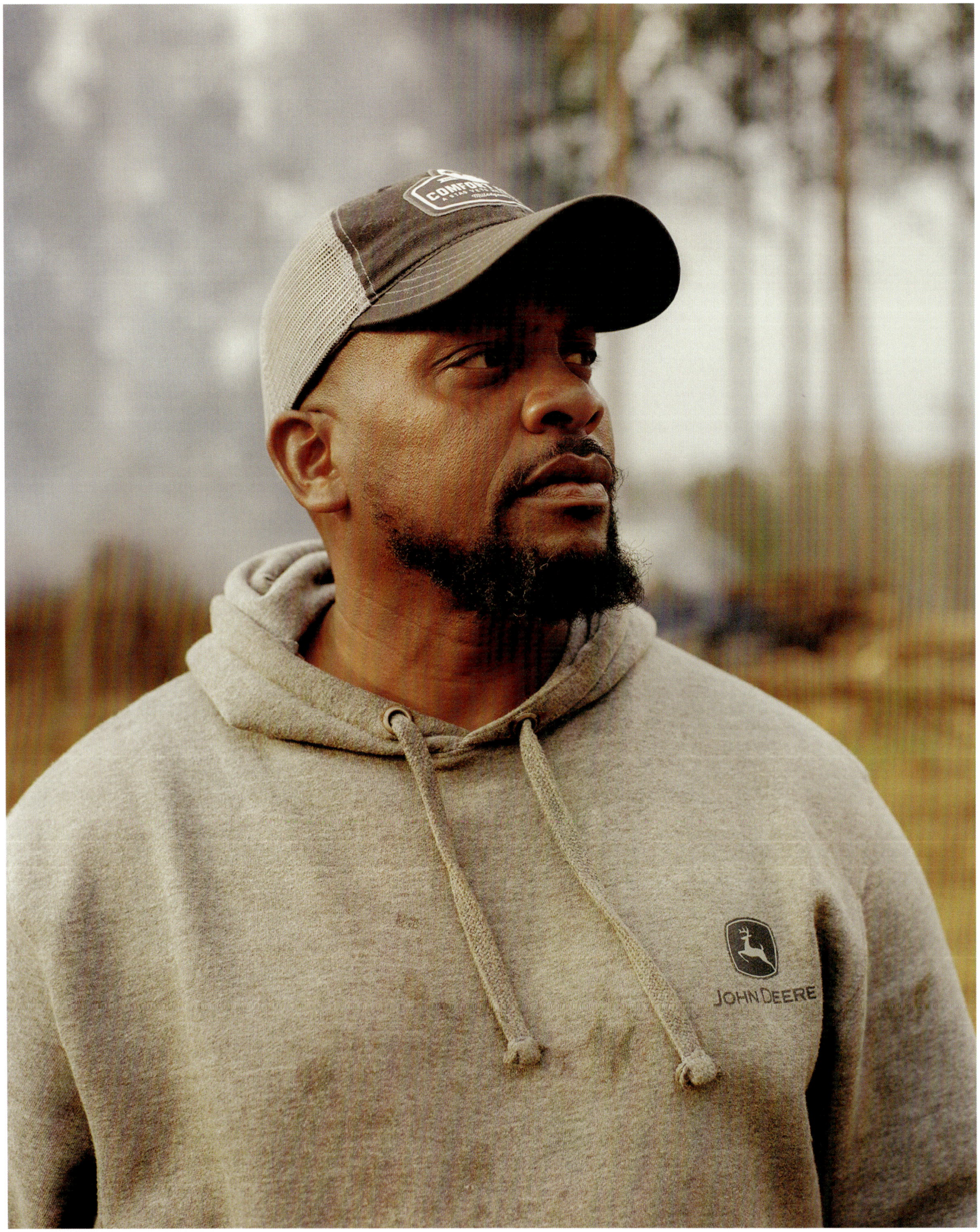

Jon Jackson, Staff Sergeant
Army - 92A / 75th Ranger Regiment
Years of service - 11 years
Deployments - 6 [Operation Iraqi Freedom & Operation Enduring Freedom]

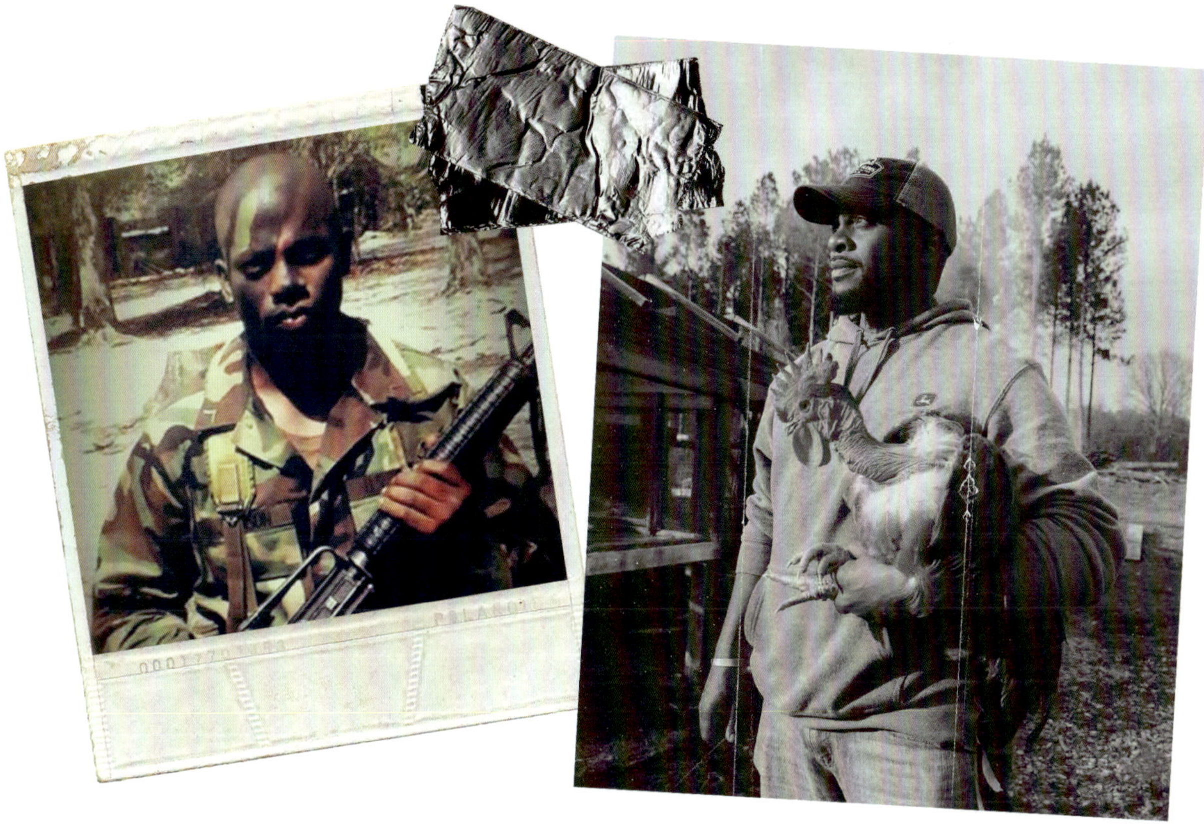

My name is Jon Jackson, and I was a 92A Supply Clerk in the United States Army. I was a member of the 75th Ranger Regiment when I separated and did six tours in support of the Global War on Terror. I was in for eleven years total and separated as a Staff Sergeant (E-6). I joined because of 9/11. I lived in Jersey City, on Garfield Ave., where my childhood view was the World Trade Center and the Statue of Liberty.

After separating from the Army, I started Comfort Farms. My Ranger buddy, Kyle Comfort, was KIA in 2010. I had this thought: "Kyle was such a cool guy and 'Comfort Farms' is a great name," and the farm was born. We're committed to helping veterans heal through learning to farm. The name can be misleading; we're not here to be comfortable. You can only grow in discomfort. We have a total farm, complete with a farmer's market. We're totally self-funded through what we produce. We were just awarded a grant to teach agriculture for the next three years. You can take classes at Central Georgia Tech, use your GI Bill, and get credit hours by learning farming from the ground up.

Separating is tough. You come from a unit like the Rangers and then you are out. You aren't deploying or jumping, you feel like you lost your edge. I went through some real serious depression ... almost committed suicide. When I clawed my way out and survived it, I felt like there's nothing I couldn't do. Even when I was having relationship issues with my wife, all the hard work on the farm had the effect of helping us repair that. You're only as good as your support. If I look like a rockstar, it's because of my family and the veterans I work with all supporting me. You need your tribe. It's never just about you!

My name is Kevin Brewington and I enlisted as an infantryman in the United States Army. My enlistment was supposed to be three years. A year and a half in, I got hurt and was medically retired. I was the radio telephone operator (RTO) on a patrol during our deployment in support of Operation Enduring Freedom in 2011. We were working to set up a COP in the Kandahar region of Afghanistan. We had just met with the elders and had our tea and talked. As we were wrapping up, suddenly the elders left, which was a red flag. As our vehicles started rolling, we saw a suspected IED on the road. Our commander made the decision to cut through a gap in an adobe wall. As I went through the hole, the IED detonated. I woke up four days later in Germany on a ventilator. I lost both my legs and getting my right arm working was the biggest challenge... besides losing my legs. Including the skin grafts, I'm guessing I had to endure fifty to seventy surgeries, depending on if you count the washouts every Monday, Wednesday, and Friday.

I got married in December 2012 and officially retired in January of 2013. I now volunteer with the Independence Fund and also the Greatest Generation Foundation. At the Greatest Generation Foundation, we take veterans back to the battlefield to heal. We have gone to Normandy, Vietnam, and are working on being able to go back to Afghanistan. It's very therapeutic for these veterans and helps bury the past. I really like helping fellow vets and I get to meet cool people, both public figures and other awesome vets. I also use my network to introduce other veterans to opportunities. Finally, I work with the local sheriff's department, showing them the importance of staying calm and using a tourniquet correctly, which is pretty cool.

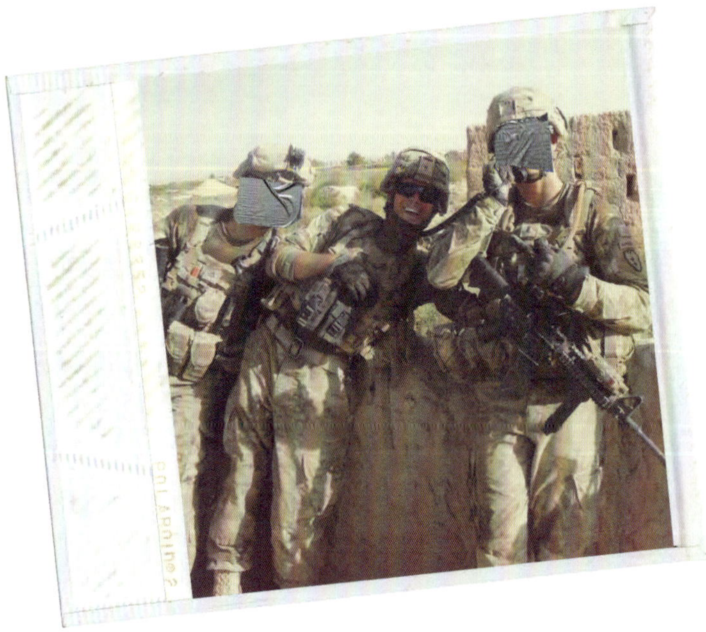

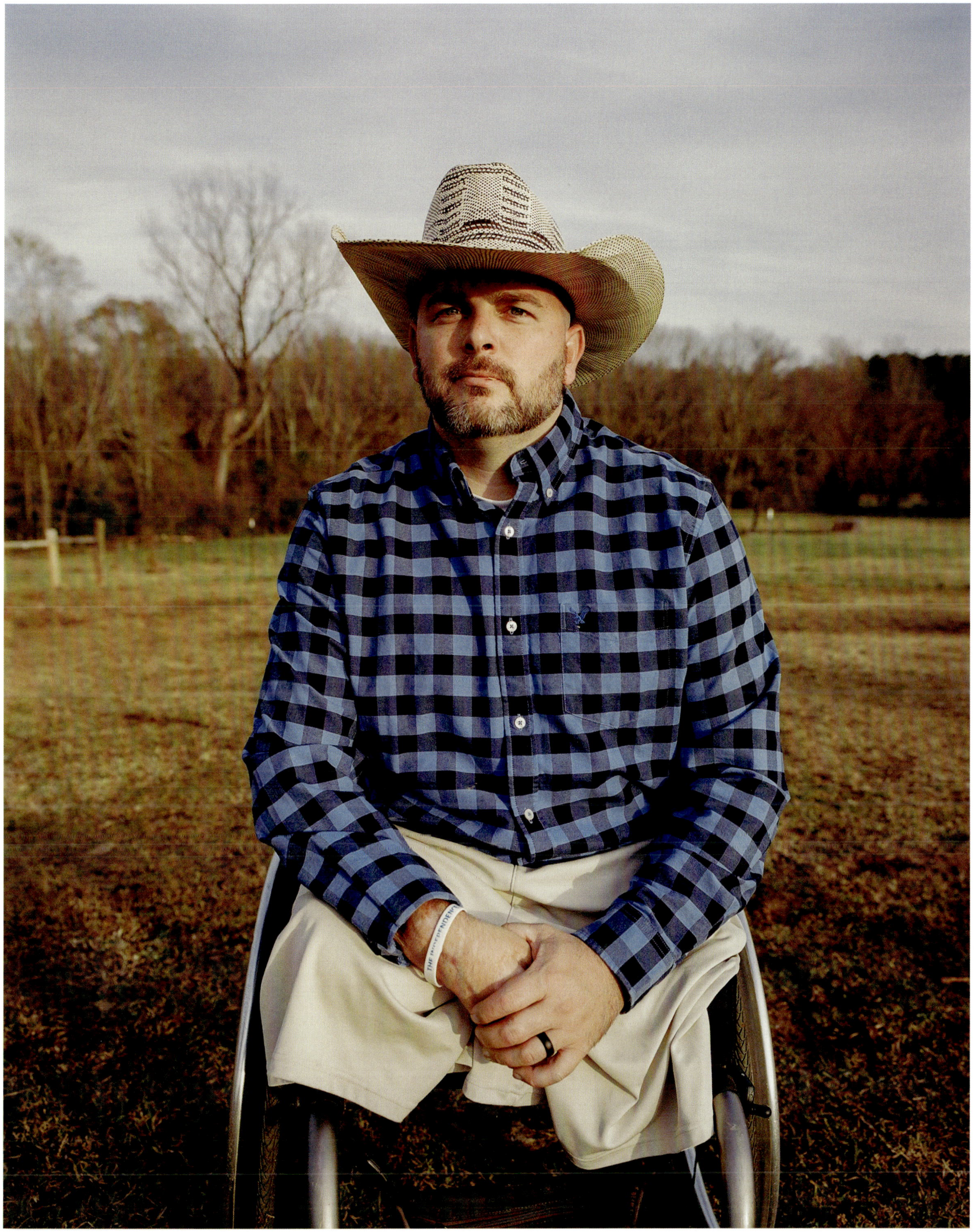

Kevin Brewington, Corporal
Army - 11B / 1-25th Stryker Brigade
Years of service - 3 years
Deployments - 1 [Operation Enduring Freedom]

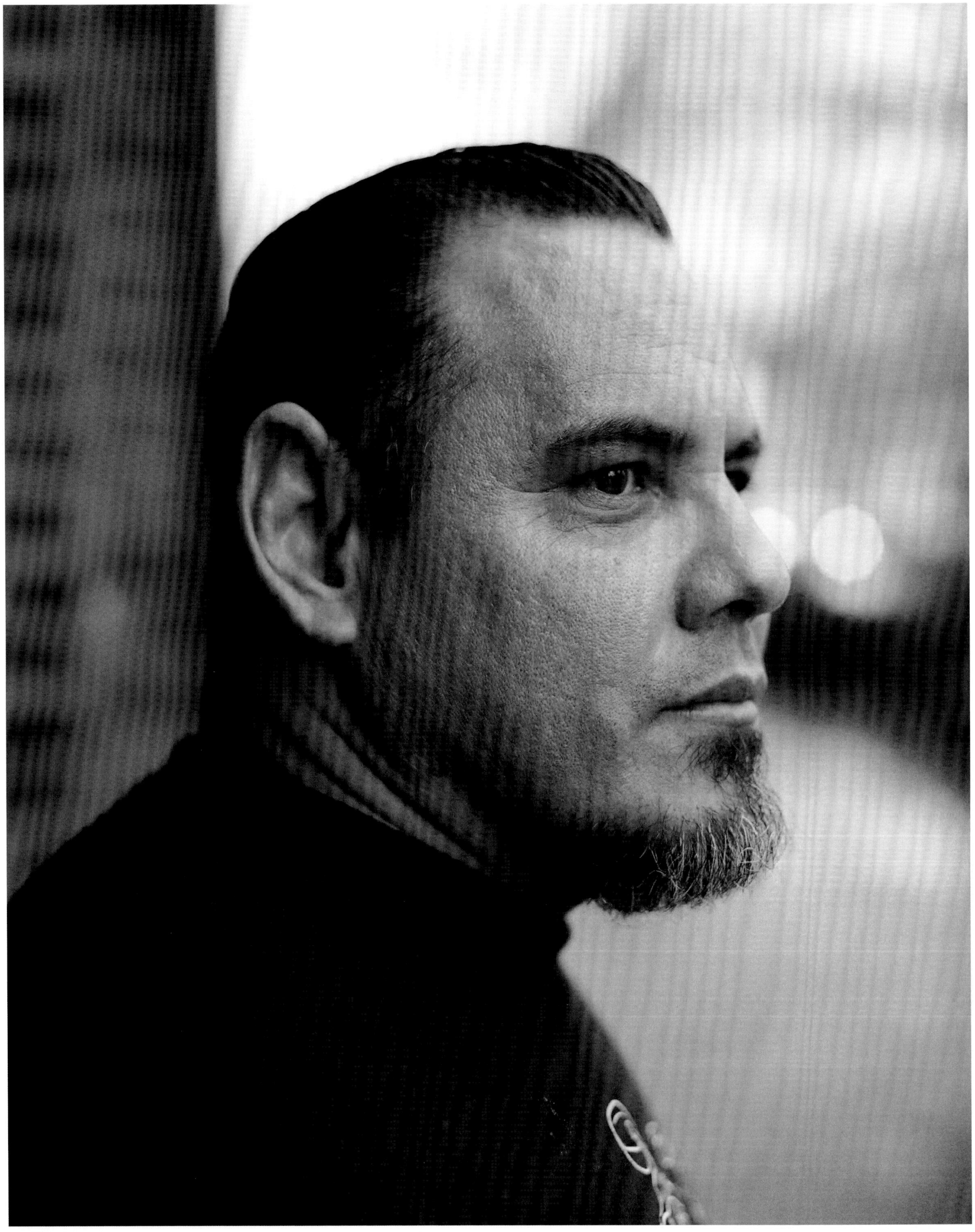

Joseph Serna, Sergeant First Class
Army - 18E / Special Forces
Years of service - 19 years
Deployments - 3 [Operation Enduring Freedom]

My name is Joseph Serna, and I was an 18E Communications Sergeant in the US Army Special Forces. When I decided to enlist, I was looking for a way to get an income for my family. I'd had my family young and needed a way to provide. I served for nineteen years before finally retiring. I deployed in support of Operation Enduring Freedom three times.

We were going out on a mission to help a sister ODA. One of the guys stepped on a landmine and they were taking the body back, but they got stuck. We took a wrecker and an MRAP, an up-armored vehicle that I was riding in, out to help them out. In the process we got stuck, too. The roads just aren't rated for a vehicle of that size. We tried to get back to the road, but it gave way; it was just a soup hole, and my MRAP flipped over and ended up in the creek. We were all fighting to get out and the water was rising. It was pitch-black, with almost no room, we're upside down and we're trying to open a six-hundred-pound door. I got buried in ammo cans, but I was able to just barely pop my head out and ask for help. I was in the back with my buddy James Treber, and he popped my seat belt and saved my life. I was in the vehicle for about forty-five minutes total. Just barely hanging on. Toward the end the fuel started to be released into the cabin. I started inhaling it, that toxic air was all I had at that point. I was hallucinating; it's like I was with the devil, bro. I think it was just the toxic fumes and the guys trying to get me out or my convulsing from that bad air, but it was nasty. It was bad, bro. I was the only one in the truck that survived. They quickly medevaced me to Kandahar and said, "You're going home."

A couple days later, I picked up the phone. I called my wife, who was pregnant at the time. We didn't have a name picked out yet, and Treber's middle name was Matthew. So, I said, "Our son's name will be Matthew," and then I hung up. I stayed in-country. It was such a hard thing, but it's where I needed to be. I was hit again a few months later in a suicide grenade attack; that took me out of the fight. I went to be an instructor for the SF pipeline. At times, I thought I was healthy enough to return to the teams, but I wasn't. I kept having hernia surgeries and right before my twenty I said I was done and was retired out.

The Wounded Warrior Project and the Green Beret Foundation have really helped me out. If you're a Green Beret, I highly recommend them. They're a great place to get help and, just like SF, they have unconventional thinking towards treatment and mental health, including the hyperbaric chambers I've used. Since I got out, I've worked to keep myself busy. Physically I had to work through my injuries. I've also needed to really take care of my mental health, and those two things didn't line up at first. I wasn't talking with anyone at the start. I started to drink a lot, which led me to see a psychiatrist. I went back to college and graduated in 2018.

Continued on next page

Looking back, I wish I could change my relationship with my wife. I put her through a lot of hardship, deployments, and my post-war transition, and we've since separated. I wish I'd appreciated her more.

MMy mantra is always "Take care of yourself." If you can take care of yourself, you can take care of others. But you have to start with you.

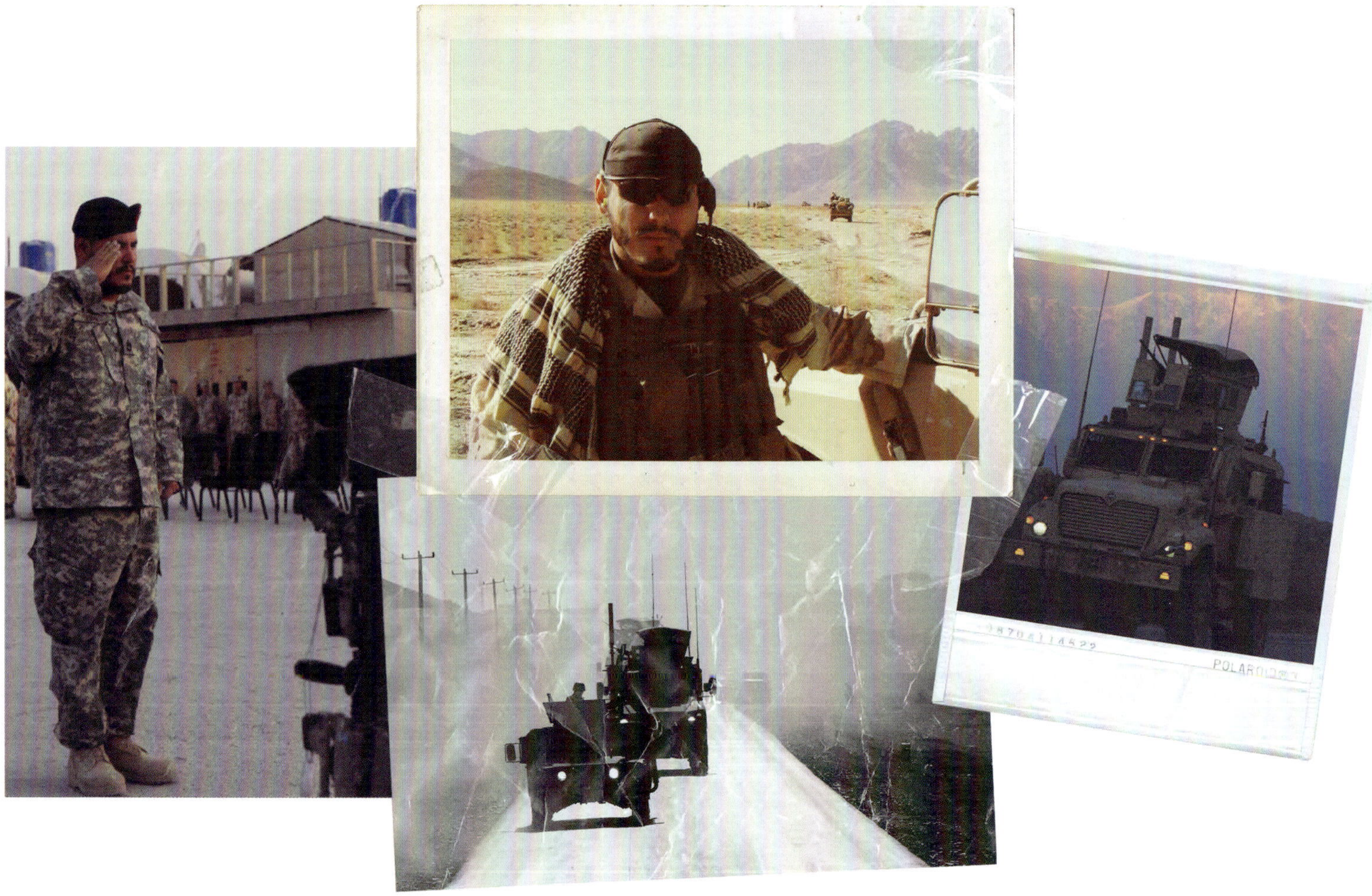

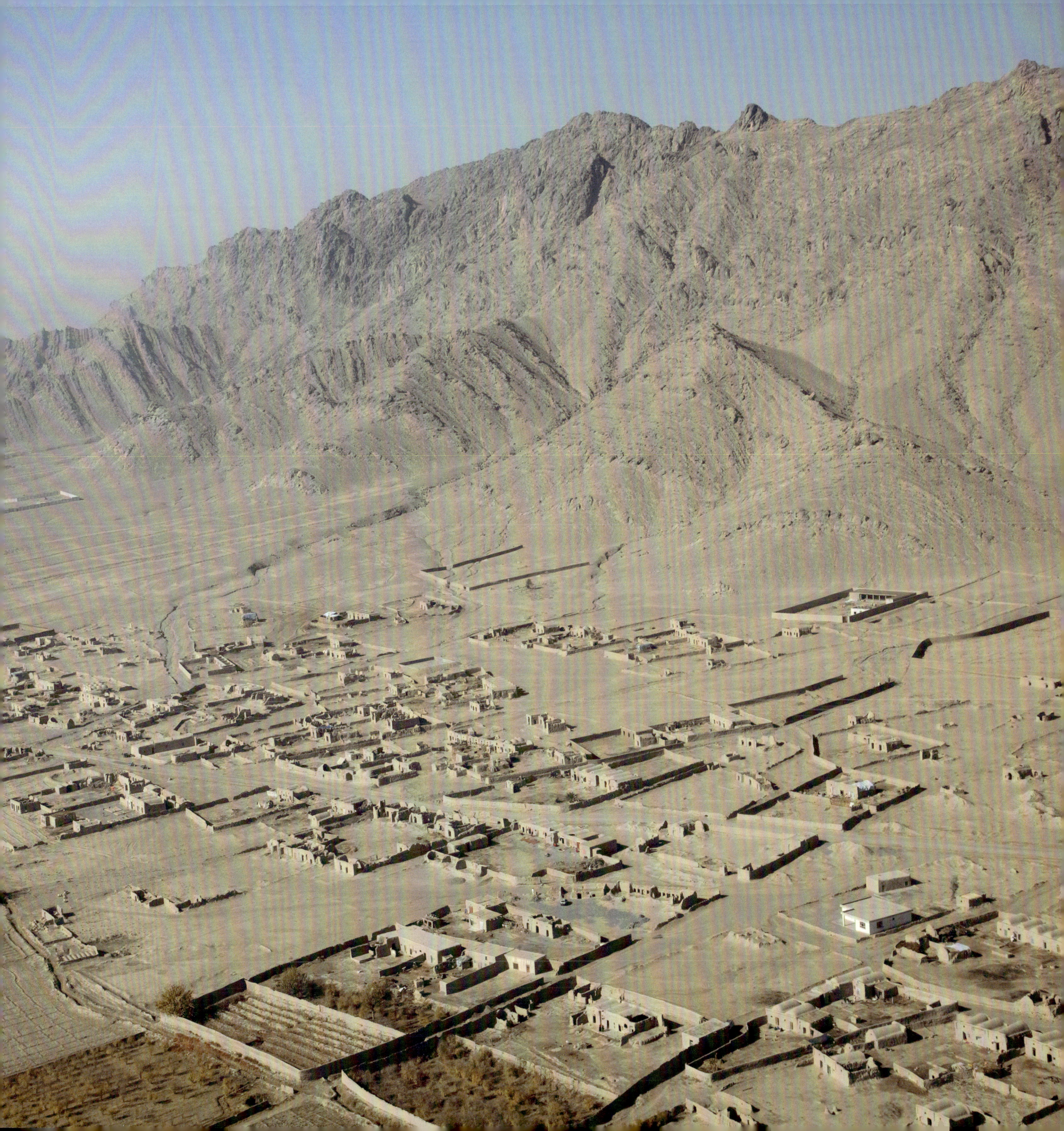

My name is Steve Bonetti, and I was an 11C mortarman in the United States Army. I was a senior in high school when terrorists attacked the World Trade Center. I could see the smoke from the windows of my high school. I deployed three times in support of Operation Iraqi Freedom and Operation Enduring Freedom before separating from the Army as a Sergeant (E-5). My last deployment was during the surge, and was a little crazy. The Iraqi police station we were working out of got hit by a two-thousand-pound car bomb. It was a really hairy few hours. Thankfully, we had no US casualties. Honestly, events like that are a really common story during that period of the war. One of my buddies who was an officer is using it as part of a case study for his Captains Career Course.

Post-military, the GI Bill and my disability rating meant I was able to focus on school. My sister's father-in-law had worked at the VA for thirty years. It really made things easier for me, to know what questions to ask and who to ask them to. After school, I opened a gym. I also worked for the VA for a while and now am back in the fitness industry. I love being in fitness because of what the Army instilled in me; I kinda got addicted to exercise once I learned how to do it properly. I work with MVP, Merging Vets and Players. I really love their mission. Also, Headstrong is a solid New York-based organization focusing on mental health. Finally, the Vet Center is great. I've heard other vets say the same thing, but I'll also say it: it's great that the VA funds them, but everything is confidential and doesn't get put in that file. The counselors are great and it's a huge asset.

I've learned that patience is the key. Be patient with people, with yourself, and really with anything.

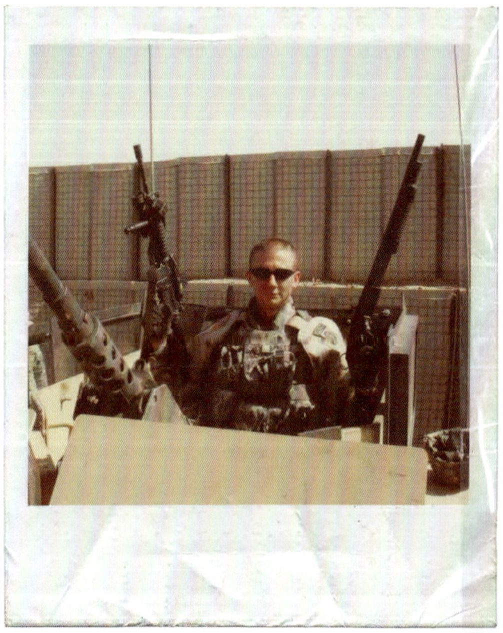

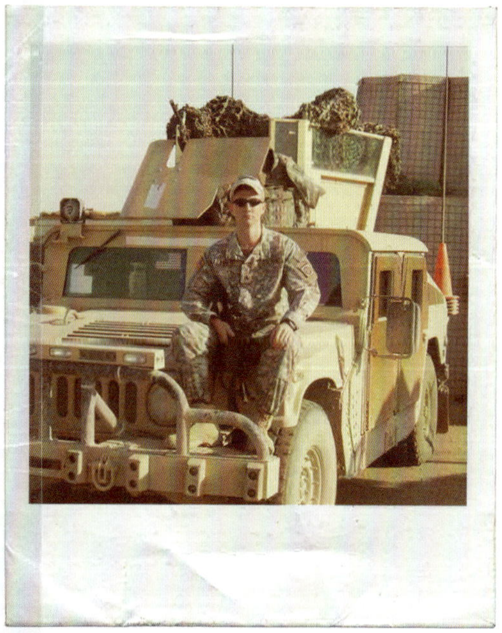

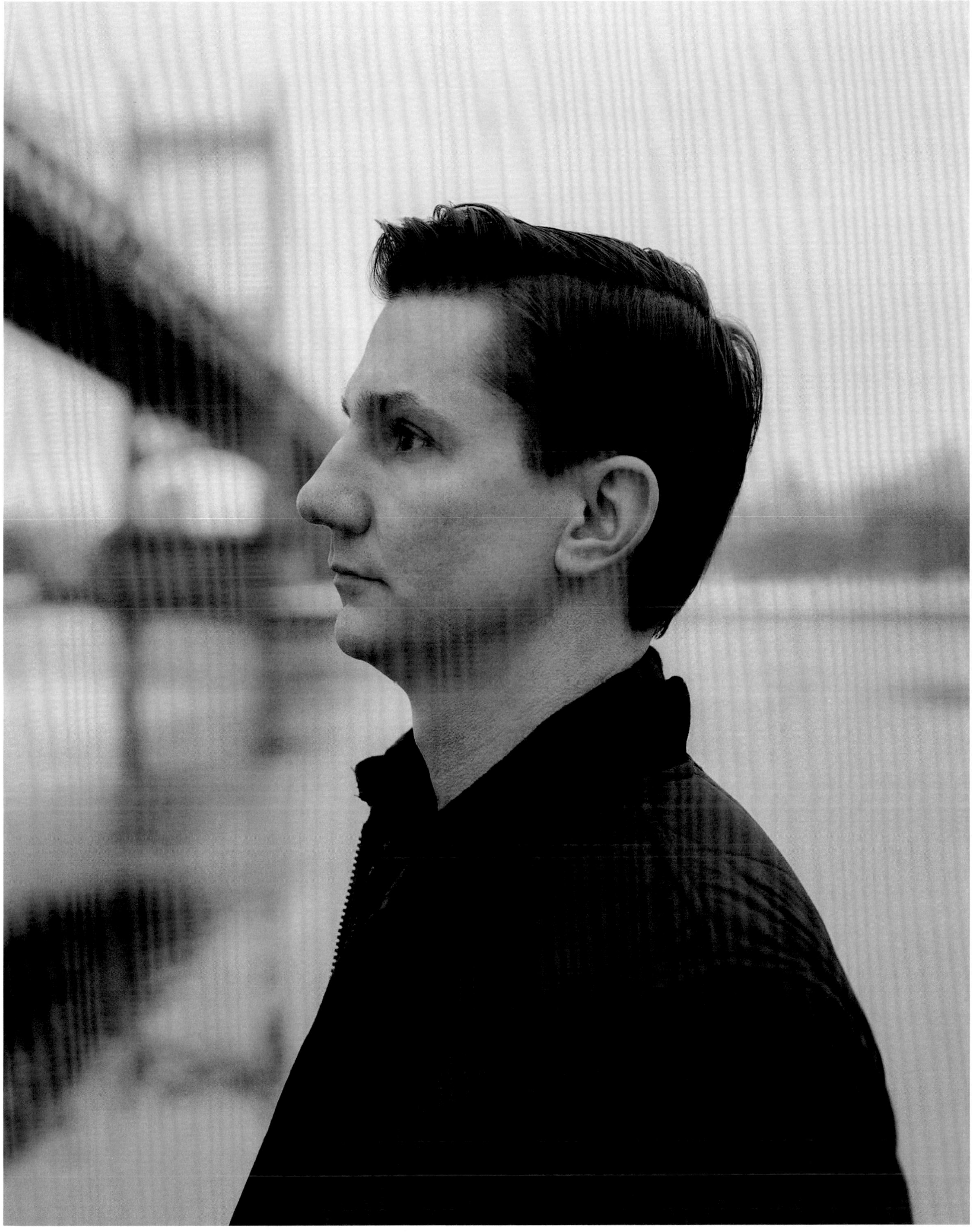

Steven Bonetti, Sergeant
Army - 11C / Mortarman
Years of service - 6 years
Deployments - 3 [Operation Iraqi Freedom & Operation Enduring Freedom]

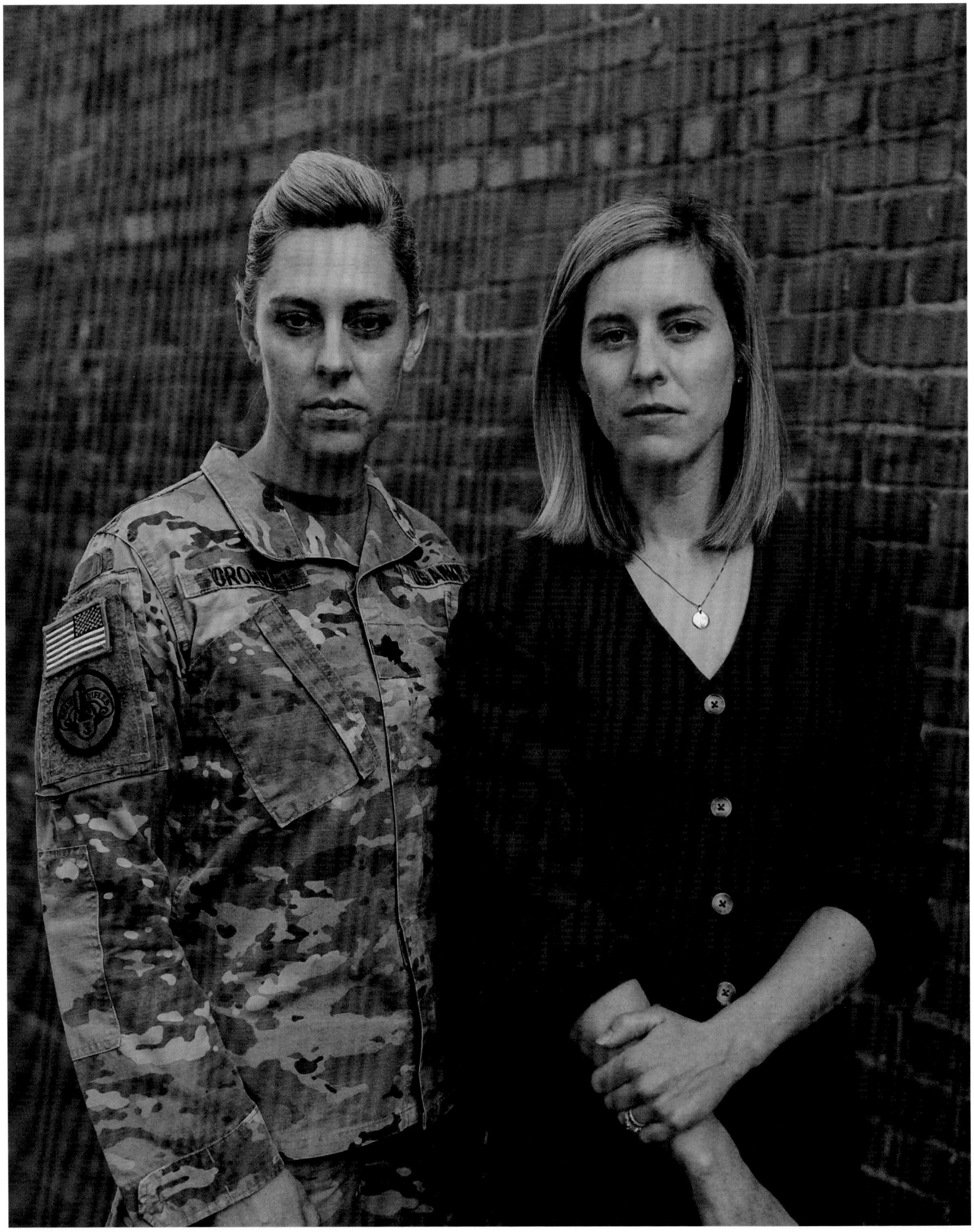

Katie Crombe, Lieutenant Colonel
Army - 59A / Strategic Plans and Policy
Years in service - Active
Deployments - 2 [Operation Iraqi Freedom]

Christine Schwartz, Captain
Army - 92A / Quartermaster Officer
Years of service - 5 years
Deployments - 1 [Operation Enduring Freedom]

I'm Katie Crombe, and I'm currently a strategic planning officer in the United States Army. I joined to pay for college. After graduating in 2003, I went right to my unit and then to Iraq. I commanded a platoon of truck drivers. Almost eighteen years later, I'm still here and I'm currently a Lieutenant Colonel (O-5). In 2018 I became the aide-de-camp to General Joseph Votel, who was the commander of Central Command (CENTCOM) at the time. I traveled with him all over the theater of operations to support the troops, both training other nations' forces and fighting the Global War on Terror. I'm the oldest of three girls and one of my sisters, Christine, here to my right, also served.

Christine: I don't know if I always wanted to join but I was fascinated when my sister first deployed. I was in high school and I could just see how much her peers and her soldiers loved each other. I was really impressed by that.

Katie: I was so happy. I mean, I was scared when she deployed to Afghanistan and I was working a desk in D.C. But we communicated a lot when we were overseas; we had that common thread where we didn't have to really explain things.

Christine: I deployed in 2011 as a 92A quartermaster officer. I was overseas as executive officer of a supply company doing re-supply runs to line units that were forward. When Katie first deployed in 2004, there wasn't a lot of communication ability. Unlike Katie, I have since separated from service and now I lead Service to School, which is a non-profit to help veterans apply and get accepted to top-tier colleges, including the Ivy League.

Katie: It's a wonderful organization.

Christine: Readjustment to civilian life wasn't really difficult in terms of taking off my uniform. That wasn't hard. But I was, and am still, a military spouse. I was just drifting with my husband for a little bit and I had a really hard time trying to find a good job when I got out. I had a hard time selling myself and felt so undervalued. I found Service to School and went to work, kept moving up. I love the mission. It's really important that vets, especially first-gen college students, understand that education can open so many doors.

Katie: Vets are such a part of the American fabric. Most of us, even the wounded, aren't the broken characters we are sometimes portrayed as. I mean, I've been through some things, divorce and normal things, but all in all I'm really happy with my choices and the support given to us soldiers. Veterans graduating from some of these schools can help change that conversation and that broken character stereotype.

Christine: Absolutely. There are so many super smart, super capable veterans. Some of the absolute smartest on the planet. I look at Kate, and she has always been a leader. It isn't something that's taught or learned, but she's always had it and it's great.

Katie: Christine is great at being true to herself. She lives the life that I tell my soldiers to: Seize the day, dare greatly! If you're willing to take risks, you can make a difference in others.

Christine: I say, "Try to put yourself in someone else's shoes and understand them and their opinions." Have a conversation before you start to judge them. The military is so important for that. It's one of those places where the most important thing is service; the rest goes out the window.

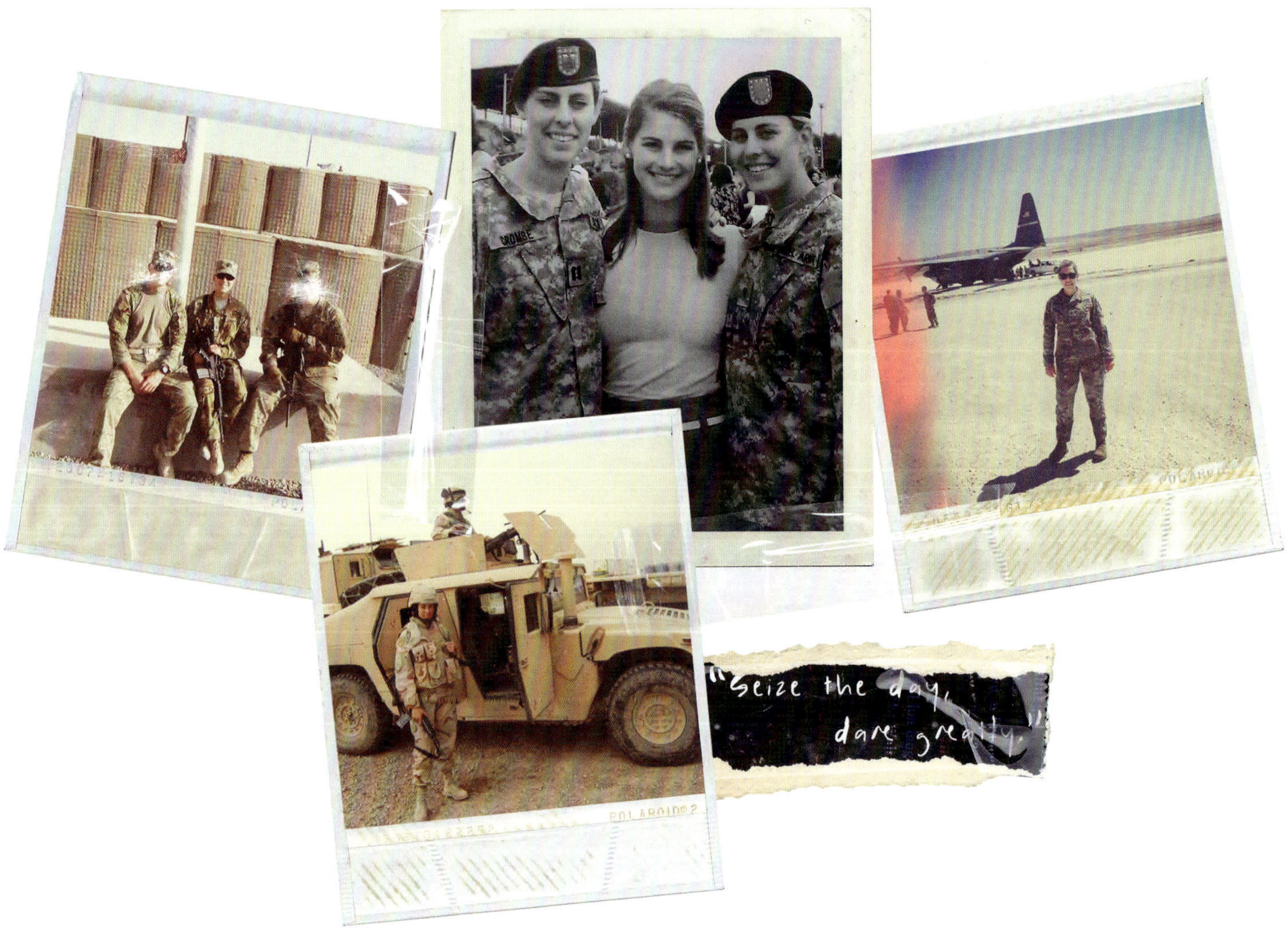

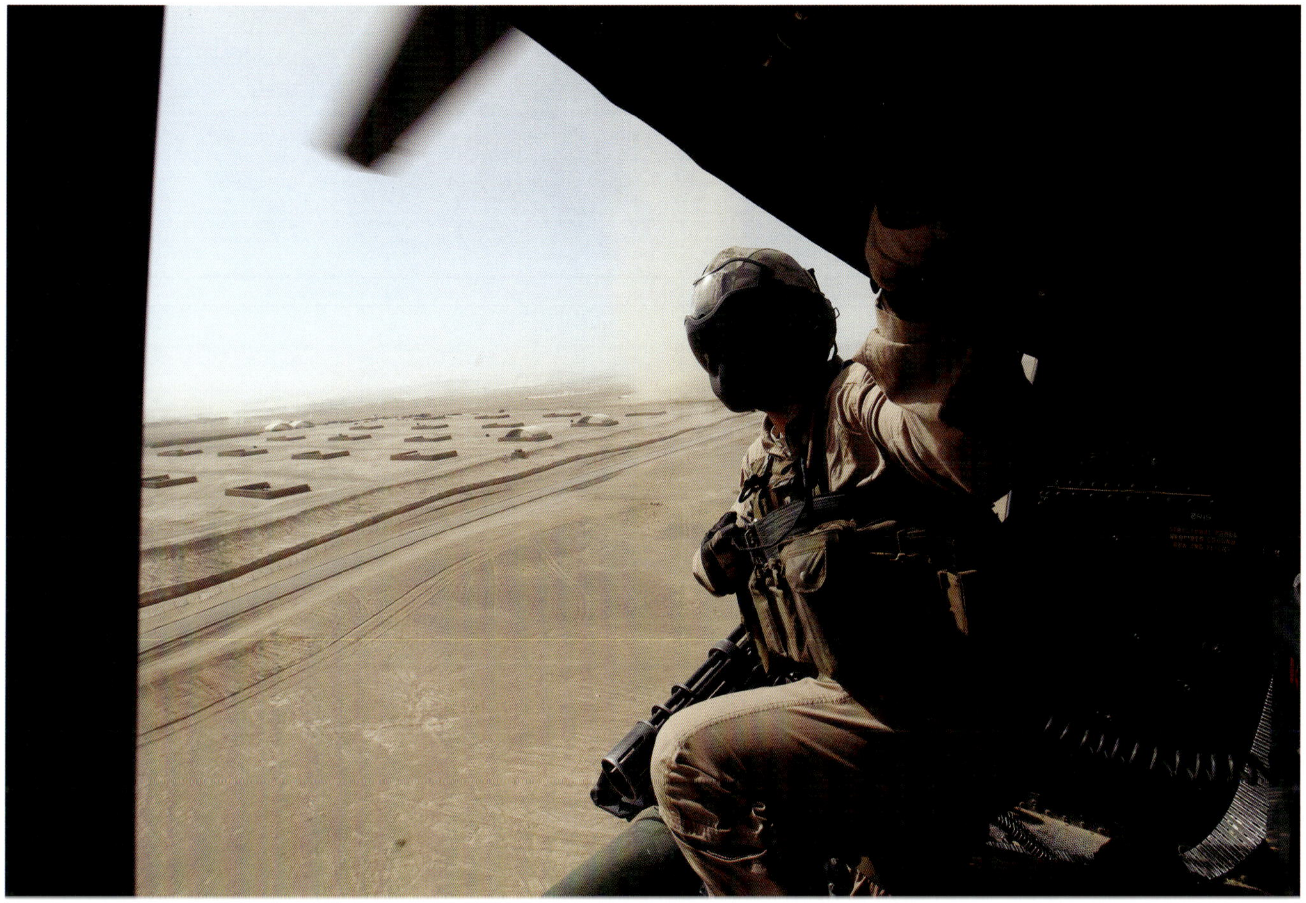

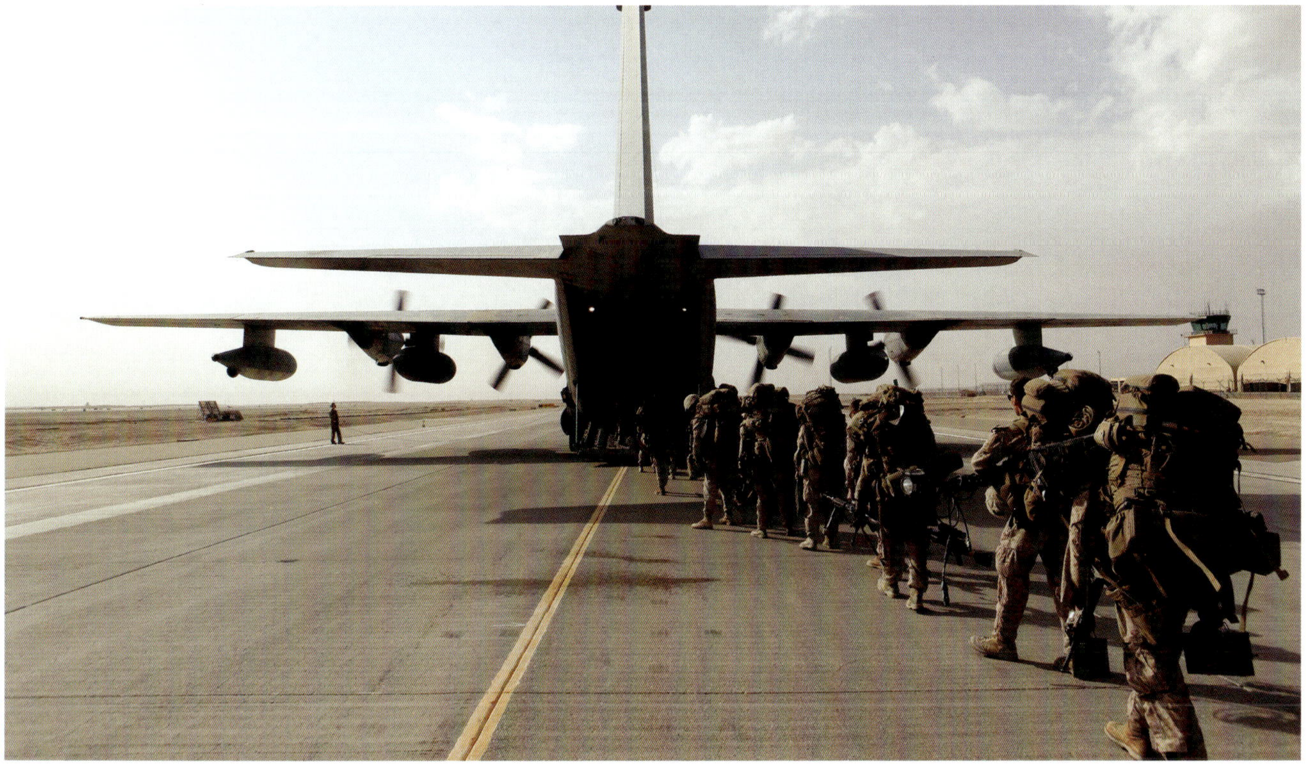

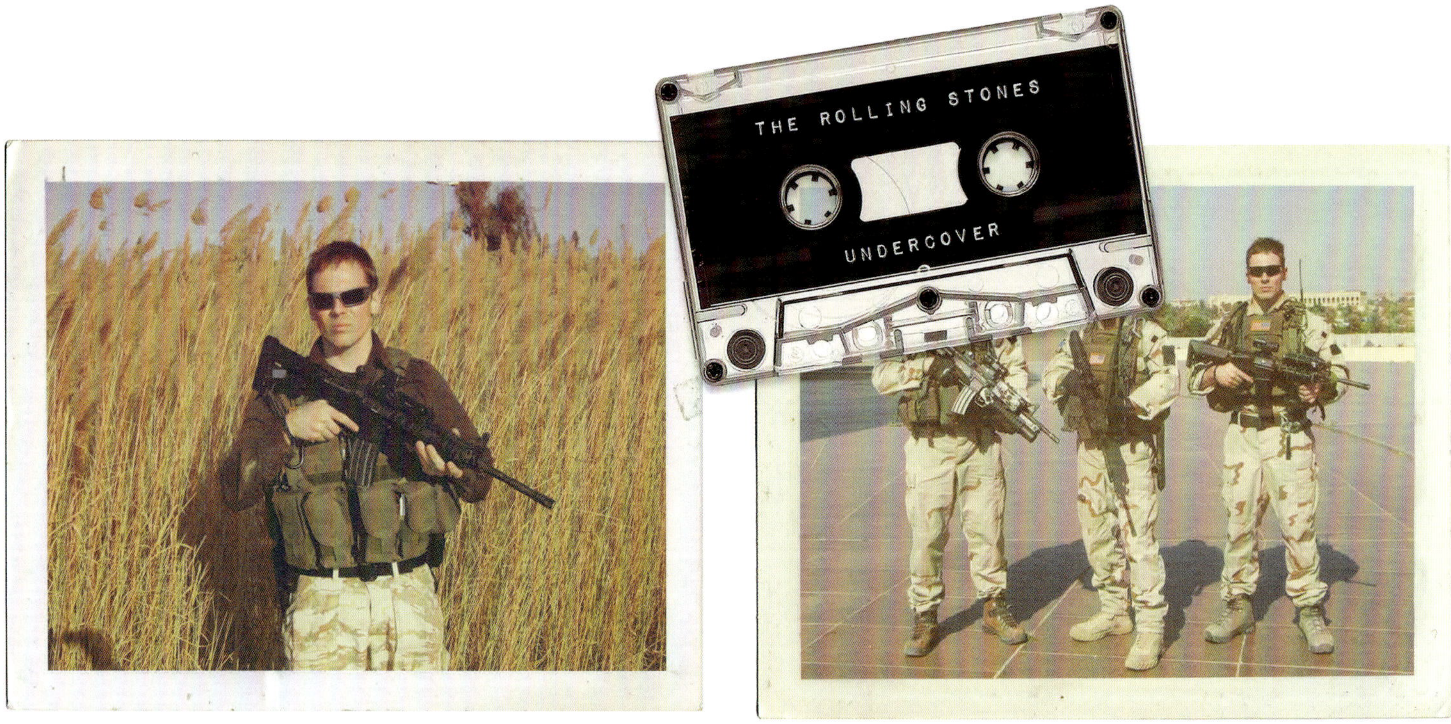

My name is Joel Carpenter, and I was an infantryman in the United States Army's First Ranger Battalion from 2002 to 2006, when I separated as an E-5 Sergeant. I joined to serve our country post-9/11. I had three deployments in support of Operation Iraqi Freedom and one in support of Operation Enduring Freedom. After I separated from active duty, I contracted with Triple Canopy from 2008 to 2013 in Iraq.

I changed a lot, post-military. I was very bought into Ranger Regiment culture and there's a learning curve when you leave that. The private sector doesn't operate as a collective like the military; you need to operate as an individual. I learned you need to learn to swim and stay afloat before you can help others. This led me to write Set Up for Success with Eric Mullen, who is a former SEAL. We saw there were a lot of success stories being shown about veterans transitioning, but not the process. A lot of guys were hiding failures and the rough parts. So, we set out to show some of our failures and embarrassing stories of our reintegration. Working in security risk management is my bread and butter, but I'm also working on a feature film development deal. I also have my real estate license and I'm getting my mortgage loan originator license, as well.

Veterans are part of an elite class of individuals and as we transition, we need to be able to leverage those experiences. Find certifications in the industries you're interested in. Figure out what drives you and how you can be an independent contractor. The military really gave us a hunter mentality to go out and hunt. It's also important to serve others. Finding out how to connect with your fellow man is critical. This led me to co-found the Aaron Grider Foundation and serve on the advisory committee for the National Veterans Memorial.

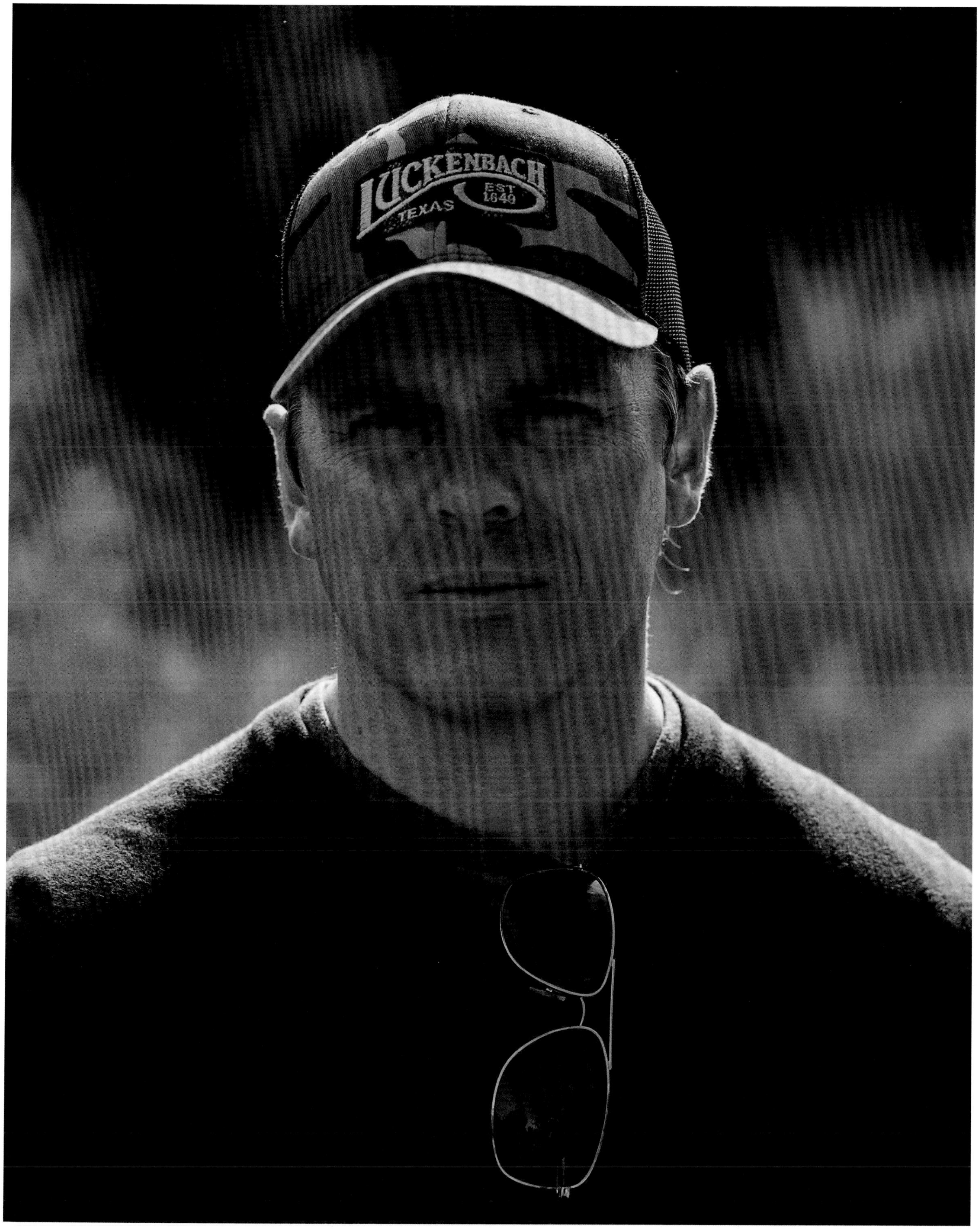

Joel Carpenter, Sergeant
Army - 11B / Infantryman / 1/75 Ranger
Years of service - 4 years
Deployments - 4 [Operation Iraqi Freedom & Operation Enduring Freedom]

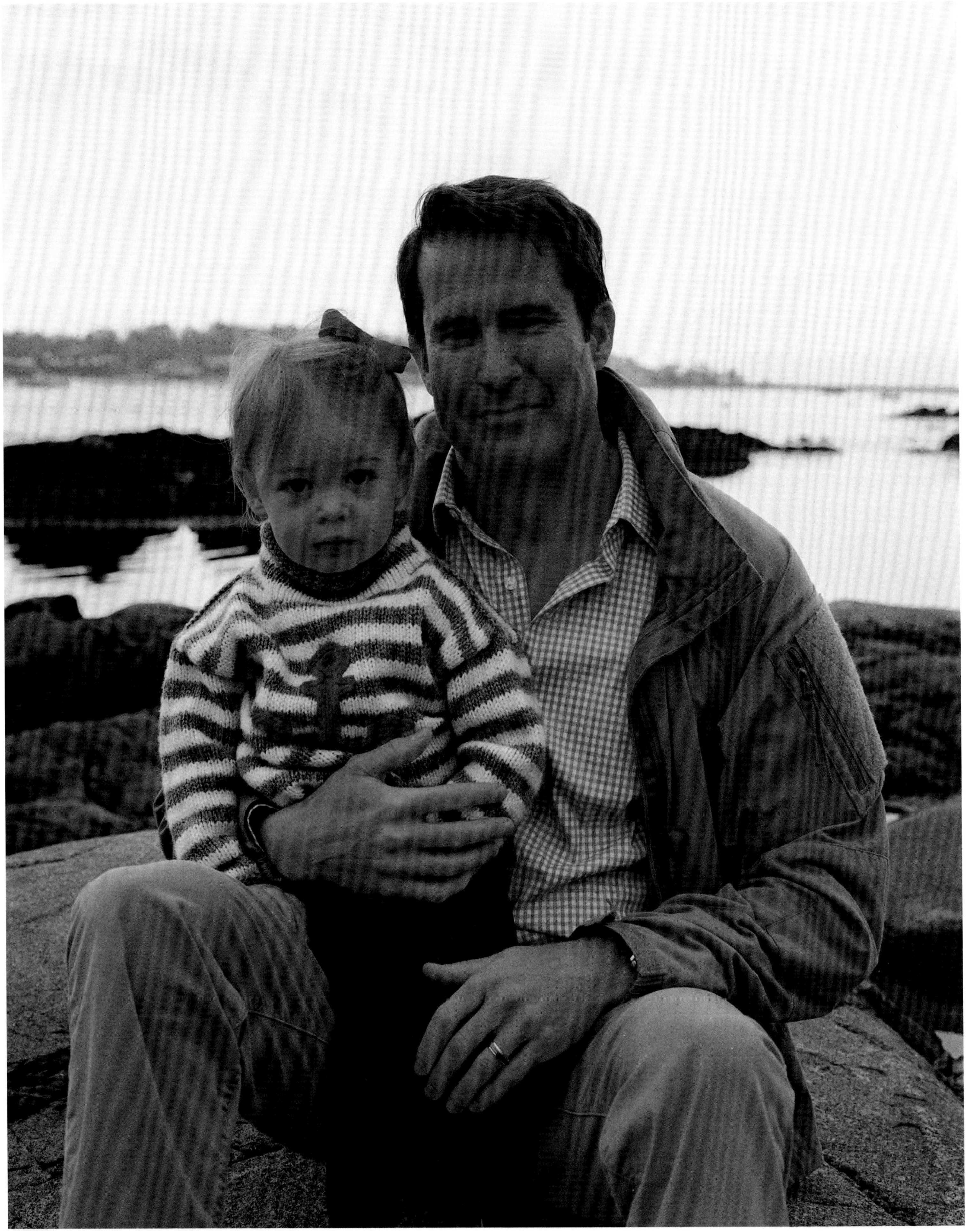

Seth Moulton, Captain
Marine Corps - 0302 / Infantry Officer
Years of service - 5 years
Deployments - 4 [Operation Iraqi Freedom]

My name is Seth Moulton, and I am a United States Marine Corps 0301 Infantry Officer veteran. I joined because I was inspired by a mentor who was my college minister. He deeply believed in public service and said we should find a way to give back. I looked at other options, but at the end of the day, I had so much respect for the 17- and 18-year-olds serving on the front lines of our nation's military, that I decided to join just before my graduation in 2001. Then 9/11 happened. I started training in January 2002 and was on active duty until the end of 2005. I spent a year in the reserves and volunteered to go back to Iraq in 2007 before I finally got out.

My transition was tough. I was lucky and had every opportunity possible—I got into Harvard Business School. But I didn't know how to sign up for the VA; I was confused about educational benefits. And sometimes it is hard to find a meaningful purpose after doing some of the most important work of your life. After business school, I had no intention of going into politics. I was a physics undergrad and honestly kinda disliked politics. I got out of B-school and was running a high-speed rail project in Texas. While I was there, I got a call from an organization that was trying to get vets to run for office. At first, I was like: "No way." But I realized that I had seen the consequences of failed leadership in Washington, and if people with my background don't step up, it will keep happening. So, I stepped up and ran, and won.

Some of the work I'm most proud of is on mental health. As one of my best Marine buddies pointed out, "After what we saw, it would be a disorder if you weren't affected by it," so I made a point of not using the term "Post-Traumatic Stress Disorder" but simply calling it "post-traumatic stress" (PTS). Myself and a lot of guys I served with have had to work through it, or still are. I was inspired to talk about my own struggle by the honesty and courage of the junior Marines in my platoon. A lot of people see veterans as victims, but that's not who we are. We have a lot to offer the world once we get home.

When I ran for president in 2018, I decided to share my story about PTS publicly, because I believe you should lead by example. Doing so made me the first candidate in history to share a personal story about mental health. I also released a three-part mental health plan:

1. Troops and vets get a mental health checkup every year.

2. Make annual mental health checkups a part of high school.

3. Create a mental health hotline so anyone can get help.

We've already made great progress on the troops, and the hotline will be up and running next year. This part of the plan was one of the few bipartisan things to get through Congress last year, and it's going to save a lot of lives.

When you're damned if you do and damned if you dont, always do. I had that in my mind when I joined, and it's been a guiding principle of my life.

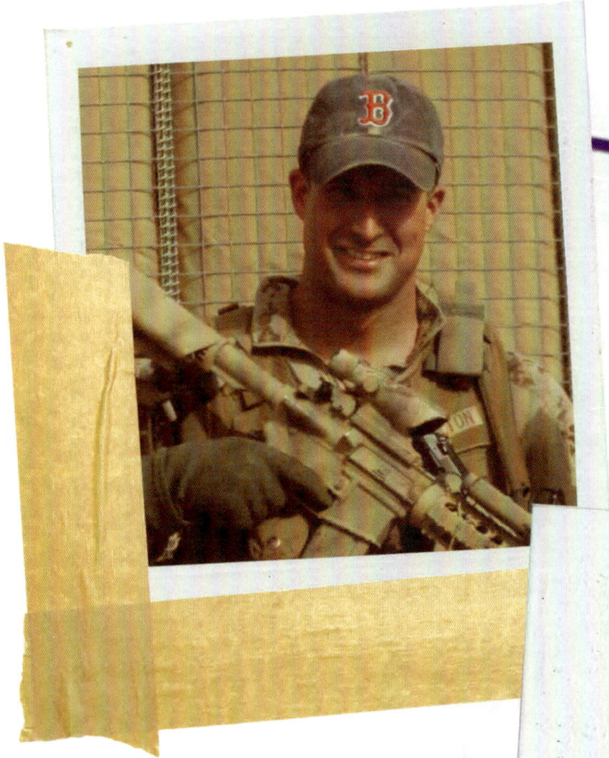

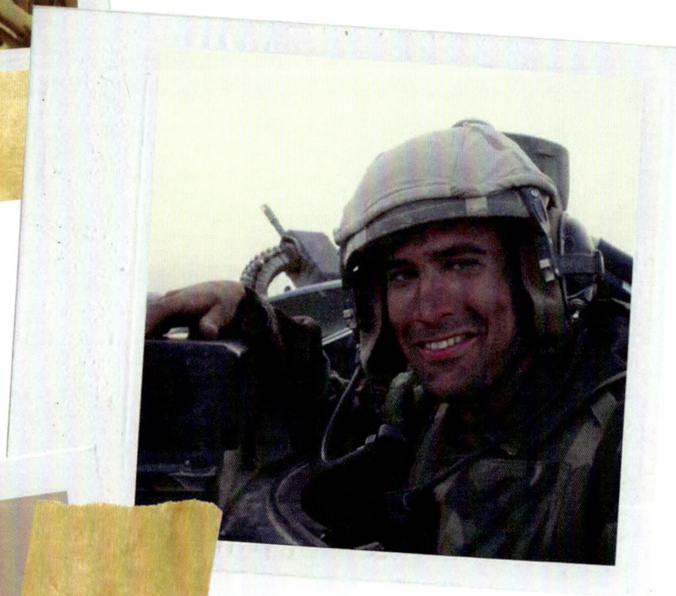

"When you're damned if you do and damned if you don't, always do."

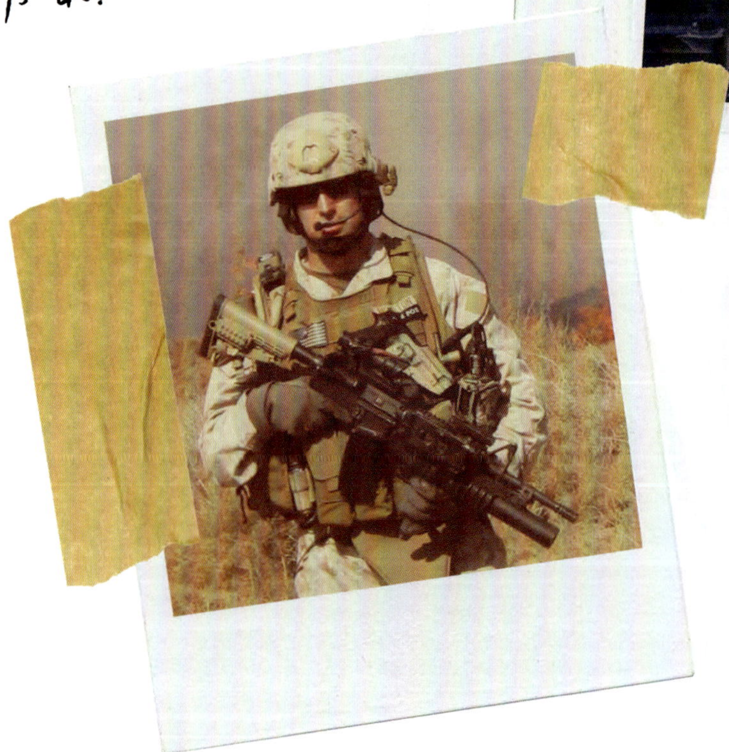

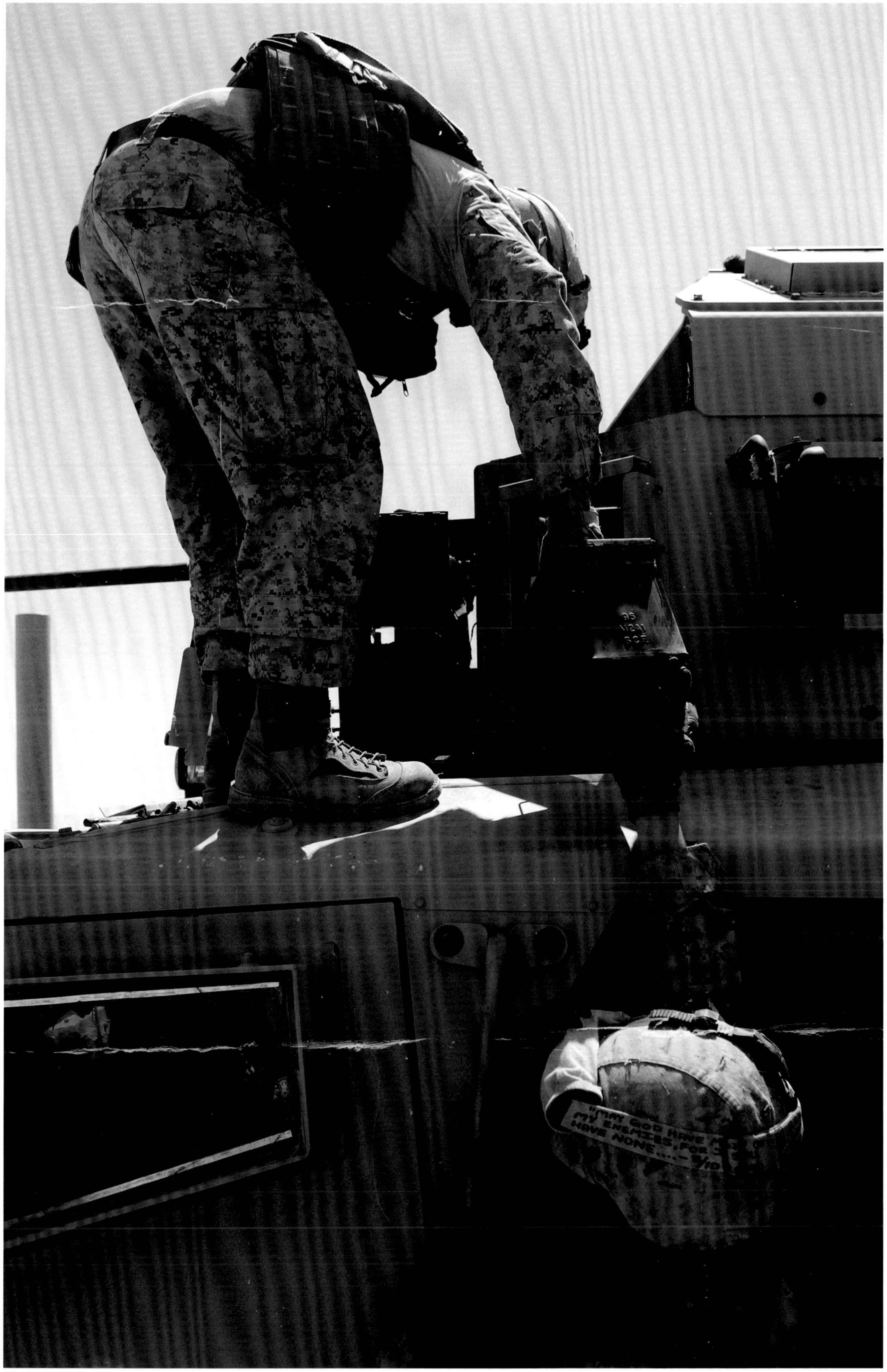
"MAY GOD HAVE
MY ENEMIES, FOR
HAVE NONE....

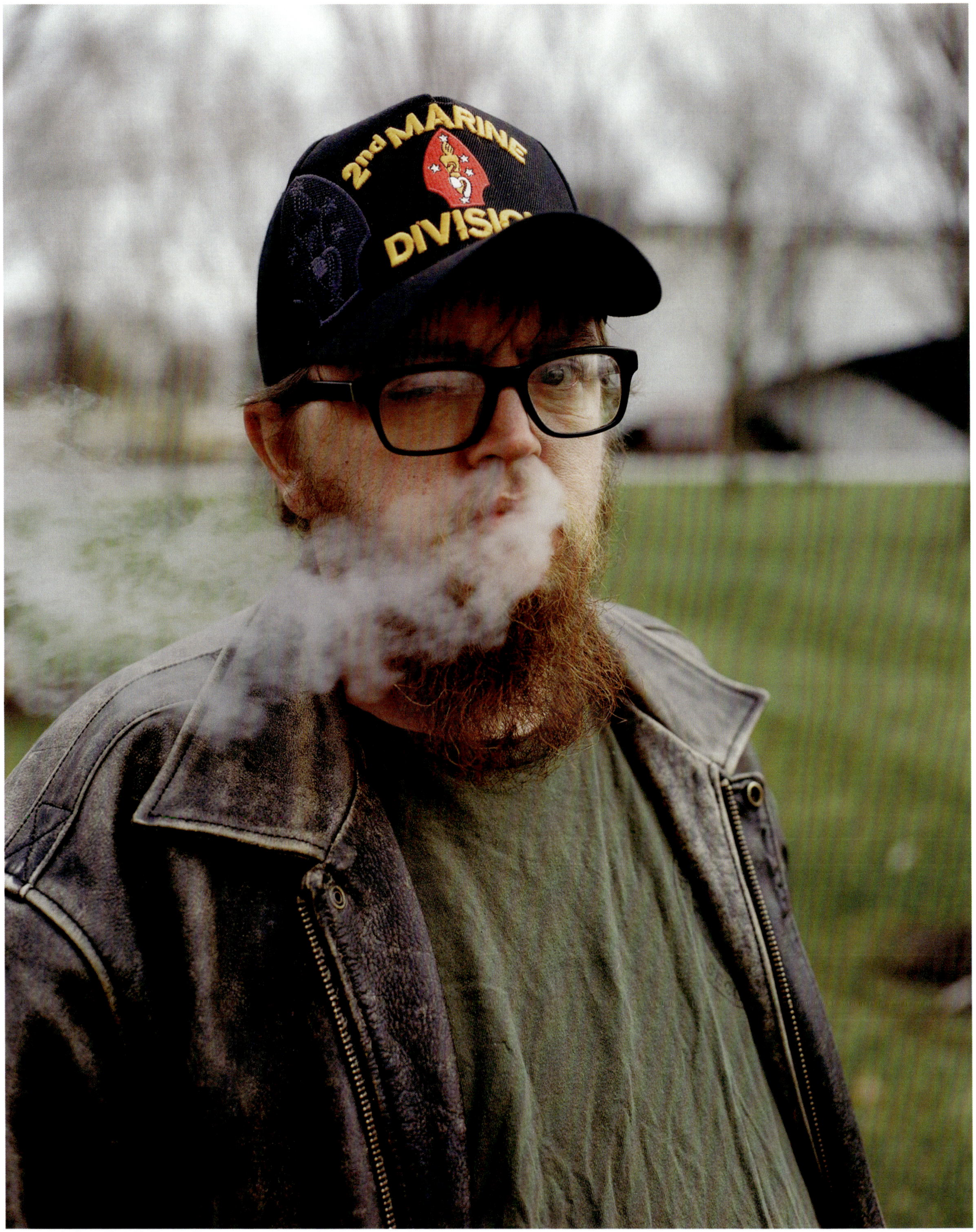

David Prince, Corporal
Marine Corps - 0311 / Infantry Rifleman
Years of service - 4 years
Deployments - 3 [Operation Iraqi Freedom]

My name is David Prince and I served in the United States Marine Corps as an 0311 Infantry Rifleman. I deployed three times: first to the Horn of Africa in support of Operation Enduring Freedom and then twice in support of Operation Iraqi Freedom. I got out as a Corporal (E-4). My first tour in Iraq was nuts. Constant gunfights and mortar attacks. We were right on the Syrian border and it made for a lot of action. Some of the attacks were also really sophisticated. There was one where they created three vehicle-borne improvised explosives (VBIDs). They put them in a fire truck, concrete truck, and SUV to try to force their way onto our base. The attack failed but it was a wild forty-five minutes after those blew up.

I have an 80 percent disabled rating due to PTSD and TBI (Traumatic Brain Injury) from my time as a Marine. I was also involved in a pretty serious motorcycle accident and am currently not working due to those issues. I volunteer with AMVETS and have since 2004. It's right in the area I grew up in and is part of the local community. They also helped me provide Christmas for my kids when I got hurt in my accident.

Being part of the early stages of the war and as an infantryman, I wish the VA had been more prepared to help guys like me transition. There are not a lot of tangible skills to take from the infantry to the civilian world. But no matter what, my mantra in life is still to adapt and overcome. You have small-ass issues or big-ass issues, and you need to figure out how to deal with them. You need to solve them to move on. You need to be able to make it work.

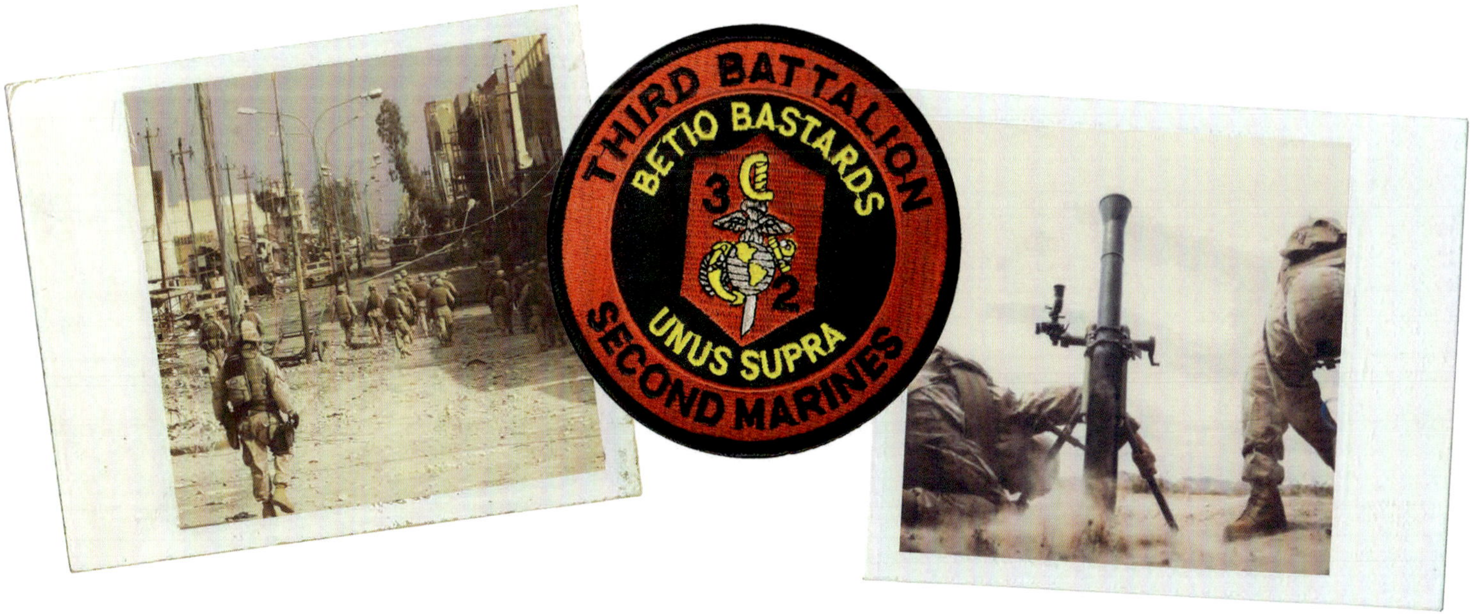

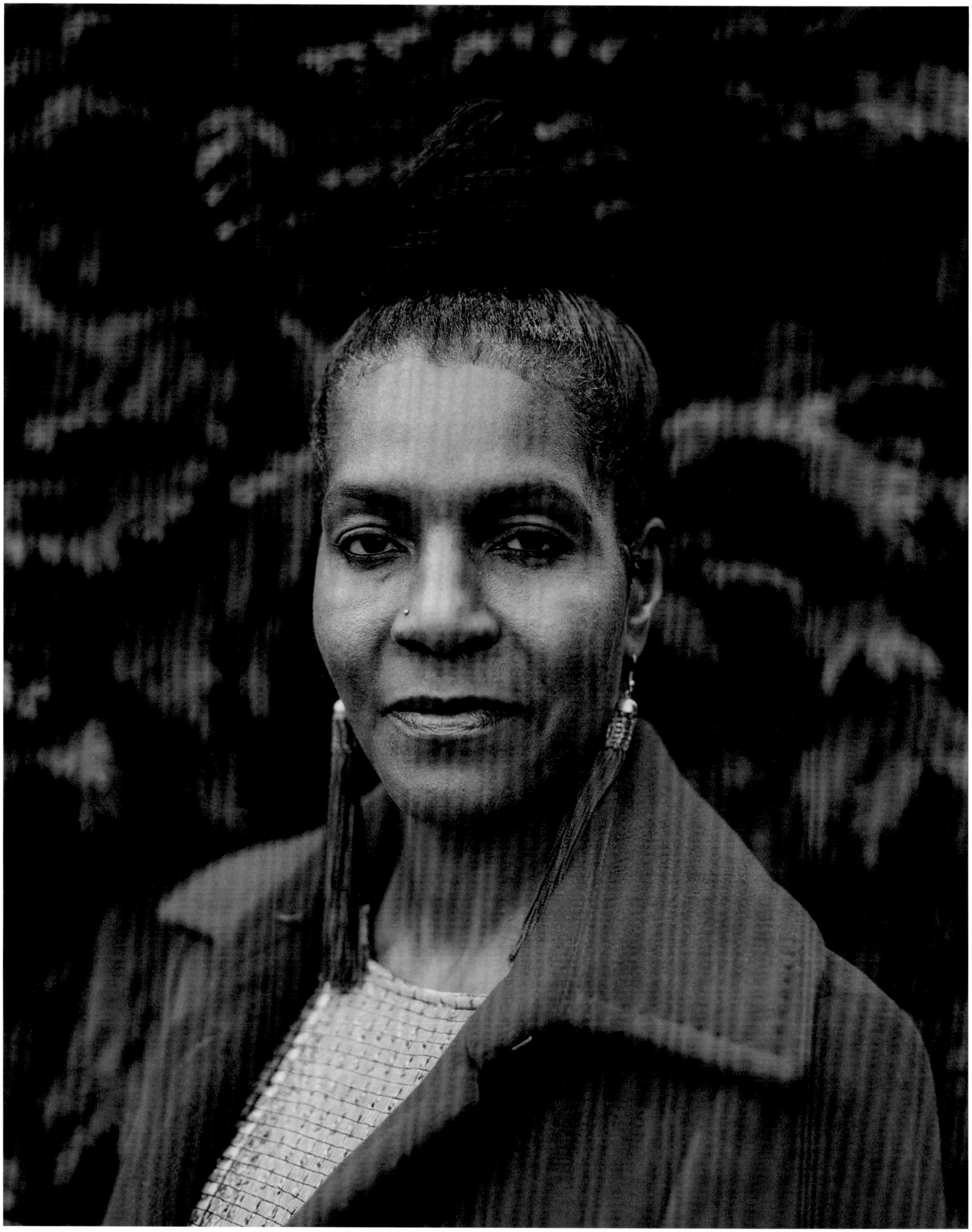

Tonya Griffin-Oxendine, Command Sergeant Major
Army - 75E / 42A / Paratrooper
Years of service - 30 years
Deployments - 2 [Operation Iraqi Freedom & Operation Enduring Freedom]

My name is Tonya Griffin Oxendine. I retired as a Command Sergeant Major (E-9) in the United States Army. I served for nearly thirty years in a variety of jobs and assignments: as a paratrooper in the 82nd Airborne Division; a drill instructor at Fort Jackson; Battalion Command Sergeant Major at Fort Knox; XVIII Airborne Corps Headquarters Battalion; 82nd Headquarters Battalion; and I deployed to both Iraq and Afghanistan and culminated my career at the Pentagon. I grew up in St. Augustine, Florida. My mother was a single parent of my brother and me, doing the best she could with what she had. We were poor, on welfare and food stamps. Often our electricity was cut off and we'd run a long, thick orange extension cord from our house to our cousin's house until my mom could scrape together enough money to pay the electric bill. There were always sounds and evidence of physical and verbal abuse, mostly toward women and children, including me and my mother. I felt there was no escape. My life would forever consist of welfare, food stamps, orange extension cords, yelling, fighting, black eyes, and bruises and pains in places no one should endure ... especially little girls. I knew deep in my heart I had to leave St. Augustine, but I couldn't afford college. In the summer of 1984, I walked by an Army recruiting station and enlisted.

As a young soldier I never talked about my upbringing because I was ashamed and embarrassed. Now, I've accepted my experiences and continue to share them with the hopes of inspiring others to know they can overcome their circumstances. I was honored to be given the responsibility of serving my country. While deployed in support of Operation Iraqi Freedom, I served as the Senior Enlisted Advisor to the commander. We made progress working behind the scenes in the local villages and schools, working with women and children. My last deployment to Afghanistan took a toll on me mentally. I was assigned to Forward Operating Base (FOB) Kandahar as the Battalion Command Sergeant Major with the 82nd Airborne Division Headquarters Battalion. The deployment was just tragedy after tragedy. I saw a lot of death and carnage. We would get the call that a hero was injured and arriving at the Kandahar hospital. One of our many priorities at that time was casualty assistance, so it was our duty and obligation to visit our heroes before and after being admitted. We would walk into the hospital with the commanding general, and a soldier who was barely conscious or severely injured would still make the effort of showing respect, rendering a salute to honor the general. But we were there to honor them. The medical staff was amazing! They saved so many lives. We also honored our fallen heroes at ramp ceremonies as they were transported to their home base.

It's lonely at the top, being a member of the senior leadership team. I suppressed a lot of emotions and struggles the more I achieved because I had to act a certain way, I had to be tough and pretend like nothing bothered me.

Continued on next page →

You tell yourself you are trying to be a good example, but I learned that although I did set the example for others, I forgot about me. There is nothing wrong with reaching out and asking for help. I carried a lot of weight, not only from my childhood but experiences throughout my service as well. I won't sugarcoat it, there are a lot of challenges being a brown woman in a male-dominated world. From the moment I got to basic training, I worked hard. I have always been a go-getter. I always tried to take the tough jobs, to blaze a trail, not only for brown girls like me but for women in general.

I was raped when I was in basic training, sexually harassed, and assaulted during my career. When other female soldiers were involved with sexual misconduct, I would see them get transferred, or even discharged from the Army. But I had a thriving career and didn't want to risk it. I wanted to keep my job. Now, though, to anybody who is being mistreated, I would say: speak up! There have been a lot of changes because of people speaking up. For it to continue to change, people need to continue to use their voice. We need to get to the place where we can have honest and hard, uncomfortable conversations in the military. Mental health is critical, but it's only one facet. Diversity and inclusion are so important when dealing with issues like this. Not only do some women endure this kind of behavior, some men do as well. Even to this day people still judge and blame the victim, and that needs to change. It was a struggle at times to be a woman, a brown woman, and succeed in the Army. It has gotten better; we are making progress, but we need to get to the point where it's one team, one fight.

It took me a long time to speak up. I loved to serve! I loved my soldiers and my mission. I am a fifty-five-year-old vet, and as bad as some of the things in my life have been, I would not change any of it. It was terrible and painful, but it helped make room for me to help other people. It may sound cliché, but I heal when I tell my story. In 2012 when I redeployed, I was serving at the Pentagon and realized I needed help and started seeing a mental health provider. I was also provided other resources like The Wounded Warrior Project and the Emory Hospital program for veterans. These were excellent resources that helped me deal with the challenges I was facing. I retired from the Army in 2014 and worked for the Department of Homeland Security but ultimately, I decided I needed a break to gather myself, so I moved from Washington, D.C. to Atlanta.

I am now a full-time employee with a non-profit organization, Merging Vets and Players (MVP). Our mission is to empower combat veterans and former athletes by connecting them after the uniforms come off. We assist them with personal development, physical fitness, and peer-to-peer support. These two groups are a part of something that deeply connects them. Our goal at MVP is to bring them together, share experiences, and help them move on and find success in their next steps. It feels good to serve others again.

During my service I achieved the highest enlisted promotion for my field and even though I experienced some hard and unaccommodating situations as a child, as a soldier, and as a veteran, the Army helped me become stronger, more independent, more reliable, and ultimately more suited for life. The Army was and still is influential in shaping my life. I have had a great life because of the Army. It changed me for the better, permanently.

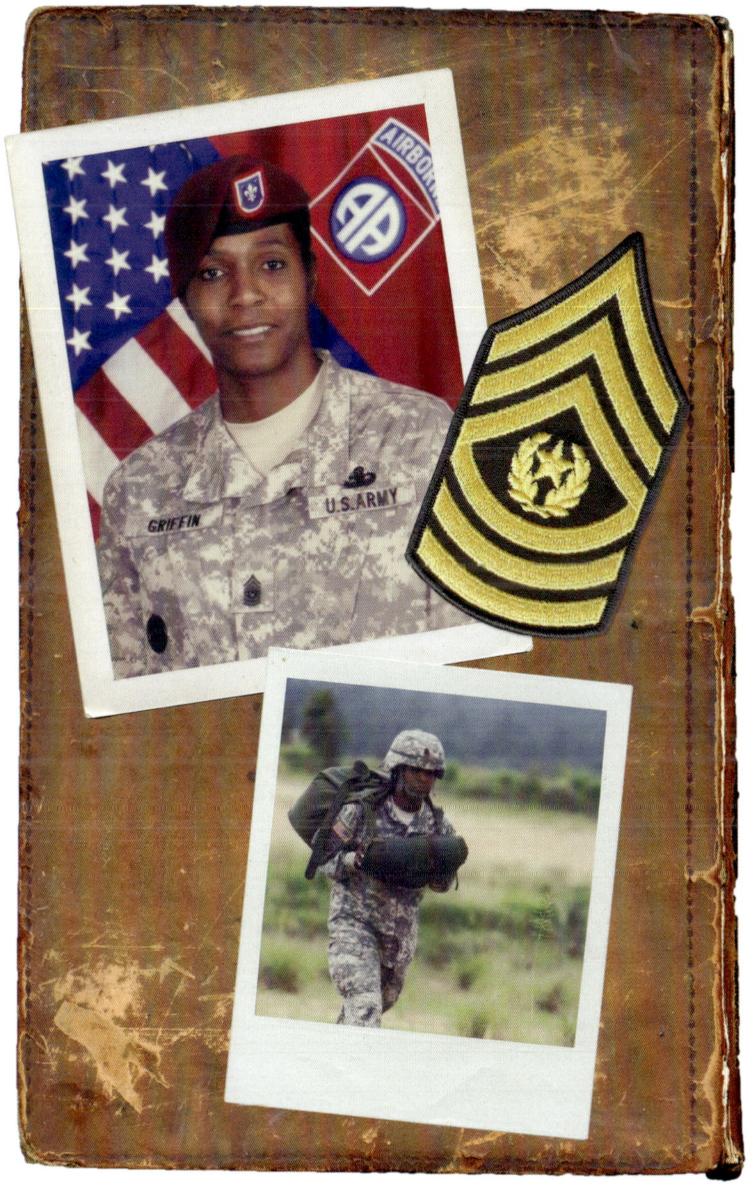

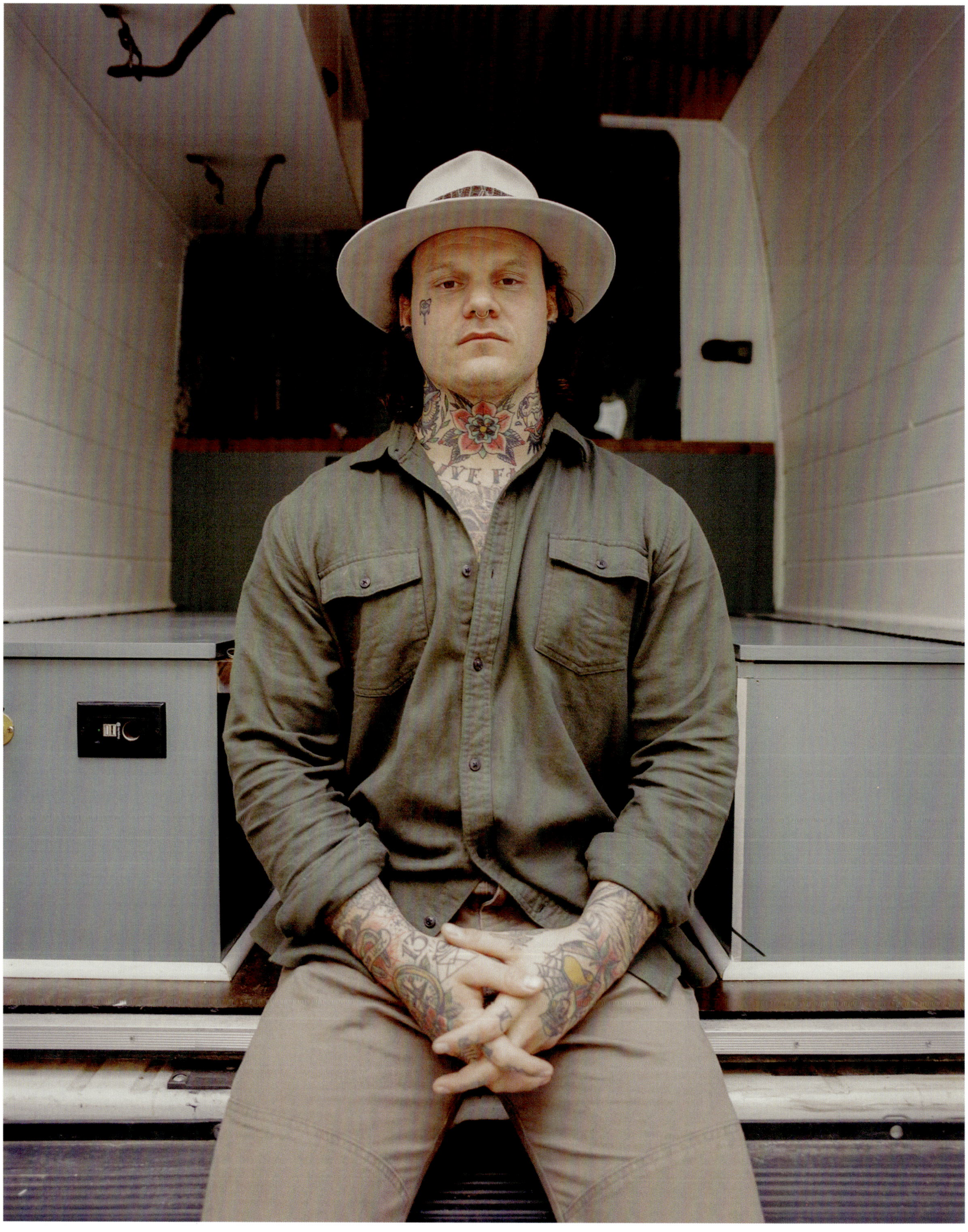

Cody Alford, Master Sergeant
Marine Corps - 0372 / Marine Raider
Years of service - 15 years
Deployments - 7 [Operation Iraqi Freedom & Operation Enduring Freedom]

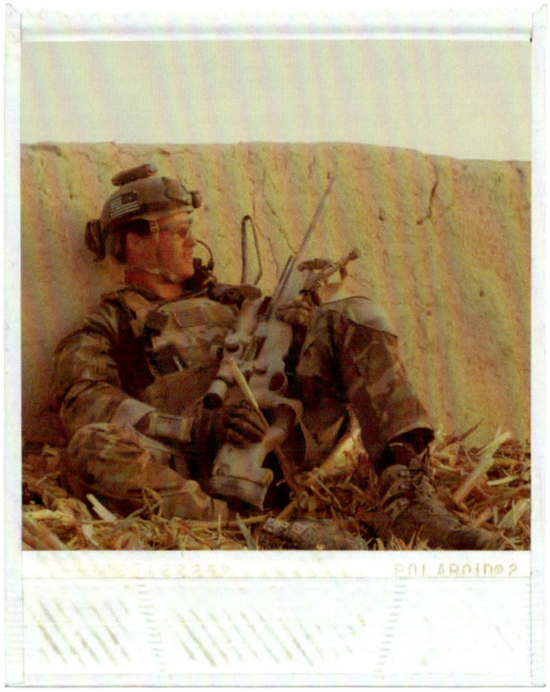 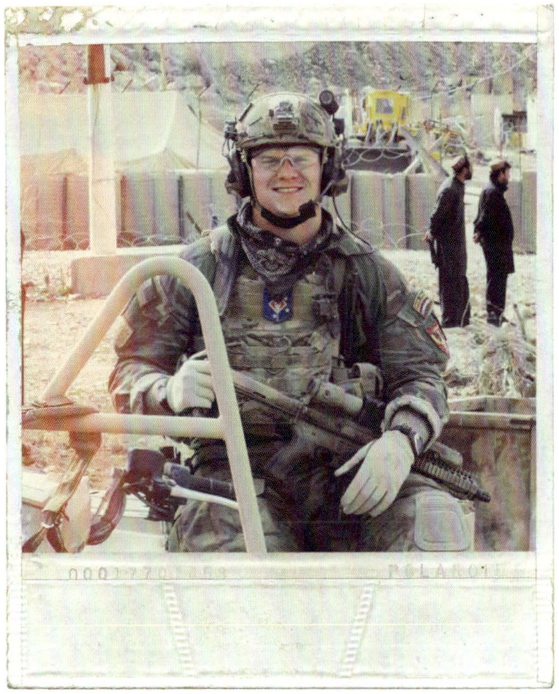

I'm Cody Alford, and I was an 0372 Marine Raider in the United States Marine Corps. I enlisted initially to become a scout sniper, accomplishing that and eventually worked my way to Special Operations. I retired as a Master Sergeant (E8), with a total of seven tours in support of the Global War on Terror. Transitioning out of the service was no easy task for me. With the help of the Warrior Care Coalition, I was able to connect to a non-profit organization called Operation Healing Forces. This non-profit allows SOF couples to disconnect from their day-to-day lives, and just be. This was huge for me and provided a perspective shift that I didn't even know I needed.

While I was in Tampa, FL doing brain therapy to restimulate my brain from numerous TBI's right before I got out, my wife asked me, "If you had six months left to live, what would you do?" I said I'd want to take photographs and travel around the U.S. "Let's do it," she said. We sold our shit, bought a van, and traveled for three years, taking time to heal and find myself. My wife is amazing. I'm so appreciative of her. Most vets don't even know who they are when they leave the military. They just don't know exactly how powerful they are. This is part of the inspiration for my clothing company, We Defy the Norm. We want them to embrace their power. Psychedelics also inspired my vision for the brand and my personal growth. It's been so healing that I get to put out how I see things in the world through the apparel. We are working to use the proceeds from the company to start a mental health retreat for veterans.

Ripples make waves. One small step can make all the difference. You might not think they mean anything at the time, but the tiny things matter.

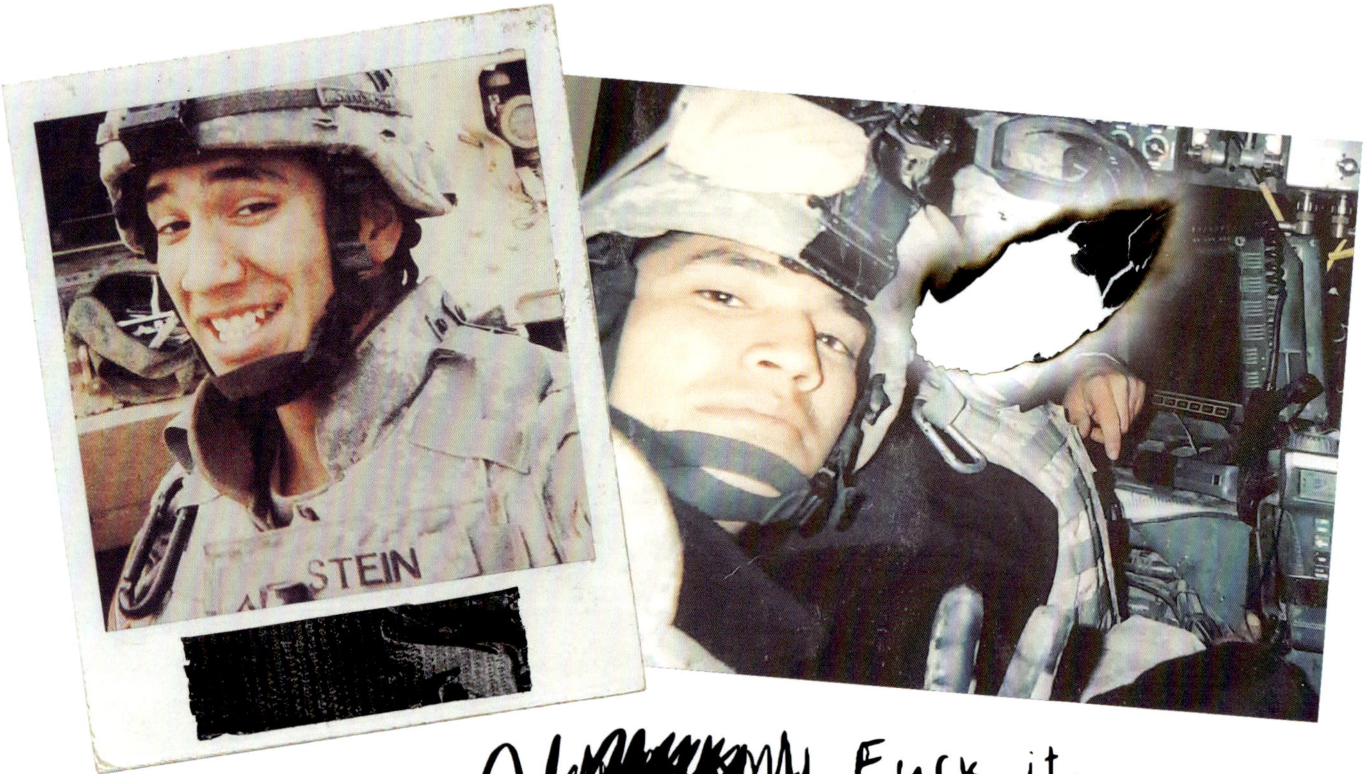

Fuck it.

My name is Josh Stein, and I was a combat engineer, 12B, in the United States Army. I joined because of 9/11 and was medically retired as a Corporal (E-4). I deployed in support of Operation Iraqi Freedom and was on the ground for four months before I got blown up. Our vehicle was hit by an EFP shape charge. It blew a basketball-sized hole straight through our Bradley Fighting Vehicle. If it hadn't blown literally straight through, the kinetic energy would've had nowhere to escape, and we'd have been in real trouble. I was in the hospital for four months, partly in a medically-induced coma. Lost both legs below the knees. My arms are also so jacked up that I'm technically a paraplegic. None of that has stopped me from being part of the Combat Veteran Motorcycle Association. I also played sled hockey with the San Antonio Rampage. We won a national championship a couple years ago. I have four daughters and coached tee ball, baseball, and football, and worked as a tattoo artist; anything I can do to keep my mind busy. Working on motors is my zen. Love working on my motorcycle and also love hunting. Here in Texas, there's a lot of hunting.

Readjustment was all ups and downs. Super high, super low all over the place, trying to figure out my body and injuries and learning everything: learning how to watch and having special prosthetics to learn how to walk again and having different legs to work and everything. You learn lessons from everything, and I don't regret hitting walls because I learned not to do things wrong again. Go after things in life, man! Take chances. If anything, I regret not going after more things while I had the chance. If I could change anything, I wish my guys had never gotten hurt. I wish I was just better. My life motto is: fuck it. You gonna be happy, you gonna be sad, fuck it. Pain, no pain, fuck it.

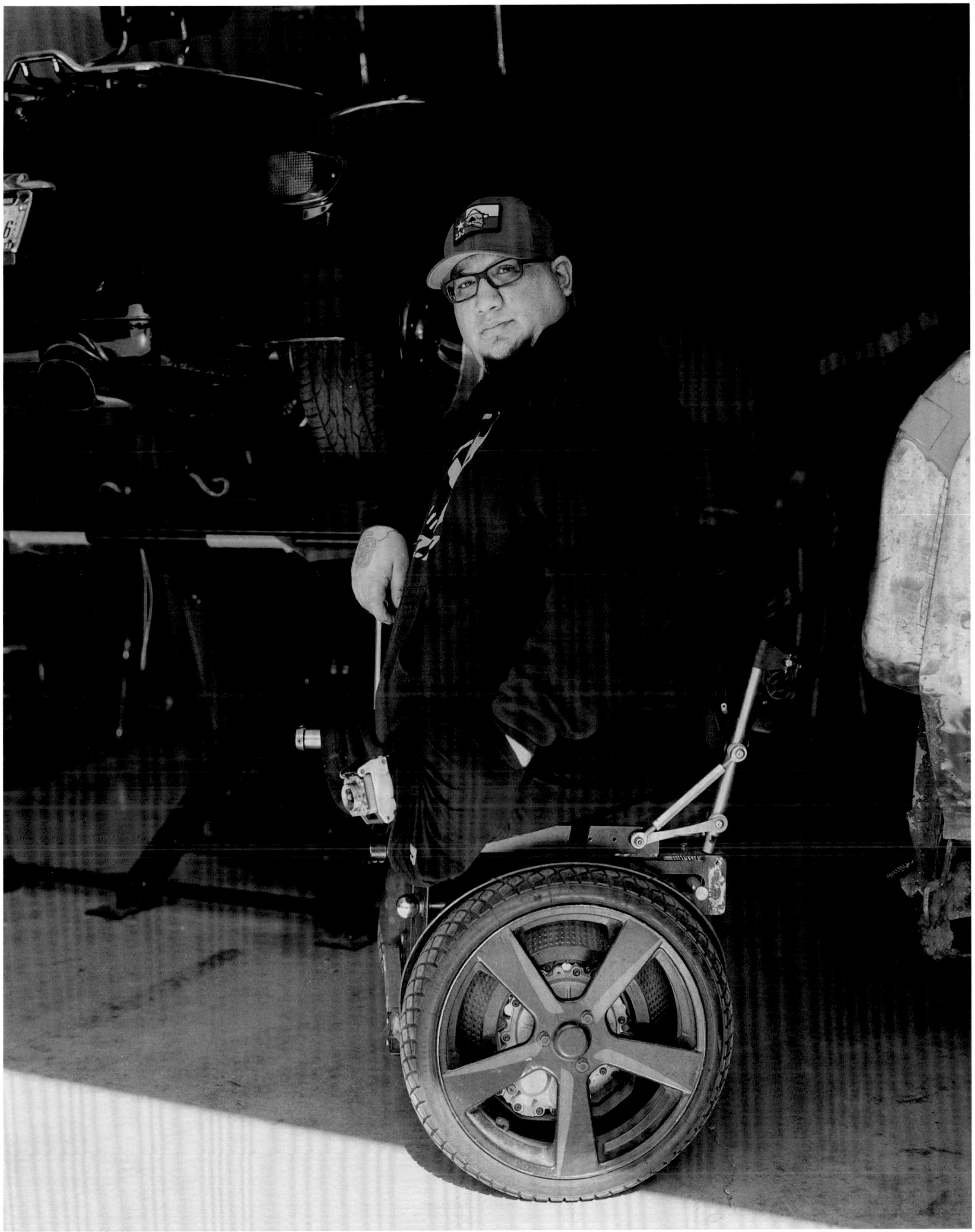

Josh Stein, Corporal
Army - 12B / Combat Engineer
Years of service - 2 years
Deployments - 1 [Operation Iraqi Freedom]

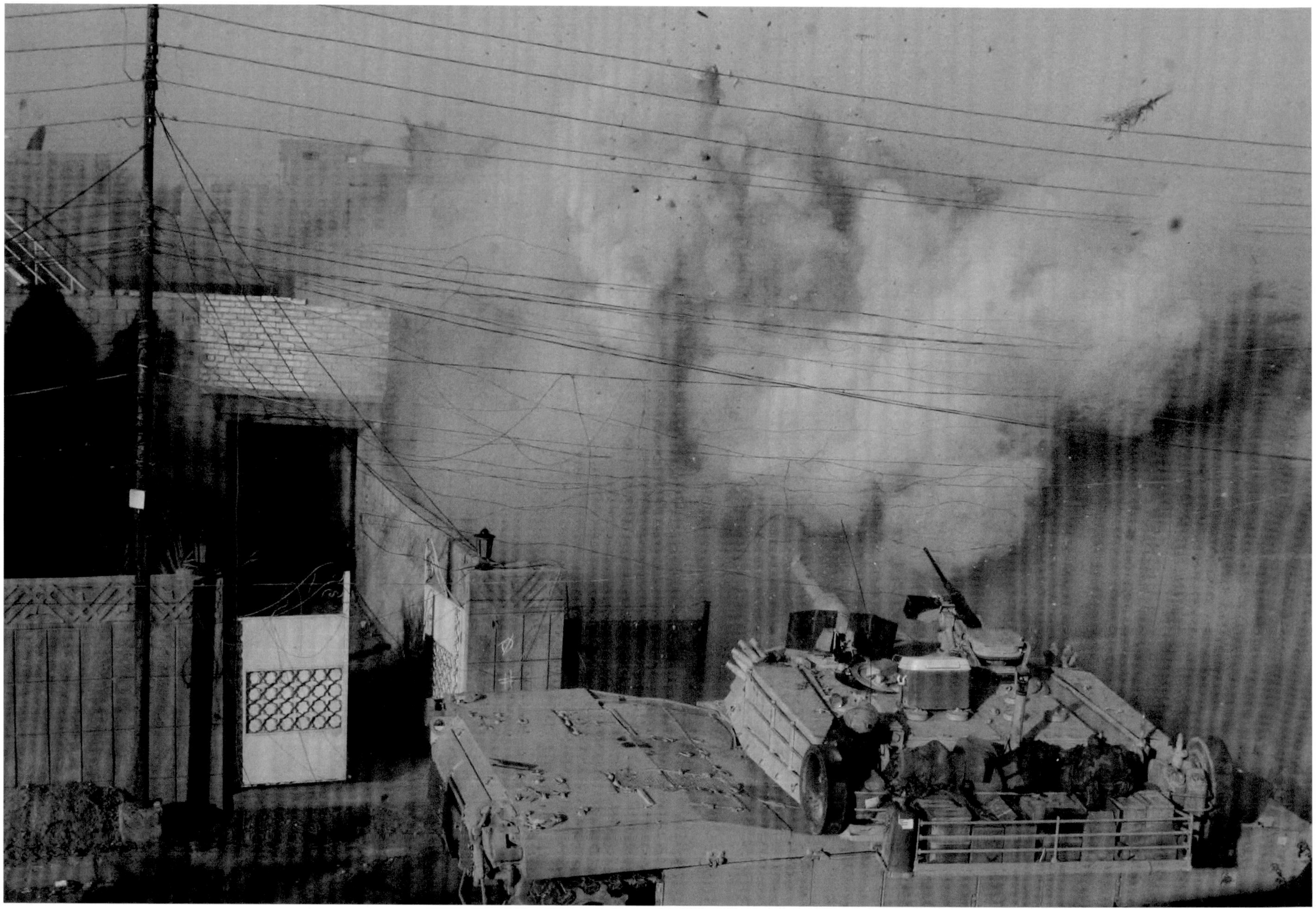

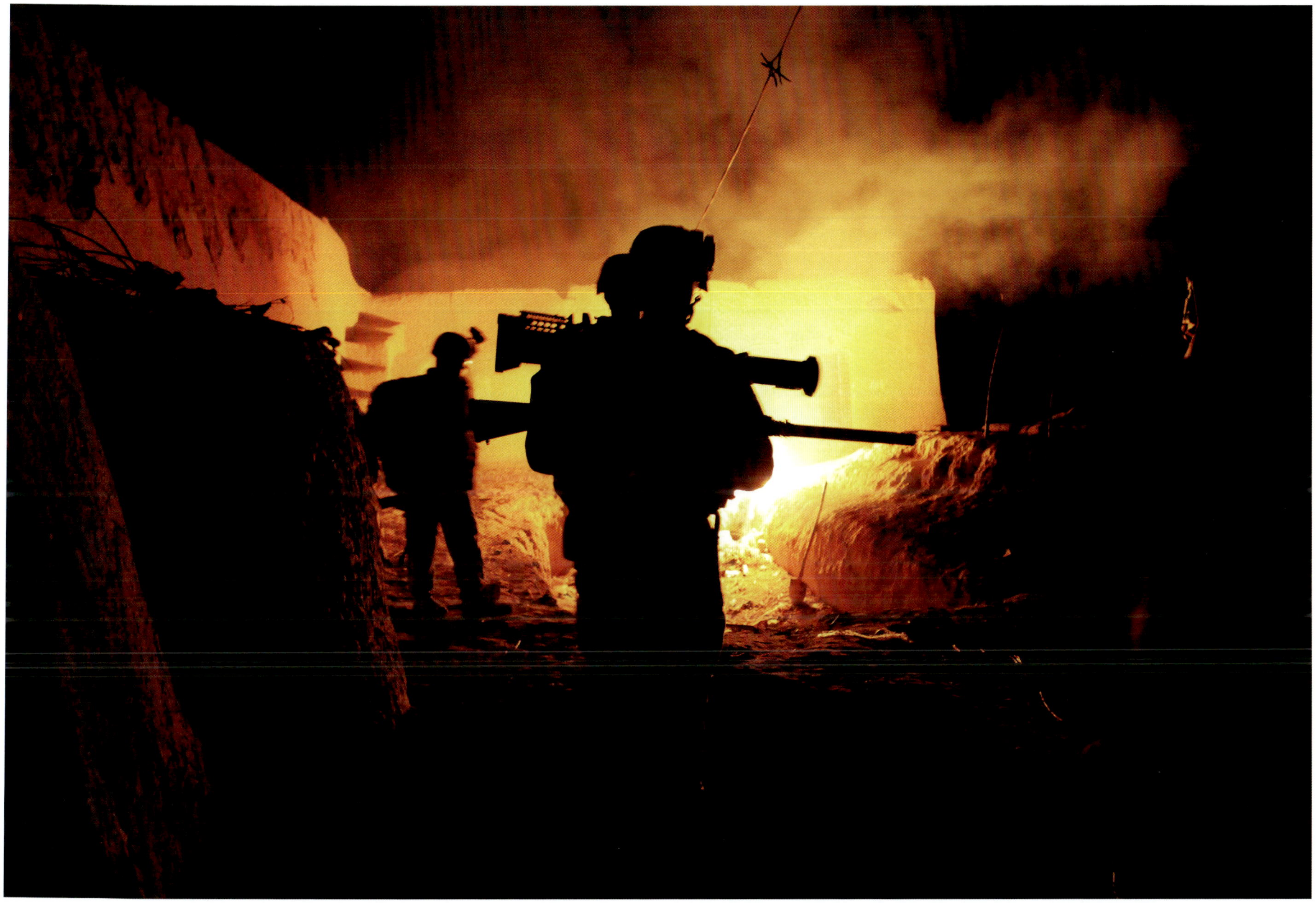

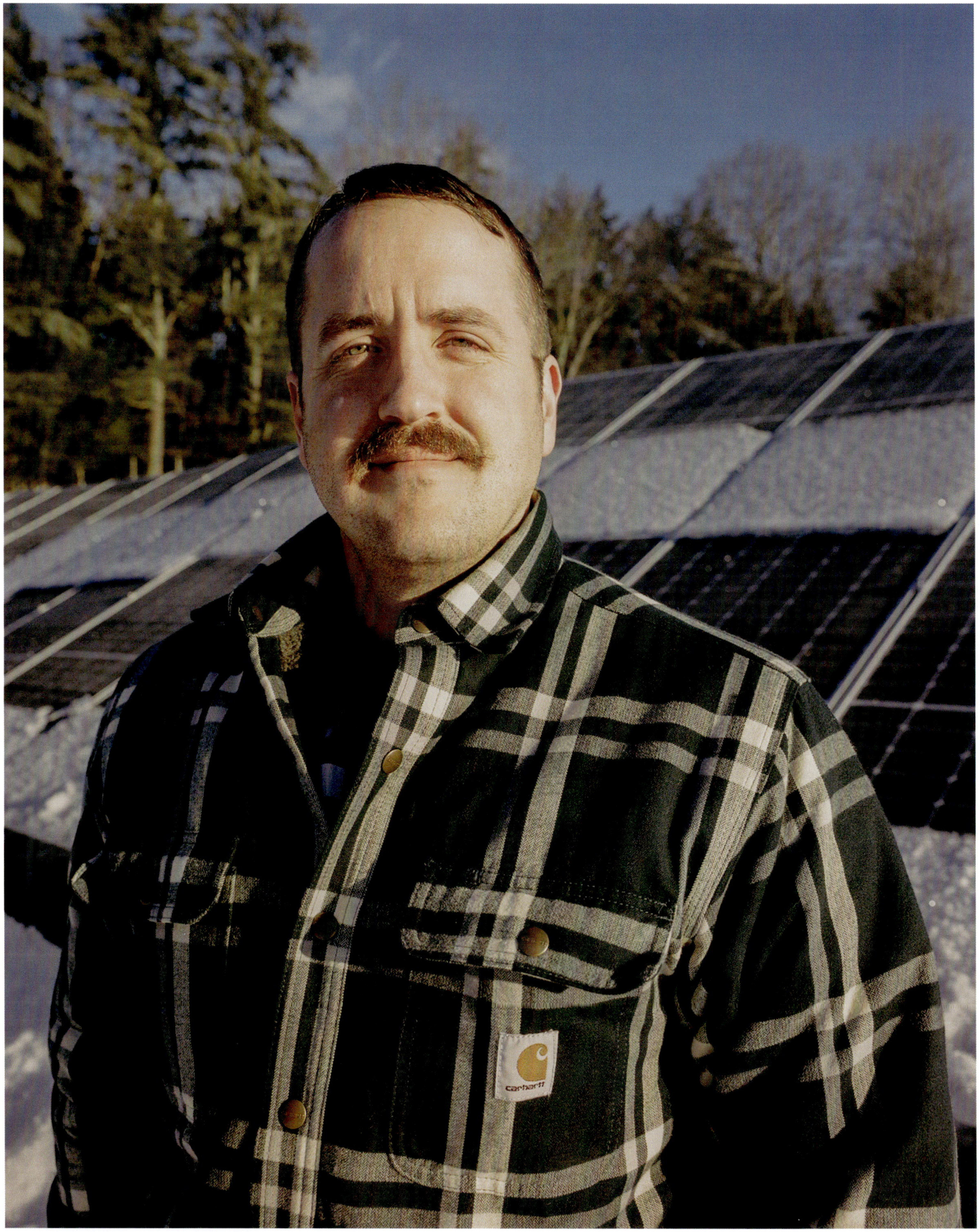

John Bumgardner, Staff Sergeant
Army National Guard - 11B / Infantryman
Years of service - 16 years
Deployments - 2 [Operation Iraqi Freedom & Operation Enduring Freedom]

I'm Johnny Bumgardner, and I was an 11B infantryman in the North Carolina Army National Guard for sixteen years. I joined the Guard kinda as an accident. The recruiters came to my high school and asked people to fill out papers, so I figured "Yeah, sure." Then I took a practice ASVAB and we were off. I deployed in support of Operation Iraqi Freedom in 2009 and in Jordan for Operation Enduring Freedom in 2014. I separated as an E-6, Staff Sergeant.

Since separating from the Guard and finishing a master's degree, I left North Carolina and started working at ReVision Energy in Maine. My job is designing solar energy systems in Maine, Massachusetts, and New Hampshire. I came up north because I wanted a job that had a mission similar to the military. Not just something for a paycheck, but something bigger. Here in the solar industry, that's a very real thing. I'm the primary estimator and preliminary designer of big, commercial solar systems. I design the base system using design software like AutoCad or HelioScope. Then I work with sales to price them out and as talks proceed, I keep improving the design based on the requirements for each client. Once we have the definite requirements, I do the final design and we hope to get it sold.

The sense of collaboration and teamwork from the military has really carried over. You see how things really work in a team and you learn so much about teamwork while serving. The success and the failure both. I have taken those lessons to help build myself and my team. My life mantra is: Have a purpose. Living with purpose, being able to wake up and have a goal and lasting meaning is so important to me.

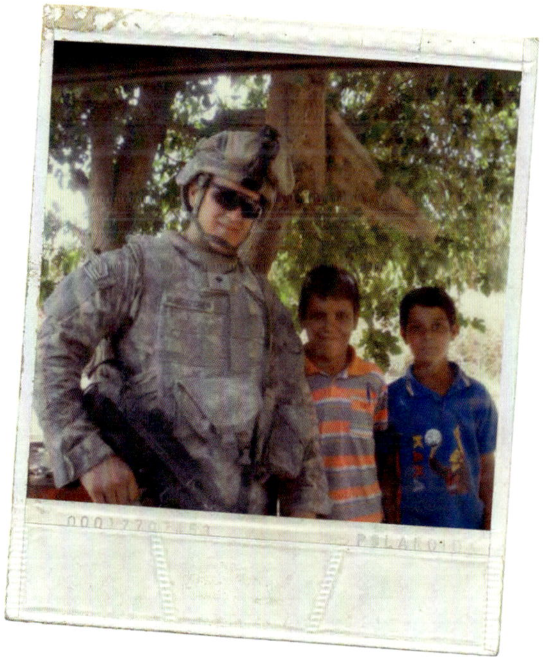

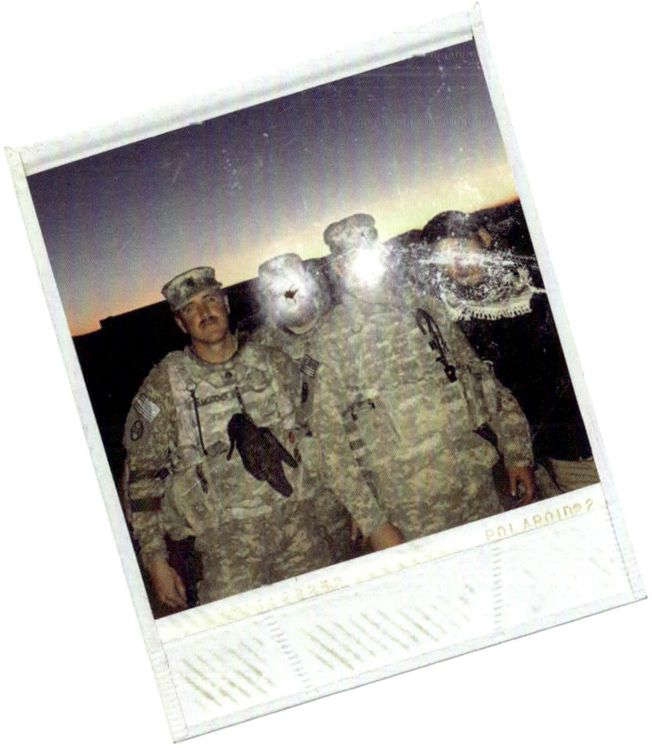

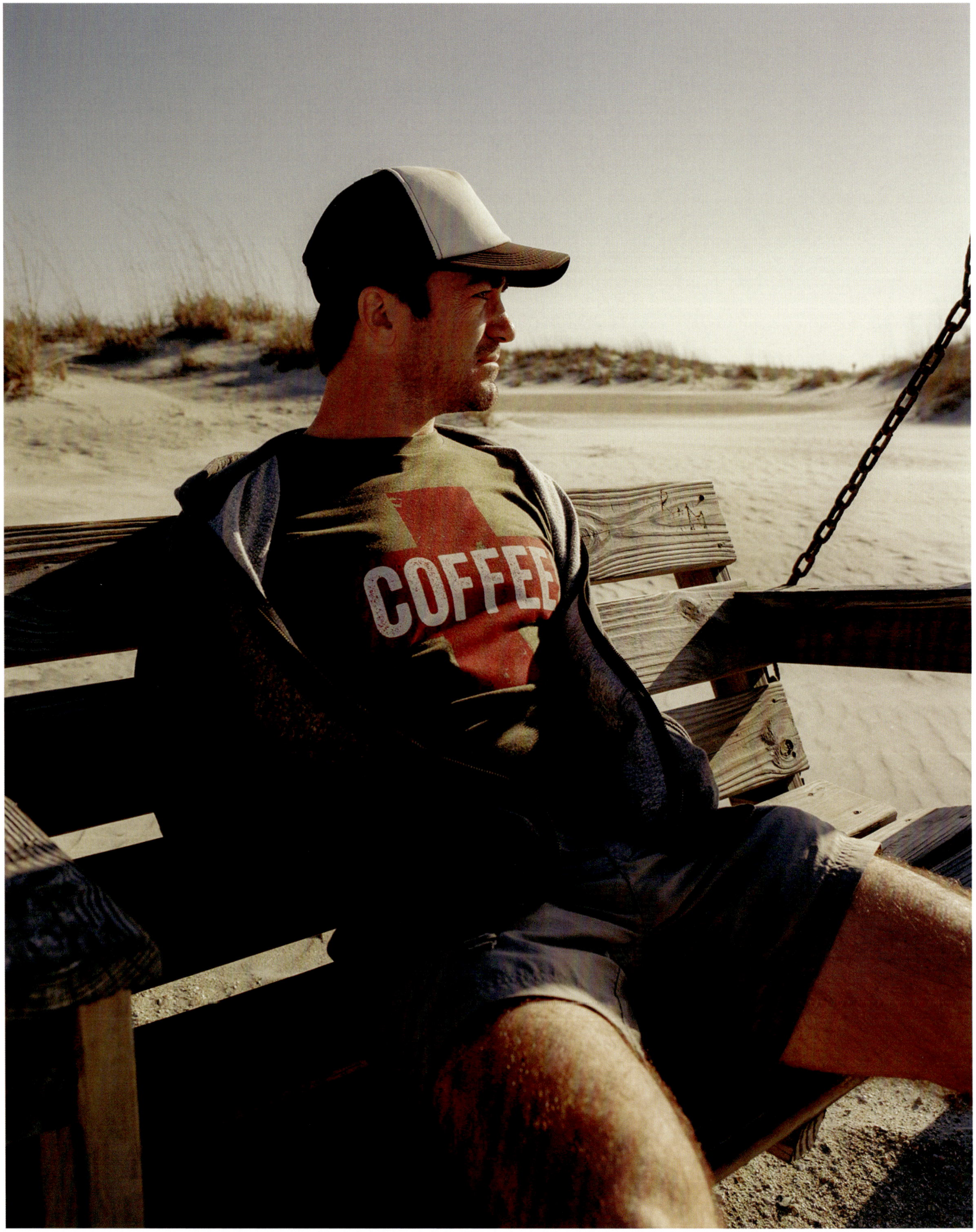

Aaron Laurich, Sergeant First Class
Army - 11B / 68W / 1/75 Ranger
Years of service - 13 years
Deployments - 10 [Operation Iraqi Freedom & Operation Enduring Freedom]

My name is Aaron Laurich. I started my career as an 11B Infantryman in the First Ranger Battalion, 75th Ranger Regiment in the United States Army. I then reclassified to a 68W Combat Medic. After I reclassed, I initially went back to 1/75. I eventually changed units and became a troop medic in the United States Army Special Operations Command (USASOC). I was on active duty when 9/11 happened. That morning my squad leader came in and made sure everyone's bags were packed. He said we're about to get busy. My first deployment was in 2002. I was pretty much continuously deployed after that. In total I deployed ten times between Operation Enduring Freedom and Operation Iraqi Freedom. In a Special Operations unit like 1/75 or USASOC, the deployments are typically shorter than conventional deployments. You have a higher off tempo and it wasn't uncommon to prosecute 100+ targets within ninety-day rotations.

I really didn't know what I wanted to do after I graduated high school, so I thought the Army would help me figure that out. I was a Sergeant First Class (E-7) when I was chaptered out with a general under-honorable discharge. I was chaptered out for opioid abuse in 2012. It was a lot like getting thrown out of an aircraft before the green light. There was a combination of things that led to it. My wife has a history of aneurysms. In 2010 she had a really bad one. She had to have multiple emergency surgeries. I missed a deployment and was already on pain medications for an injury. It was the perfect storm to abuse drugs. Worse, I had 24/7 access to the drugs as a medic. It was a slippery slope and I ended up on a slip and slide right off the cliff. Addiction in a unit like USASOC or the Rangers is especially hard. You try to fight the battle on your own. You tell yourself, "Hey you're a Ranger, you need to be better!" You're also worried about judgment, being seen as a failure. You worry about someone finding out, they won't let you deploy, and then getting kicked out of the unit. So, you hide it. Addiction is a really hard battle to fight alone. It was thirteen or fourteen months, but eventually mine came to light and that was it.

After separating, I took a job in New England, specifically because it was so far (900 miles) from Fort Bragg, NC. I had a lot of guilt and I was full of shame. I felt I had let my family and my teammates down. There aren't a lot of medics in USASOC and when they lose one you can't help but feel you fucked everybody. In New England, you aren't going to run into a teammate at the grocery store or run into a guy you knew from that life at Lowe's. So, I exiled myself. I believed I didn't deserve the benefits from any veteran support organizations that most people receive when they leave the military because of my addiction. In my eyes, at that time, those were only for people who served honorably.

I met my wife in Key West when I was doing a military school. When we saw each other we both instantly knew there was something there. I deployed shortly after we met, and she wrote me letters every day. That communication really set the tone of our partnership. Then in 2016 my wife had another series of aneurysms and additional health complications. I was in a really dark place with my own addictions.

Continued on next page →

My PTSD was really manifesting itself. So, there I was in the hospital with her, I'm a pretty smart special ops medic, I am a board-certified flight paramedic ... and I can't do anything for her. Ultimately, all of this stress led me to realize we weren't happy living in New England. We decided we wanted to go to a place where we could be happy. Early in our relationship I was living on Tybee Island, near where I was stationed with 1/75. The surrounding area was always a happy place for us. So, we moved back. Here we've found peace and community. We found happiness. It was the best thing we could have done. Being here has saved my life. My mental health has drastically improved, and her health has improved, as well. Being out on the water, whether it's paddle boarding or kayaking has been so soothing. There is just something awesome about the ocean and the tides. It's never the same water twice. There is something so cleansing about that tide coming in and out. It's a great way to sum up issues so many people are going through.

I've become part of some veterans' groups. I support the Three Ranger Foundation, GallantFew, and Fight the War Within, a local veteran anti-suicide group. When I first got out, I didn't have the desire to do anything with a veterans' organization. Now I try to be involved. For anyone struggling, remember: You're not the only one going through it. Everyone is struggling in their own way. Everyone's military experience is unique, but some of the root problems and challenges are similar. Special Operations needs to continue to do a better job of encouraging soldiers to come forward, to seek help. As a member of Special Operations, we volunteered multiple times to serve at the tip of the spear, and there is a lot that goes with that. When you hide your struggles, when you keep it in, many start to self-medicate. Then everybody loses out when that soldier falls off.

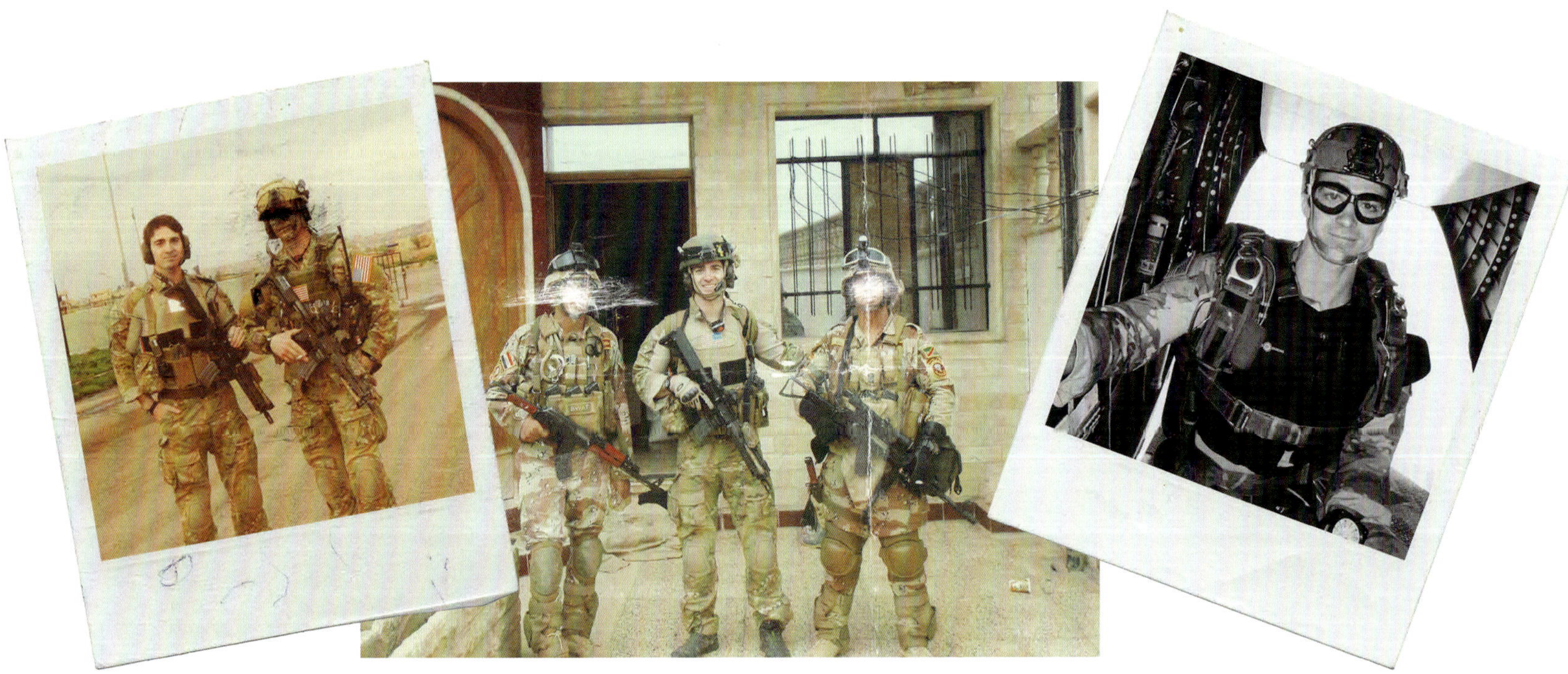

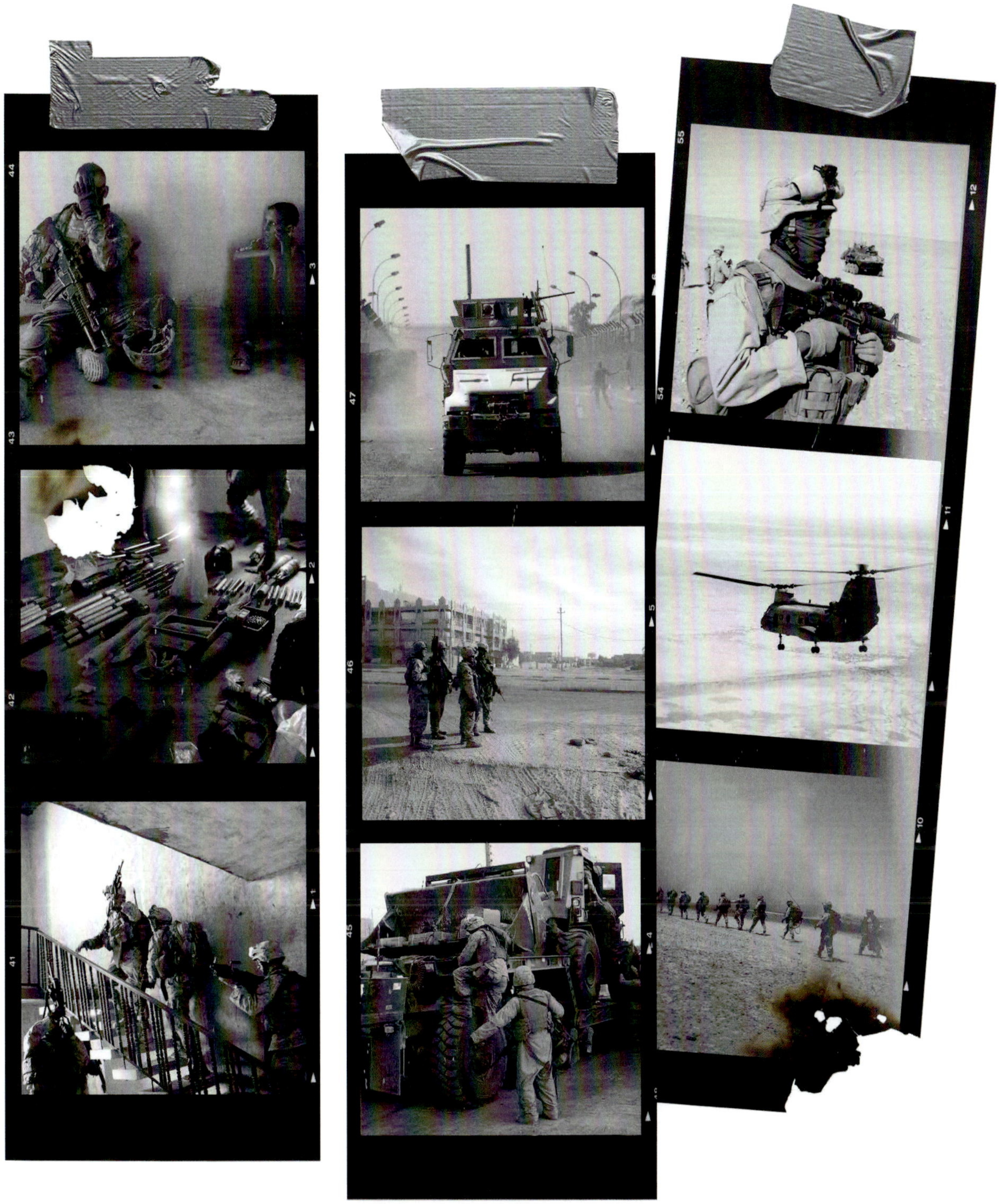

My name is Troy Tomlinson, and I was a tank mechanic in the United States Army. I joined to get money for college and to get my family off to a good start. I retired as an SFC after twenty-two years in the service and two combat deployments in support of Operation Iraqi Freedom.

After retirement, I came to Florida to learn the golf industry. My transition was difficult. I was offered some nice jobs doing similar things to what I had done while I was in uniform, but they didn't feel like a fit. I wanted to continue to be a leader. Golf really helped me separate. I ended up getting what I thought was the job of a lifetime. I was working at a high-end resort for a year as an assistant golf pro and then for another four years as a caddie master, but it didn't quite fit. I started working with Fairways for Warriors to be back around soldiers and to help them transition. I'm now the director of golf for the organization and absolutely love it. Just being around the troops again, and the vision of the organization to help combat vets from all eras, is amazing. It's so great to see the generations from Vietnam and Desert Storm get together with some of our younger veterans. They all take care of each other.

I'm also working to figure out how to help vets who are struggling. When you're transitioning out, you have to go through TAPS and that's good. The American Legion has a lot of influence, and if they can't help you, they know the person who can. Also, the VA has a lot of resources, and it's getting better. It's more than just the hospital. You need to invest a little time, but it's absolutely worth it as you transition. You only got one life. Live it the best that you can, whatever your definition of that is. Know you are never alone.

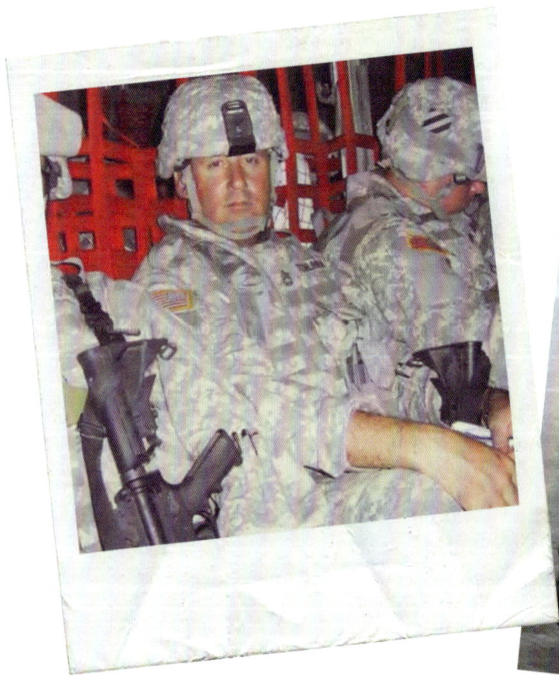

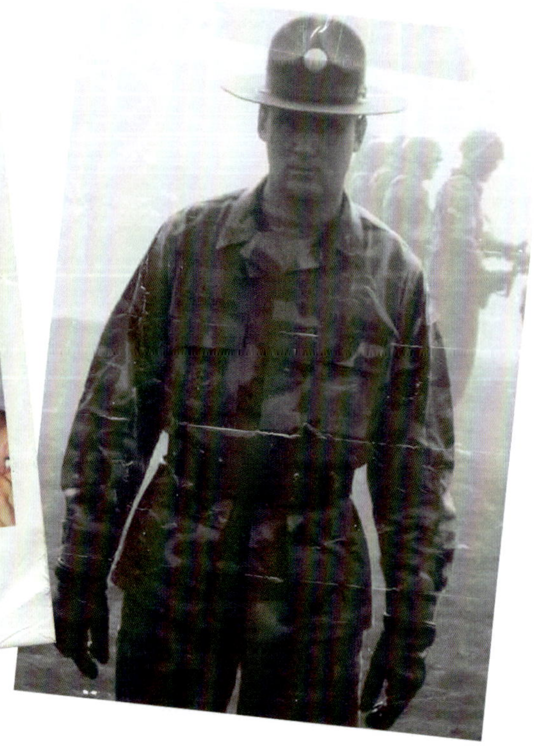

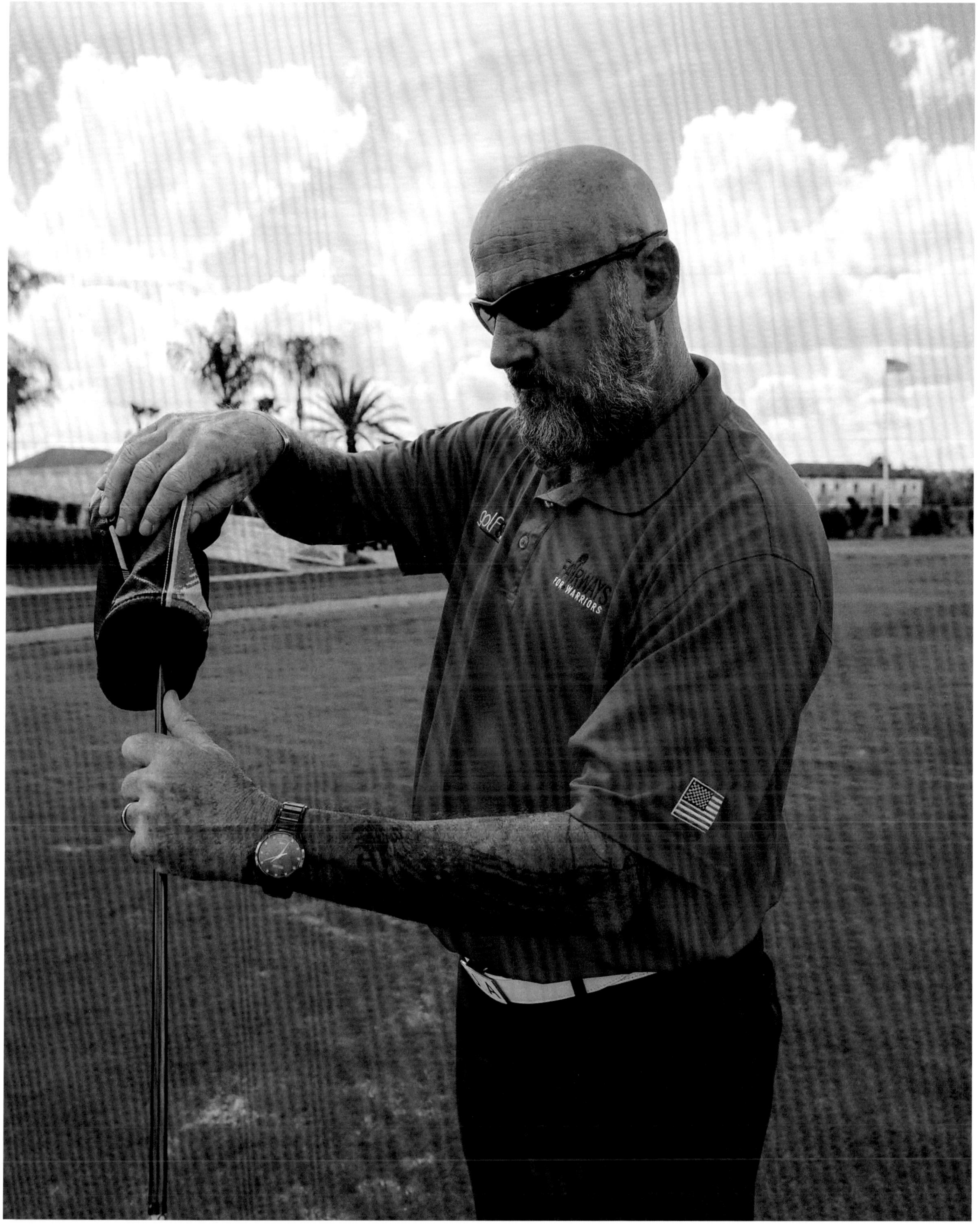

Troy Tomlinson, Sergeant First Class
Army - 91A / Tank Mechanic
Years of service - 22 years
Deployments - 2 [Operation Iraqi Freedom]

My name is Richard Graham, and I was a member of SEAL Team 10 and SDV Team 2 in the US Navy. I enlisted because I wanted to hunt drug dealers. It was just before 9/11. I had a scholarship for architecture, which is what my father's business was, and I got cold feet. I had some friends talking about SEALs and Escobar and it was intriguing to me. Like, these Special Ops dudes were mysterious, and I asked myself, "Man, how do I do that?" I was in boot camp when the USS Cole was bombed, and I was in phase three of BUDS on 9/11. It was a little crazy because I still had another year and a half of training to complete before I was able to deploy. I eventually did deploy twice in service to the Global War on Terror and medically separated as an E-5 after six years in.

I had a major blood clot and had to be medevaced. They tried to fix it at Portsmouth Naval Hospital, but the procedures gave me nerve damage in my arm. The clot meant I wasn't allowed to jump or dive. Eventually, I was medically discharged. Technically, I should have been medically retired, but there weren't a lot of organizations helping service members separate when I first got out. Worse, the military hadn't really figured out how to take care of guys like me back then. I went through the VA and got a 30 percent rating — and mostly not for the same things preventing me from serving in the SEAL teams. It was infuriating. I had been thinking, "Man, I'm in for at least twelve years," and suddenly the military was no longer an option. It was like, "What am I going to do?" SEAL culture didn't really help either. In the SEALs guys are always jockeying for deployments. They're always hiding injuries or downplaying things. No one wants to go to the doctor. There are now so many more advocates and people that are able to explain what SEALs and other Special Operations guys go through. The system understands better and understands more.

Readjustment was difficult, because it came before I was ready for it to be there. It wasn't my choice. It was this hidden thing I couldn't see in my body that stopped me. Meanwhile, I'm looking at a guy who lost a leg going back overseas. At the time I was really mad about it. Now I see it as a chapter in my life. I'm out doing so many other good things! I have a training company where we do firearms instruction. Right when I got out, I started to work at the gym as a personal trainer. Some of the guys I trained were cops. They remembered I was a sniper and I'd share some knowledge and that evolved to teaching shooting full-time. We're evolving the courses now. You need to be strong and eat right and work out right. You need the situational awareness and education to be able to do those things. It's cool, because in many of my classes I have police and civilians training side by side; the experience helps them relate to each other. Do what you know is right. In the SEALs they say: earn your trident every day. Just because you got it once doesn't mean you deserve to keep it.

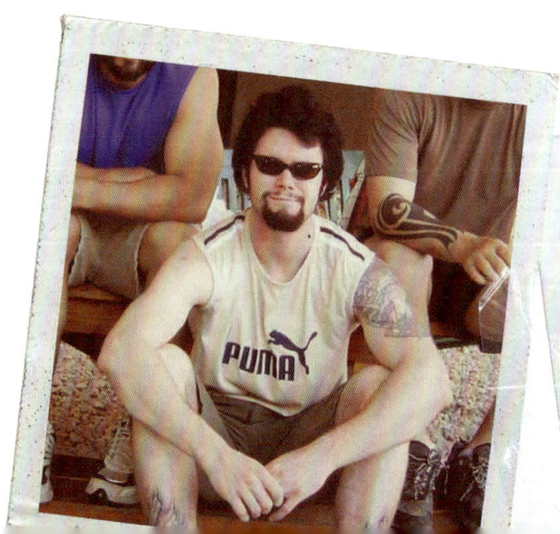

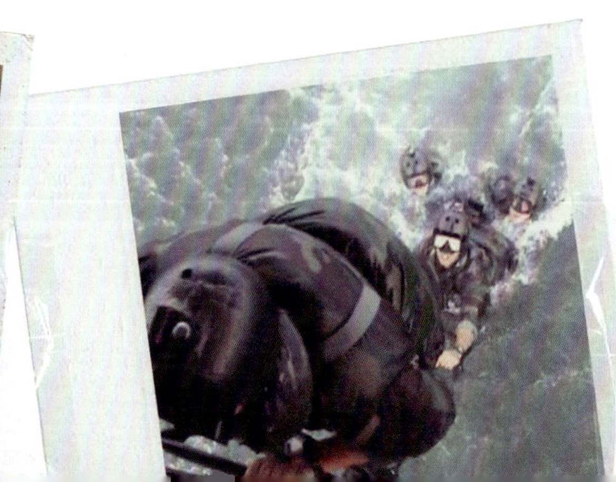

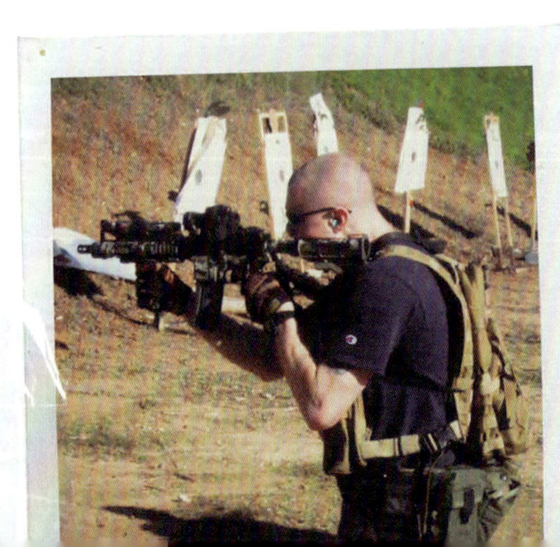

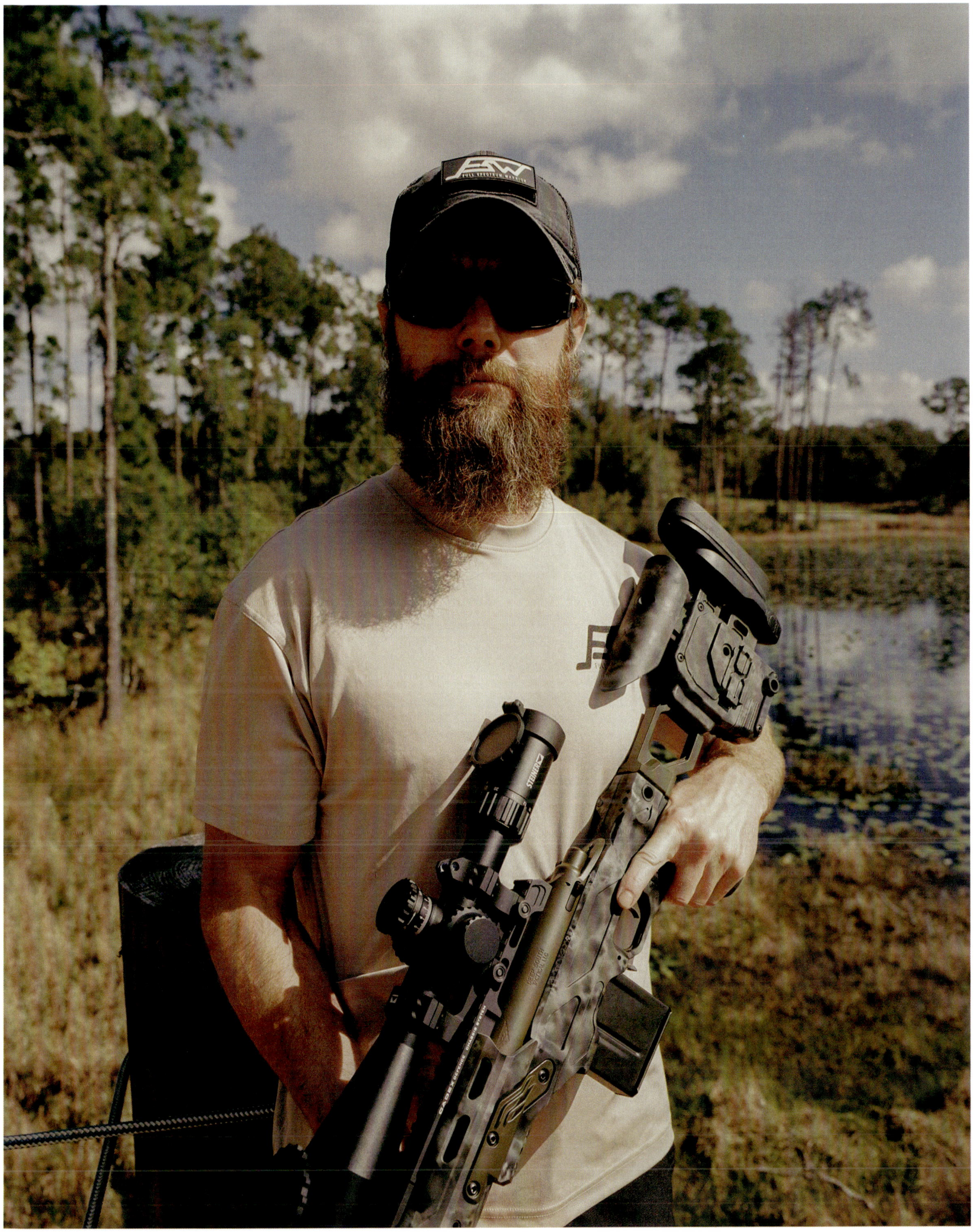

Richard Graham, Petty Officer Second Class
Navy - Seal Team 10 / SDV Team 2
Years of service - 6 years
Deployments - 2 [GWOT & Bosnia]

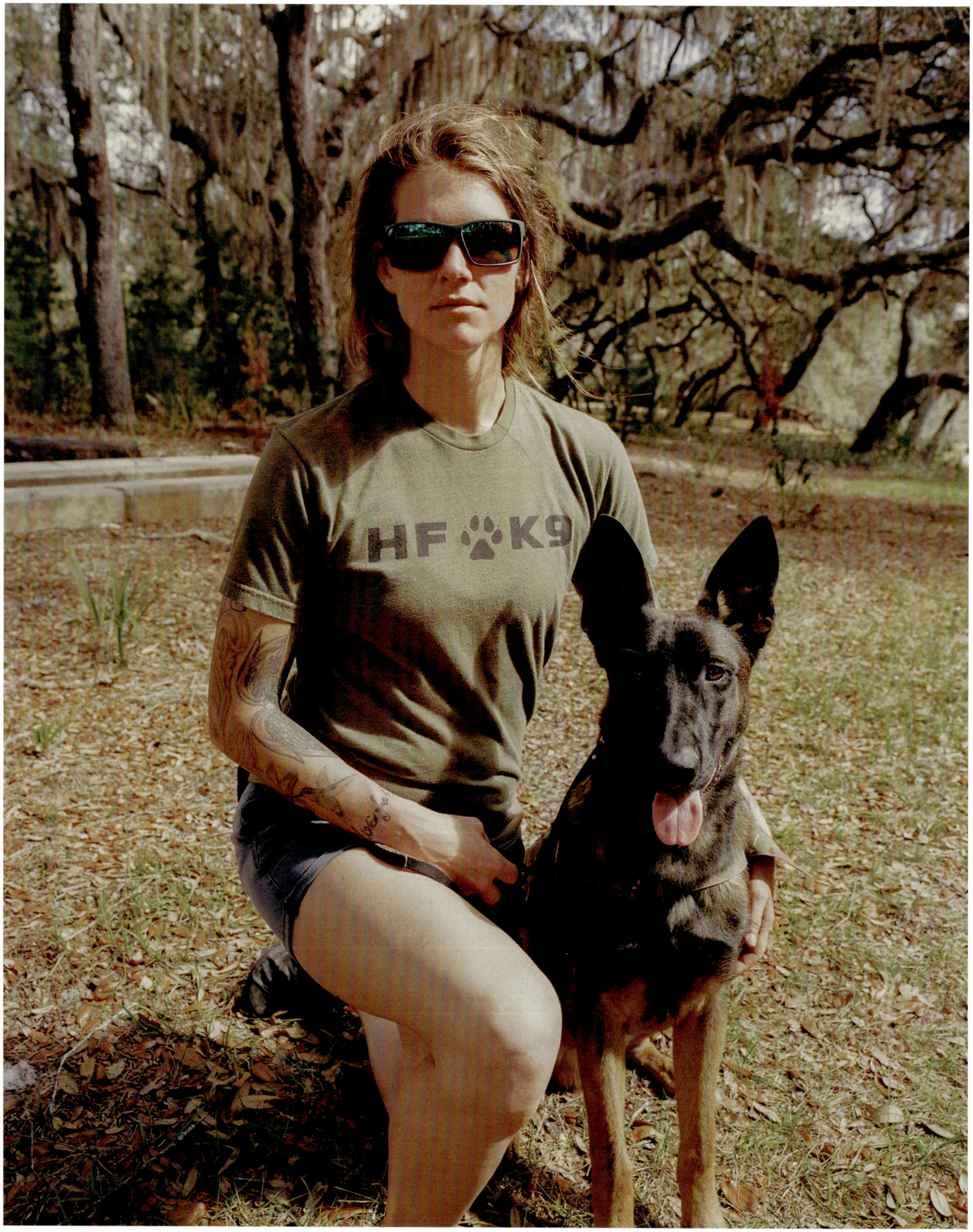

Callie Farill, Senior Airman
Air Force - E3751 / Firefighter
Years of service - 6 years
Deployments - 2 [Operation Iraqi Freedom & Operation Enduring Freedom]

My name is Callie Farill, and I was a firefighter in the US Air Force. I left the Air Force as a Senior Airman (E-4). I deployed in support of Operation Iraqi Freedom in 2010 and Operation Enduring Freedom in 2015. I also went back to Iraq as a contractor post-service. I was one of those kids that lived in a small town and didn't want to stay in a small town. I had gotten my firefighter level one and two certifications while in high school. My dad had been in the military and he suggested it.

When I separated, we had some pretty good guidance; they were starting to get TAPs (Transition Assistance Programs) right. The DAV was a great resource for disability. They'll assign you a person and walk you through your entire disability process. Purple Heart Association will, as well. After contracting I had some money but was tired of being a firefighter. Eleven years into the gig, my body was starting to break down. I moved to Utah and TacGas offered me a job as an event coordinator. I was doing a logistics job, but it was not fulfilling. I have loved dogs since I was a kid, they just gravitate to me. So, I googled "K9 GI Bill," found a program, and moved across the country to take a thirteen-week course. Once I was certified I started my career as a K9 handler. I worked the NBA All-Star Game in 2019, the "Hamilton" touring show, and eventually picked up a job in Vegas where I helped get a K9 program set up. Gigs in the K9 industry ebb and flow. I met Rich at Home Front K9 and now I'm working on training dogs for them.

I've gotten into some hairy situations, but if you stop, you get stuck. Keep going forward!

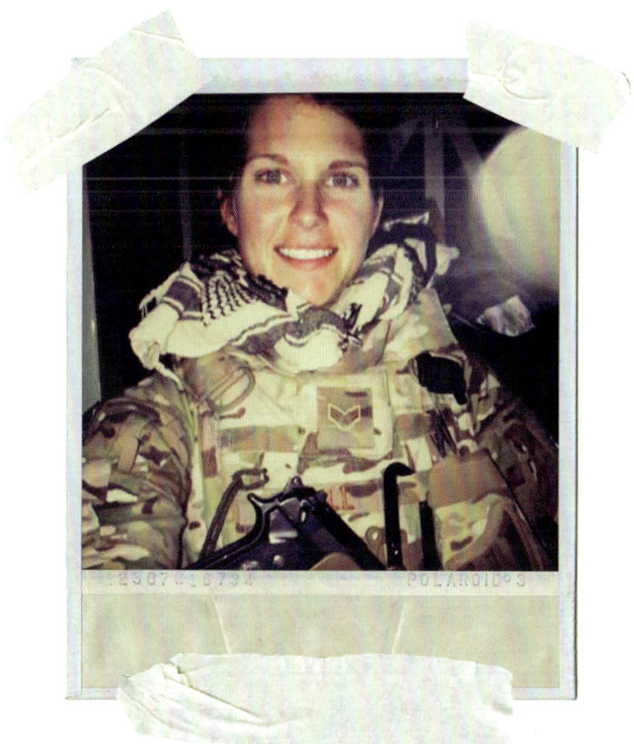

My name is Andy Gamberzky and I was a Combat Controller in the United States Air Force. Combat Controllers, also known as "CCTs," are the smallest special operations force within USSOCOM. Military Free Fall and Combat Dive School are just some of the standard courses we attend. We kept tabs on everything that's in the air during missions. We also surveyed airfields as a part of an airfield seizure mission. I grew up in New England, and when 9/11 happened, my dad was in NYC. I was in middle school and when he got home, we started talking about what happened. The conversation left a never-ending impression on me. It was such a moment in time. I left active duty in 2017, but I'm still in the Air National Guard as a Tech Sergeant (E-6). I deployed in 2011 and 2014 in support of Operation Enduring Freedom. In my tour in 2014 I was shot in the leg and had to be medevaced. I went back to Operation Iraqi Freedom in 2016, and it was a wild tour. All our leadership in the Special Operations Task Force had been ground-level commanders of the same battle space in 2006 and 2007. The guys we were pitching missions to had been in our shoes a decade ago; you just couldn't bullshit them. They had been there; they even remembered grids! They knew what was up and what questions to ask. You always had to be dialed in and totally prepared. When you're in a Special Operations unit, you're over seas three to six months and home three to six months.

For me it took a long time to adjust after I got off active duty. My wife was a huge influence on me as I transitioned. She was also in the Air Force and we met while we were both serving, so we had a lot of shared experiences. But the most frustrating thing about transitioning was just getting frustrated with people for shit that doesn't really matter. Time is the best healer. You need to talk, but everyone is different and needs to deal with it their own way, on their own timeline. As a veteran is transitioning, I'd say: care about the right things. Figure out the things that matter most and spend your time doing them. Too many people care about shit that doesn't matter. Don't fall into that trap. Force Blue is one of the things I really care about. It's a non-profit that takes former combat divers and has them go through the process for scientific diver certification, so they can eventually go on trips and work with scientists. You often find that scientists are so smart and have ideas but just don't have the manpower. But with Force Blue, when you bring in military guys, we're so into mission-planning and we can get in there and help. It brings in a lot of camaraderie, too. Coral reefs are dying, and they produce a ton of the world's oxygen, so I try to show up and do my best.

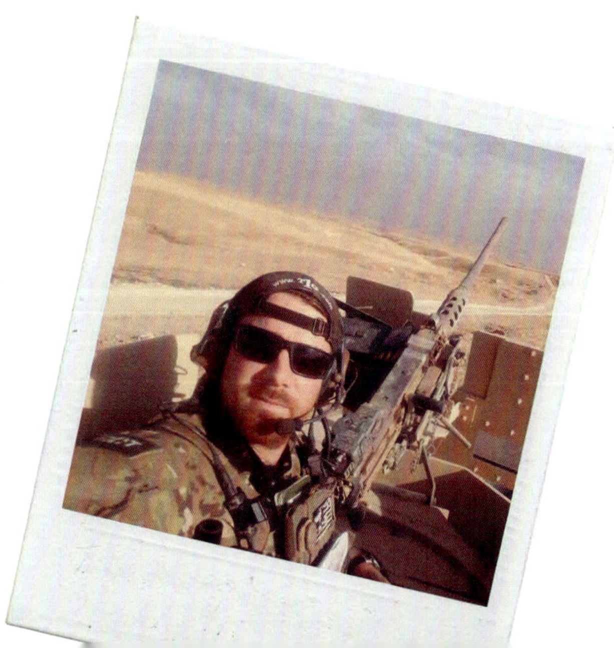

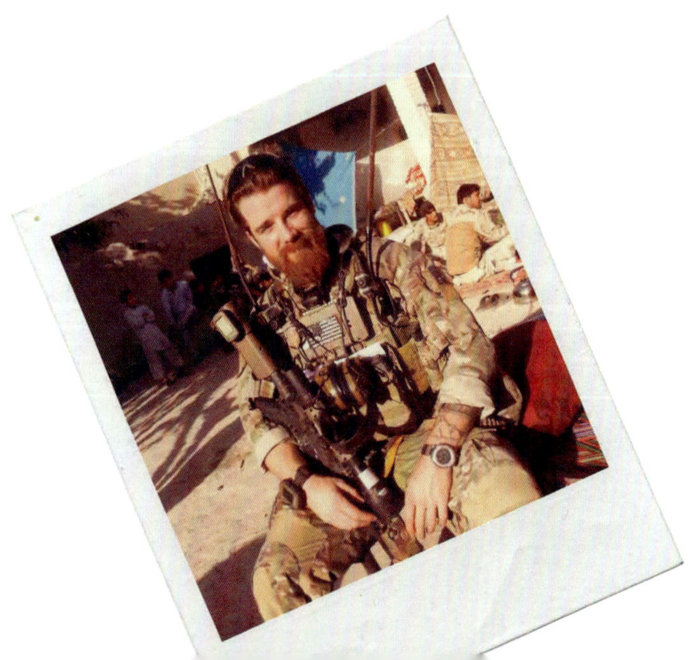

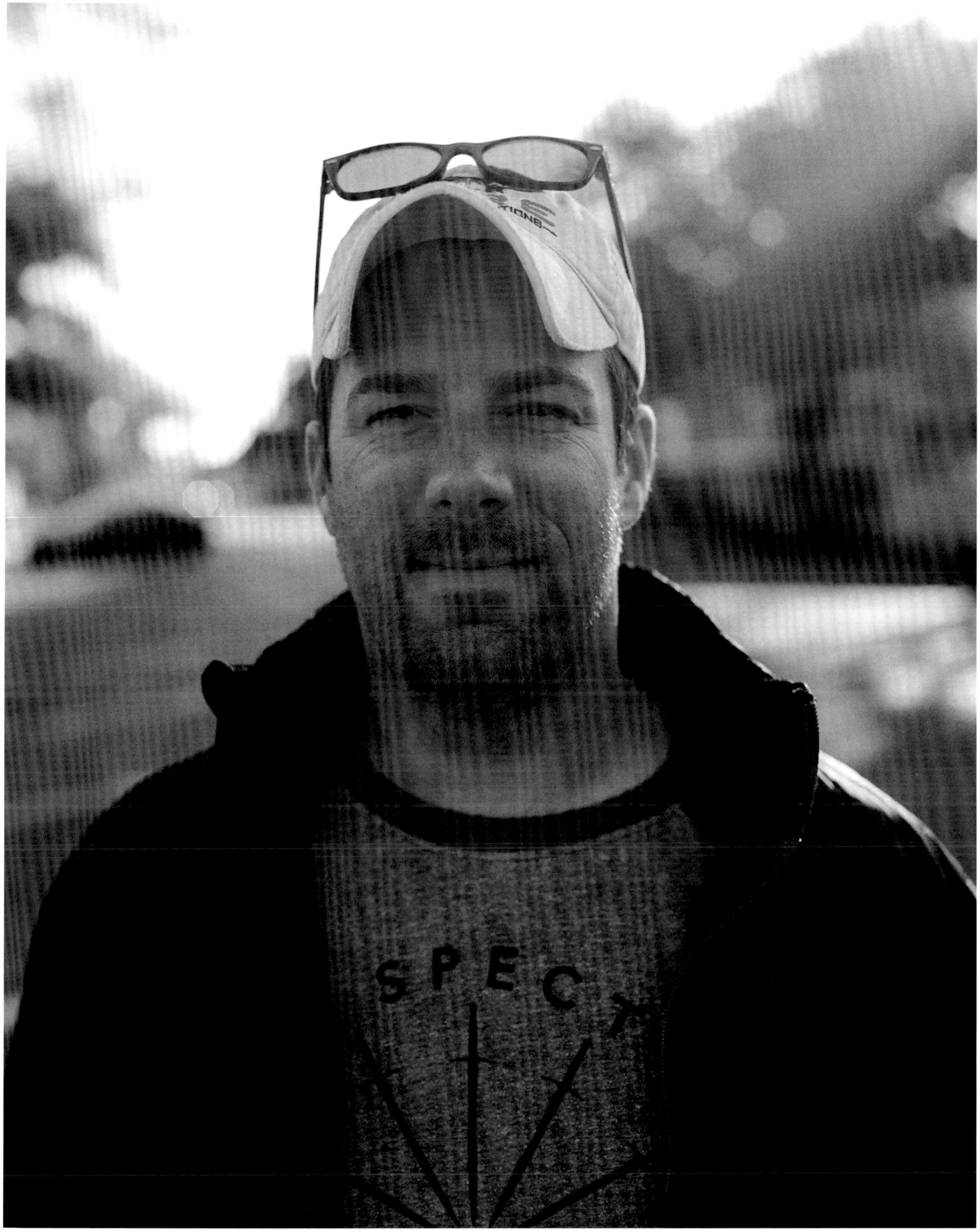

Andy Gamberzky, Technical Sergeant
Air Force - Combat Controller (CCT)
Years in service - Active
Deployments - 5 [Operation Iraqi Freedom & Operation Enduring Freedom]

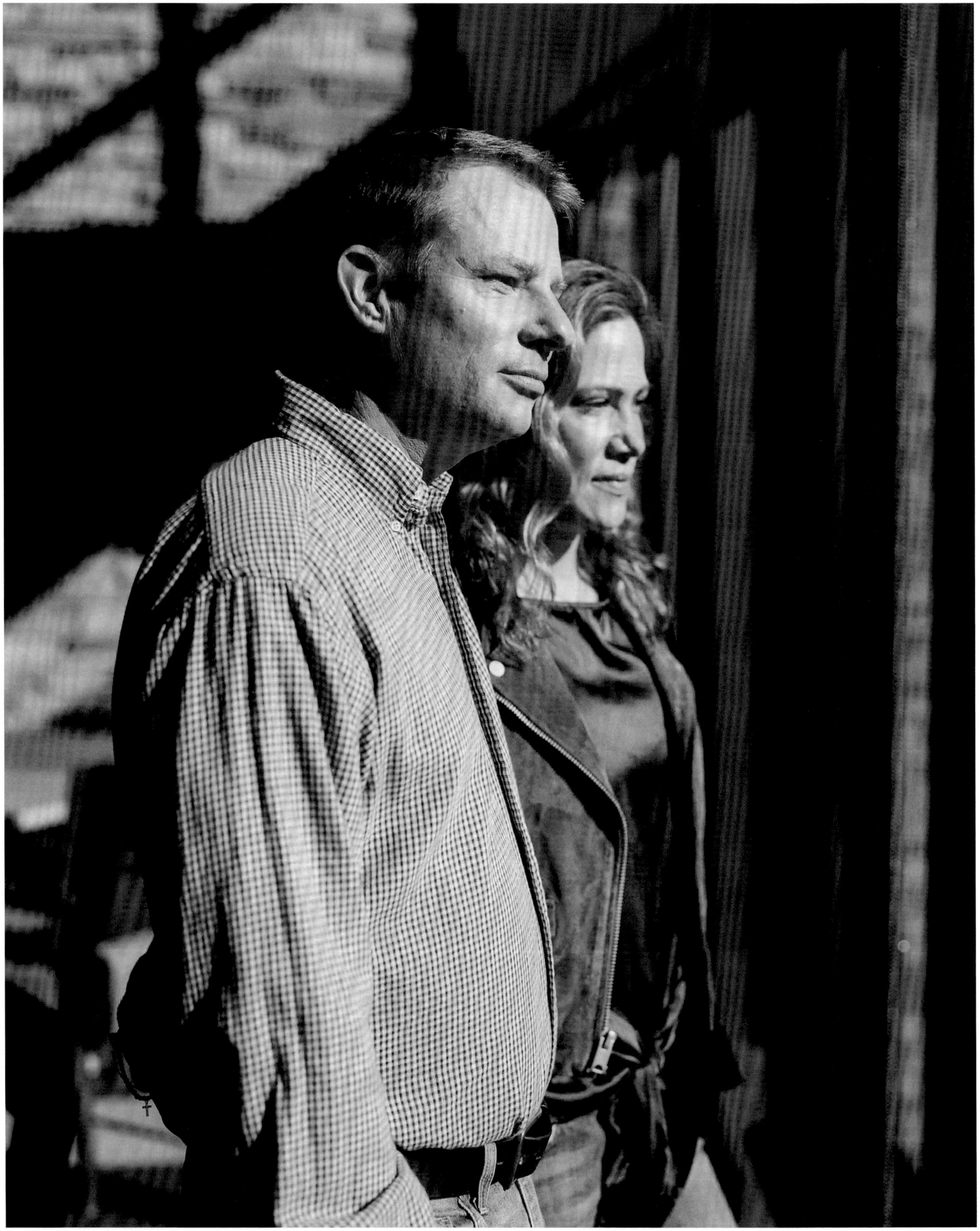

Chris & Angie Baker, Lieutenant Commanders
Navy - 1310 / Naval Aviators
Years of service - 8 years
Deployments - 1 each [Operation Anaconda & Operation Enduring Freedom]

We're Chris and Angie Baker, and we were both pilots in the United States Navy.

Chris: "Angie's dad was an Air Force Academy grad and wanted her to be an aviator. Then Top Gun happened and, well, that was it."

Angie: "Chris, on the other hand, was talked into joining the military after he was already a consultant in Chicago. He was looking for a new thing and found the Navy. We both left as Lieutenant Commanders (O-4)."

Chris: "Angie flew the P-3 Orion during Operation Anaconda on recon missions during the early phases of Operation Enduring Freedom."

Angie: "Chris flew EP-3's on various deployments in support of Operation Enduring Freedom."

The Semper Fi Fund has a special place in our hearts. One of our children is special needs and they really came through and helped support us with specialized care. Since then, we've been supporting them every year with their combined federal campaign donation.

Chris: "Since separating from the service, Angie has been focused on being mom to our five kids, doing approach coding for Boeing, looking to starting a franchise business and managing rental properties."

Angie: "Chris transitioned right into commercial flying for Southwest Airlines after nine years on active duty. The timing was perfect. Just crazy luck to be able to go right to the airline. He has been a captain now for five years. The move to Southwest made the transition much easier."

Chris: "Angie's transition was harder than she thought it would be. In the military, there are like-minded people all around you, but in the civilian world, there isn't as much in common with the people you meet and are around."

Angie: "With me and Chris still flying, there's a lot of military people in the industry, but also there's the camaraderie of being pilots."

Continued on next page →

We hope our fellow veterans appreciate how amazing they are and how much their service matters. There is a small percentage of Americans who have served and it shows how amazing the United States is. We served with such quality people. To take on the responsibility of defending the US, in your early twenties, in a time of war, is an awesome responsibility.

Chris: "Angie's life mantra is to constantly improve, work to be the best you can be every day."

Angie: "Chris would tell you that you can't take anything for granted. Tomorrow isn't promised. You go overseas and friends don't come back; you need to remember not to take life for granted."

We know it's the biggest challenge, but we say: Work to find a new purpose. Connect with other veterans. Don't be afraid to work with a headhunter or recruiter to help you translate your experiences as you separate from service. It's ok to ask for help! And start looking before you leave. Look at what you want to do and realize all the cool things you've done might not sync up. So think hard about the new qualifications or skills you might need to have nine months or so before you leave.

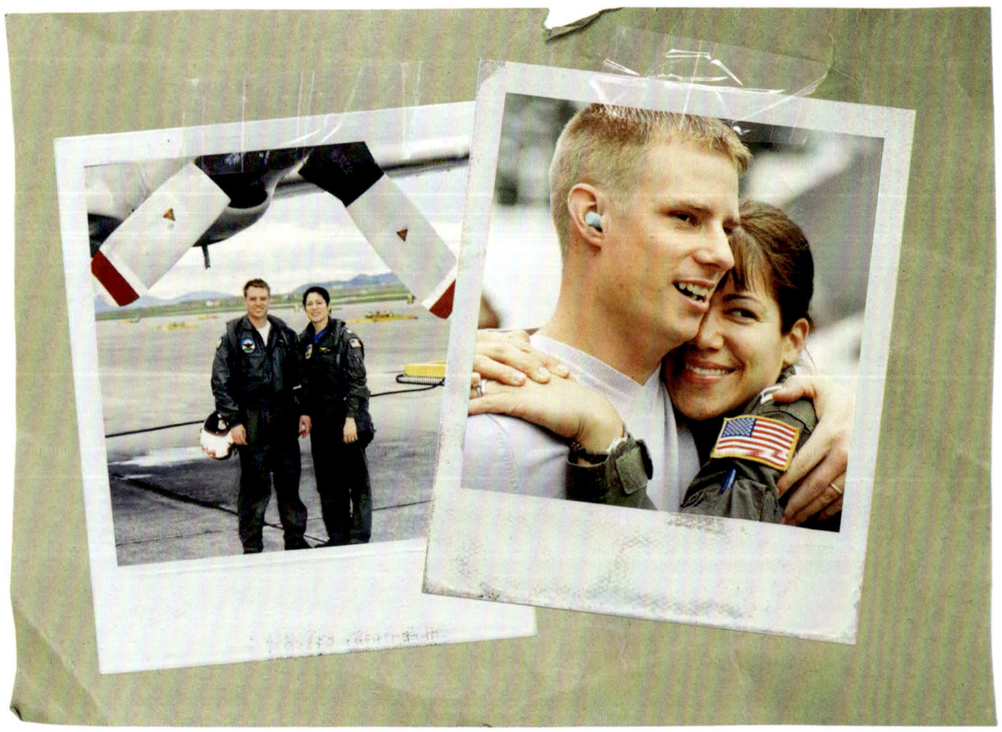

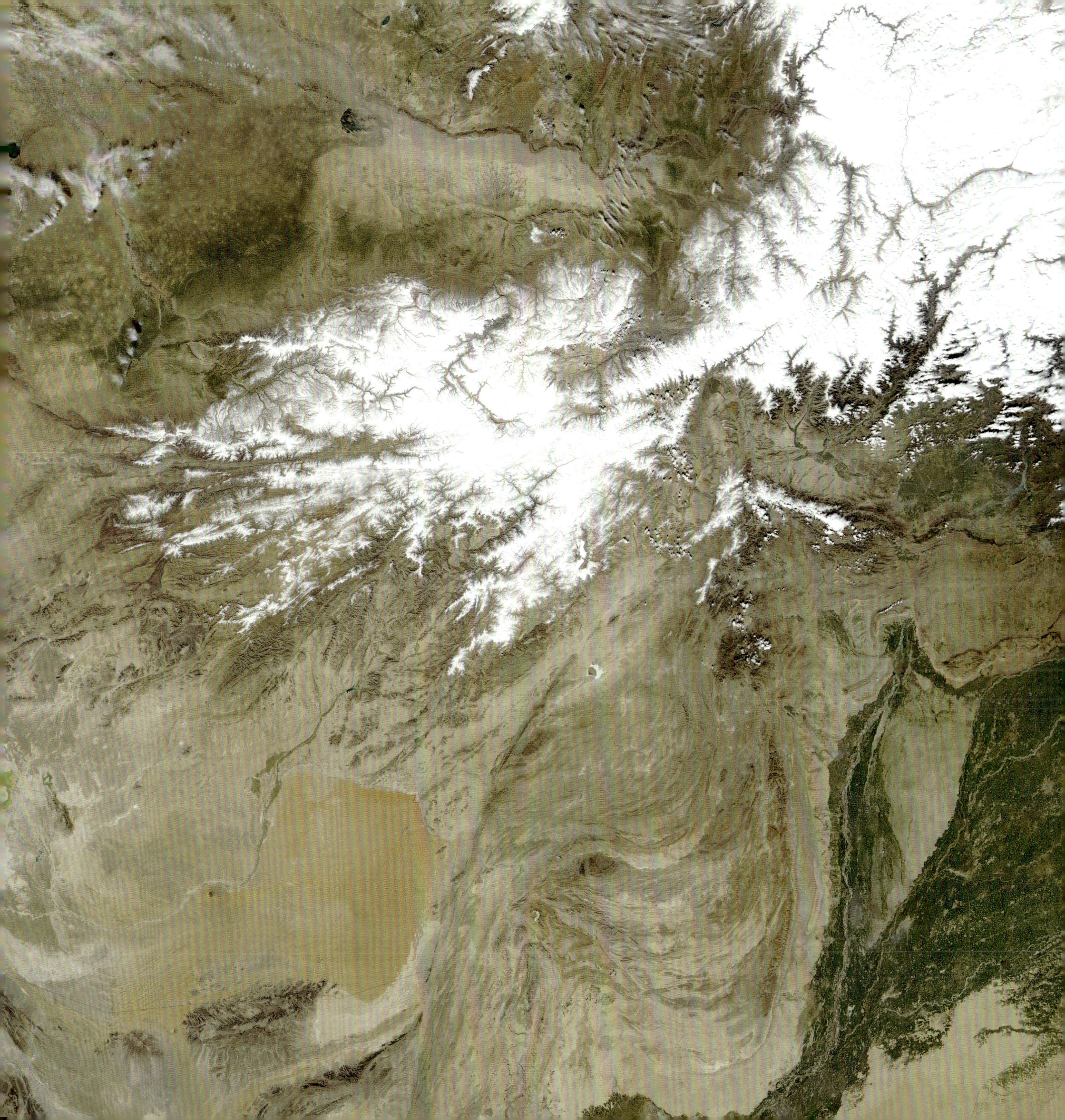

I'm Brian Laverentz, and I deployed as a Hospital Corpsman in the United States Navy in 2013 as part of the 26th Marine Expeditionary Unit and participated in Operation Eager Lion. I had initially wanted to go to medical school, but med school is really expensive. My Dad had been in the Navy and suggested being a corpsman to decide if I really dug medicine. Grandfather and uncles on my dad's side served too. After I separated, I started working as an ER tech part-time and I was a full-time pre-med biology student. Currently I'm in my third year of medical school. I also really missed the service and sense of camaraderie, so I commissioned into the Army as an officer in October 2019.

I really enjoy using my Instagram page to share medical education content. It has also allowed me to connect with other military members, which kickstarted a program I am developing to help establish mentorship for enlisted who want to become physicians. There's not really a streamlined process for doing something like that, and I want to help these individuals find their own path to success and connect with those who've already walked it. I cover how to use the GI Bill, manage family stress, and explain ways the VA can help and what programs are available to veterans. I also support the HunterSeven Foundation, which does a lot of great work with veterans.

A life mantra I've thought about a lot lately is, "The process of getting better isn't something that happens overnight." I heard it at a conference, and it really stuck with me. The accumulation of marginal gains leads to greatness. In the military there is always that "Shut the fuck up and get your shit done, no excuses" mentality. But sometimes you have a large goal, and you get frustrated and angry and upset because you haven't gotten it done yet. Focusing on making tiny improvements has helped ease some of that frustration.

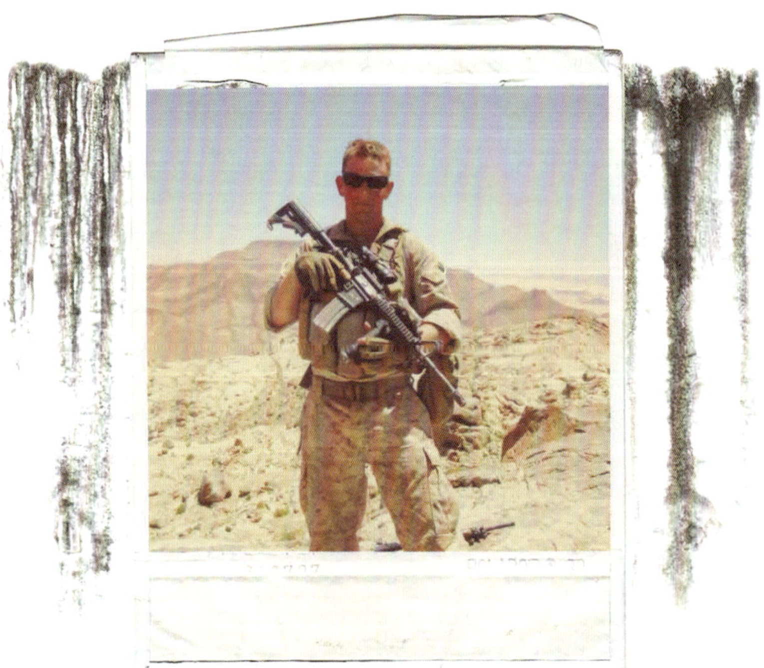

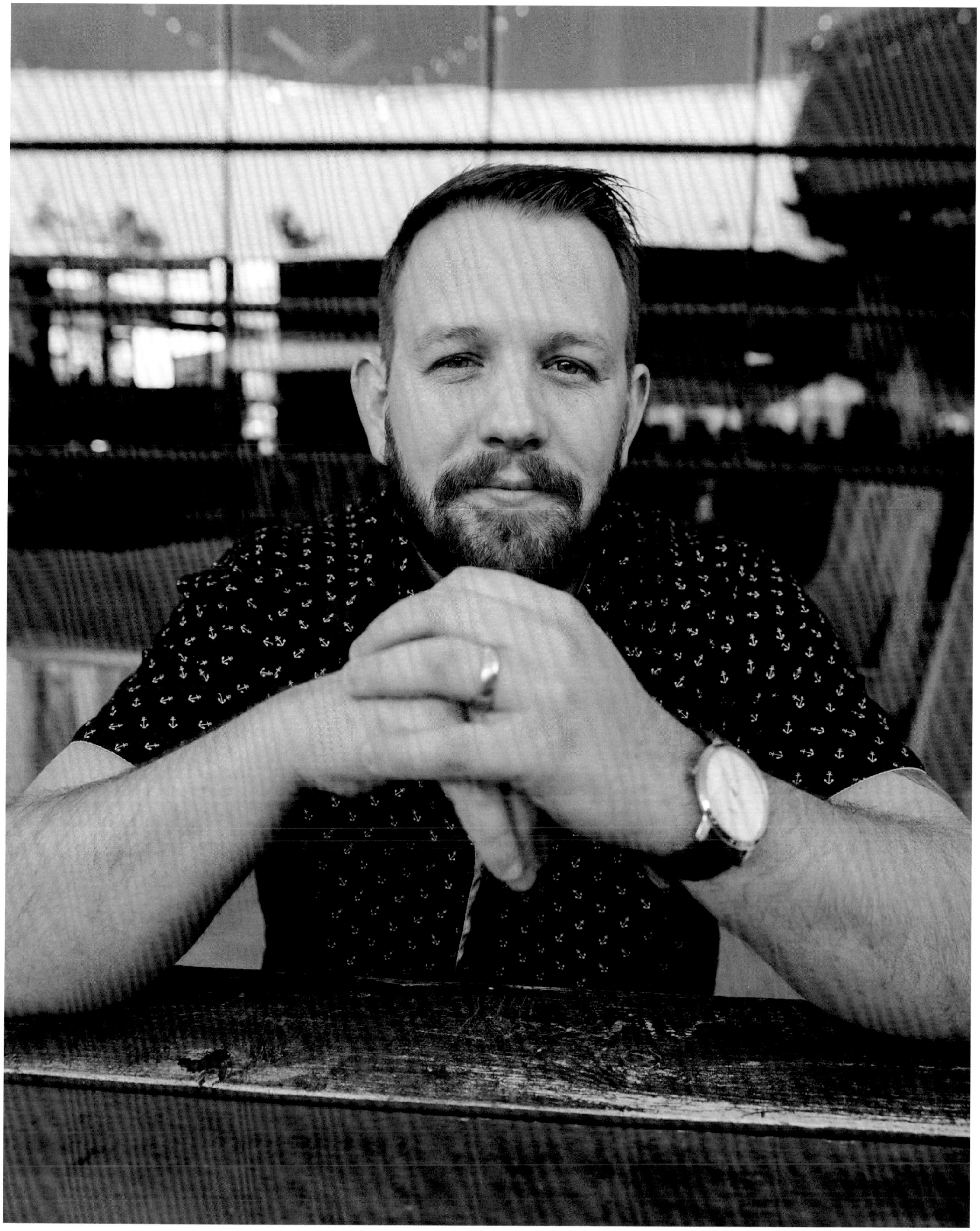

Brian Laverentz, Petty Officer Third Class
Navy - Hospital Corpsman / 26th Marine Expeditionary Unit
Years of service - 5 years
Deployments - 1 [Operation Eager Lion]

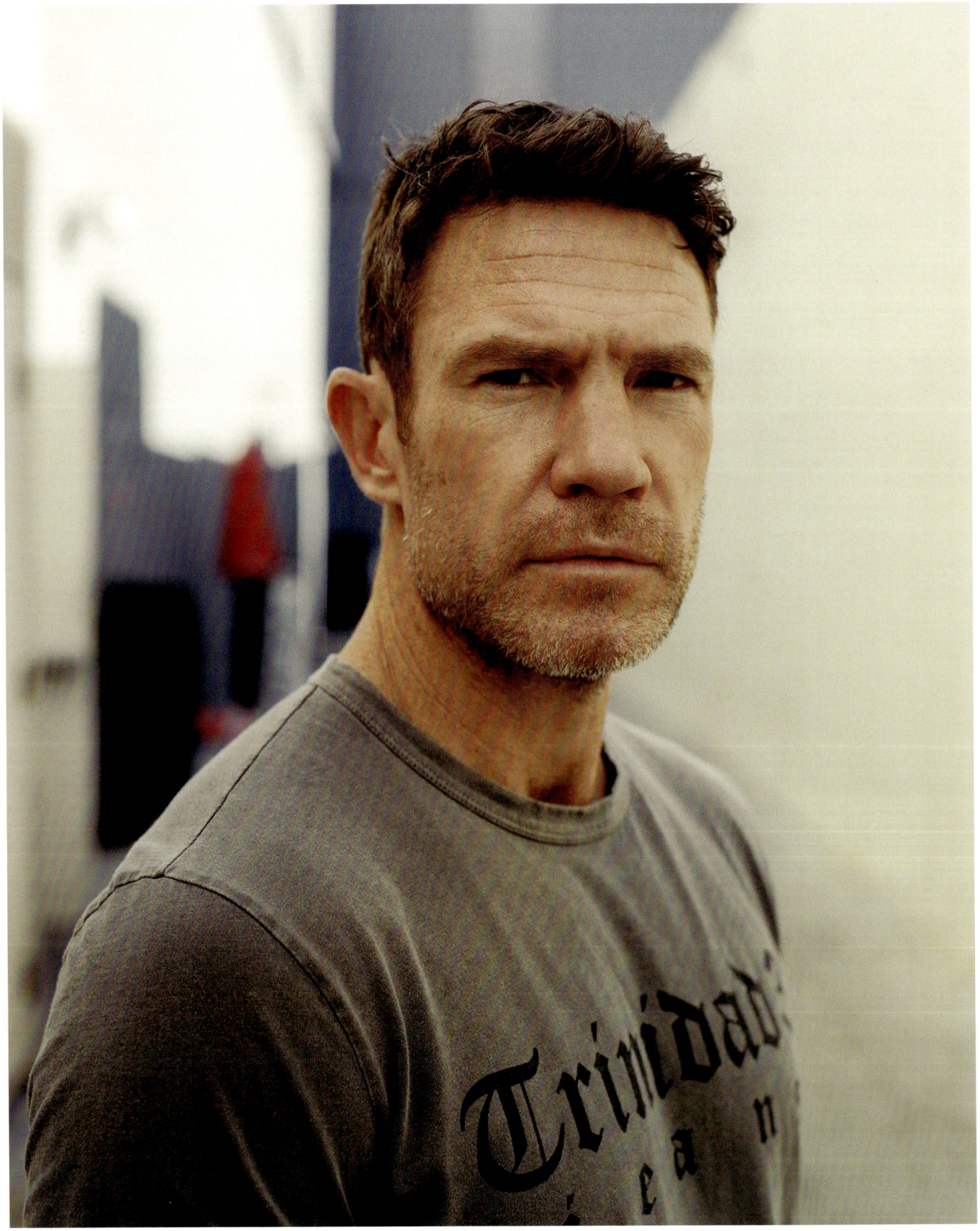

Nate Boyer, Staff Sergeant
Army - 18E / Special Forces
Years of service - 10 years
Deployments - 3 [Operation Iraqi Freedom & Operation Enduring Freedom]

My name is Nate Boyer, and I was an 18E Communications Sergeant in the United States Army Special Forces. I joined because I had been a volunteer, doing relief work near the Darfur region of Sudan in 2004. After that experience, I wanted to fight for those who can't fight for themselves. I separated from the Army after ten years of service as a Staff Sergeant (E-6). I deployed in support of Operation Iraqi Freedom in 2008 and Operation Enduring Freedom in 2013 and 2014. I was really fortunate to have football as part of my transition out of the military. I re-enlisted in the Texas National Guard when I was in college. It was an interesting experience being an older student who was half in college, half in the Army, always on the go. In 2015, I lost both my college football uniform at the University of Texas-Austin and my Army uniform. It was rough. I had so much of my identity tied up in those uniforms. That led me to co-found Merging Vets and Players (MVP), a non-profit that brings former professional athletes and combat veterans together to help them find purpose when they leave the playing field or battlefield. I'm also working in film and television and am currently in post-production on my first feature film, where I made my directorial debut. The film is about the genesis of MVP.

My advice as far as readjusting to civilian life is that if you don't know what you want, what you love, or what you're passionate about — get out there and just try things. Veterans are very capable and what we've learned and survived can be applied to almost anything. I barely made my college football team, then I barely made it on the field as a long snapper and worked all the way up to barely making it to training camp with the Seattle Seahawks in the NFL. But I made it. I kept hustling and investing time, making sacrifices, etc. The work ethic that was instilled in us is something that will always be with you as a service member. Embrace the suck. It doesn't mean just accepting a shitty situation; it means being ok with who you are and where you are in this moment but understanding that you can improve your fighting position if you want to. It's up to you. Obstacles are just opportunities to work hard, to grow, to be great.

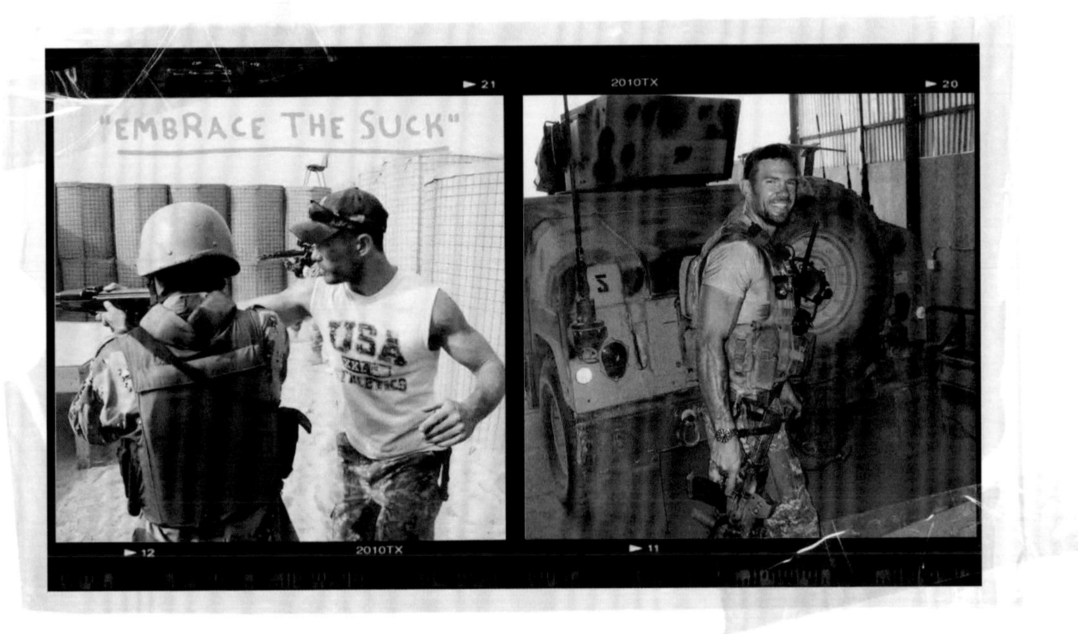

My name is Ryan Bunge and I started as an electronics specialist, and eventually cross-rated into military police in the United States Navy. I joined the military because my grandfather had served in World War II; he was in one of the LST's during Normandy. I wanted to be like my granddad and enlisted in 2000. Being in boot camp during 9/11 was crazy. We had our normal morning classes and around 9:00 a.m. they called us to form up. They said, "The nation is under attack, that's all we have. If you're from New York, you get a phone call. If not, go back to work." Over the next couple of days, we started to see all the threat conditions elevate. Everything started changing, but we never got any more information until we got out of basic. I deployed twice in support of Operation Enduring Freedom and left the Navy in 2006. The first couple of years post-separation were rough. I didn't have a lot of marketable skills on paper, and I was kinda aimless. I was also angry a lot. I'd do something as minor as locking my keys in the car and I'd flip out. Looking back, I wish I could tell my younger self to find a really talented therapist. I told myself I was fine, but I wish I had gotten help sooner.

In 2011 I got my first job at Saft Batteries as a battery systems technician, and I've been there for ten years now. I work on big batteries that work with wind farms and things like that. Batteries are really complicated devices. I was always asking our engineers how they worked. I just kept learning about them and the systems they power. In 2012 there was a job as a field tech that opened up. I didn't have the formal education, I was in over my head, but I worked really hard. I pushed myself to learn on the job and in 2013 became an application engineer, applying the batteries to specific projects. I'm still an engineer and it's taken me all over the world to install these systems. The whole microgrid world is an emerging technology field.

Unending, rigorous self-reflection is so important. It's difficult to practice and it's something that I fail at all the time, but my life is so much more peaceful and robust when I do it. When someone's driving me nuts, instead of thinking, "They're driving me nuts" I think, "What can I do better ... what am I doing that might be pissing them off?" Look, there are just ass-holes out there. If I find you really are just dead weight, so be it, but if I came to that realization after asking myself a series of questions and trying to see things from your perspective, I think it makes things better.

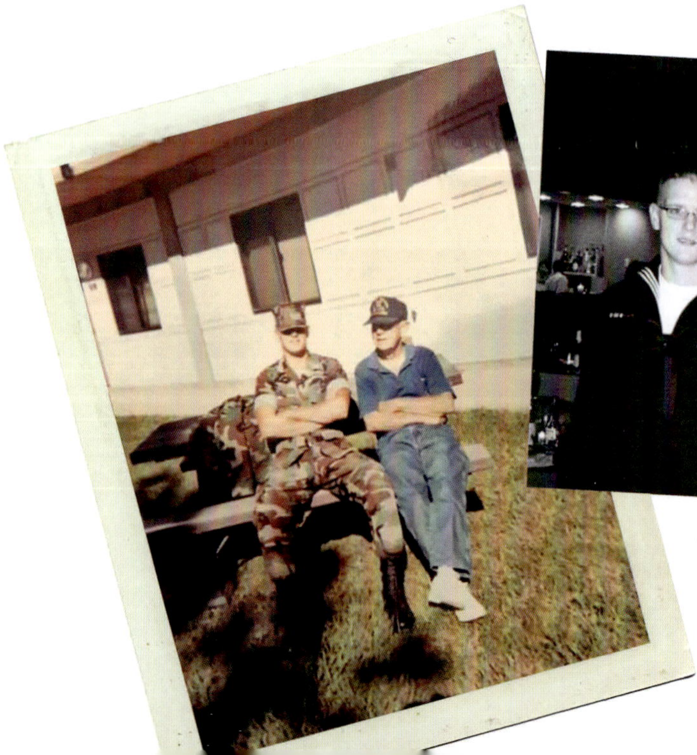

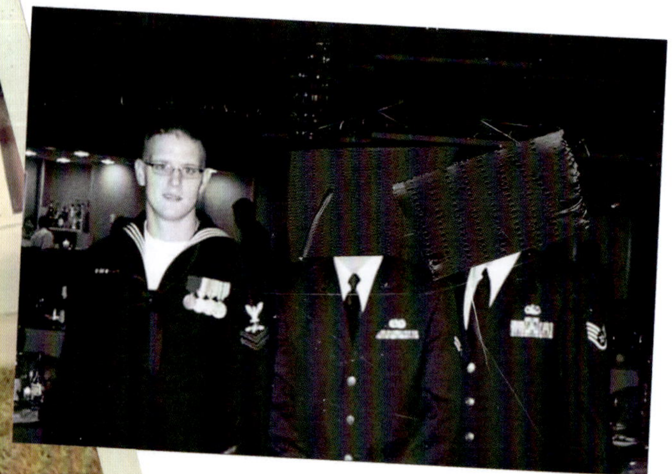

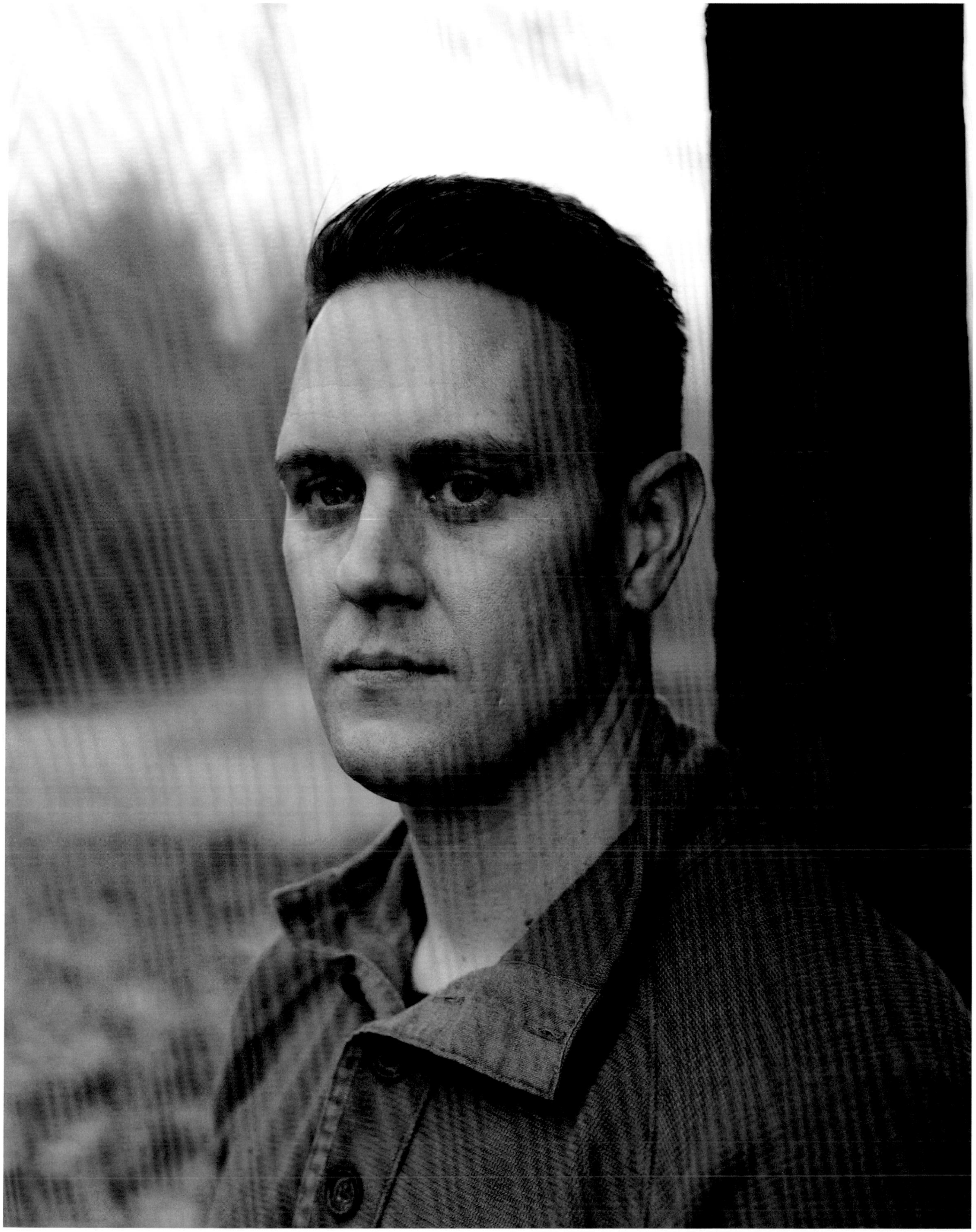

Ryan Bunge, Petty Officer Second Class
Navy - 6615 / 9545
Years of service - 5 years
Deployments - 2 [Operation Enduring Freedom]

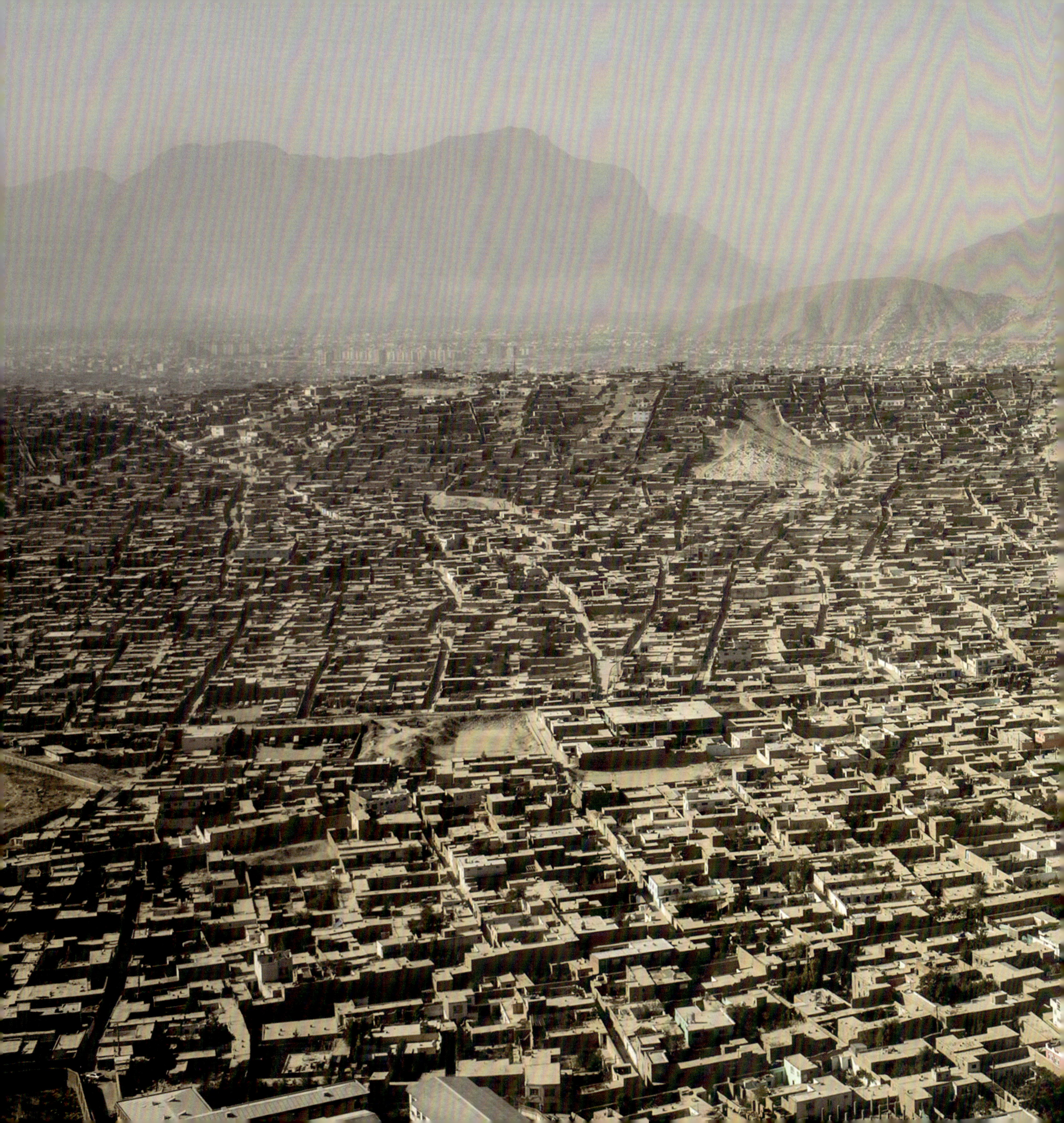

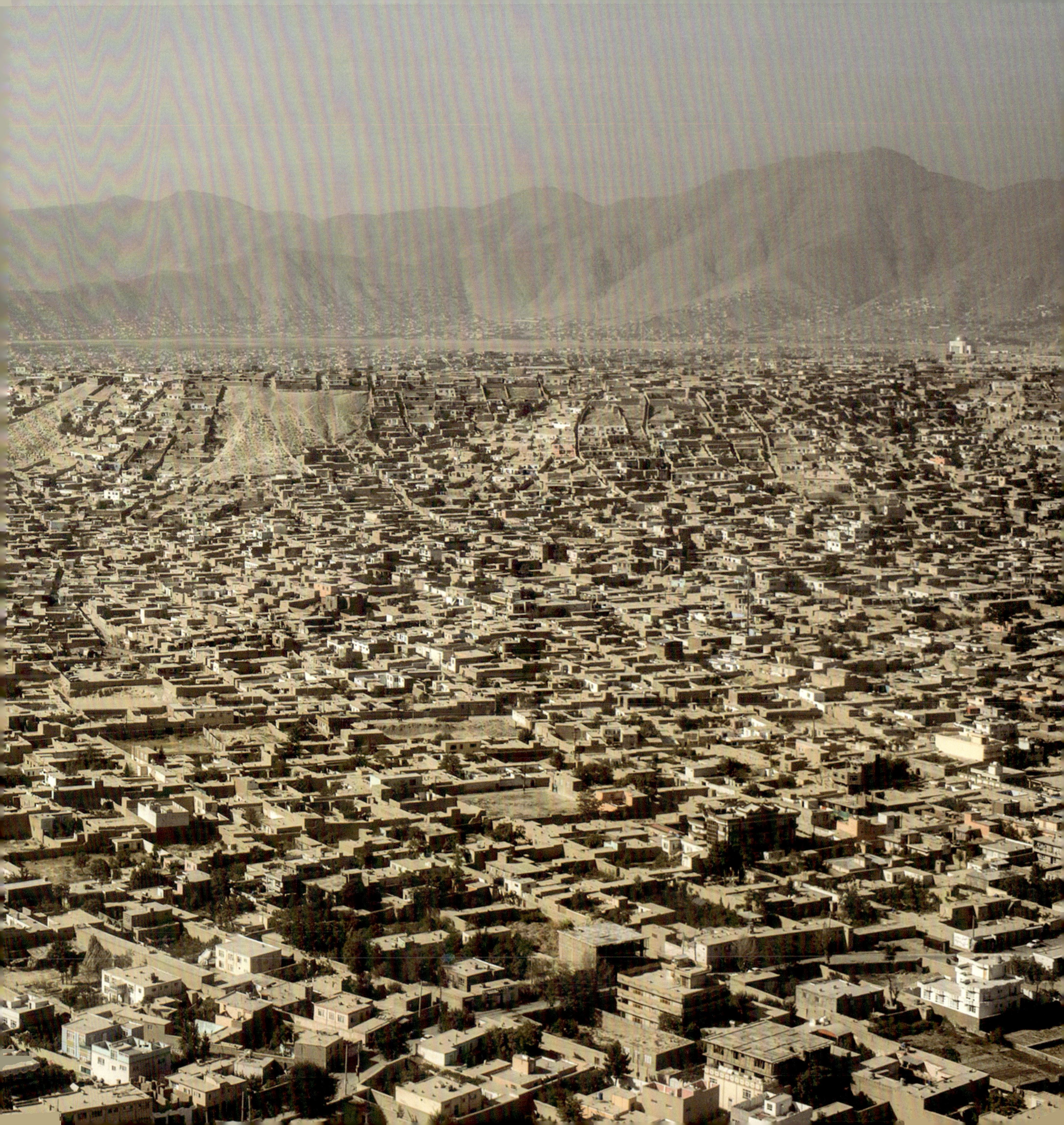

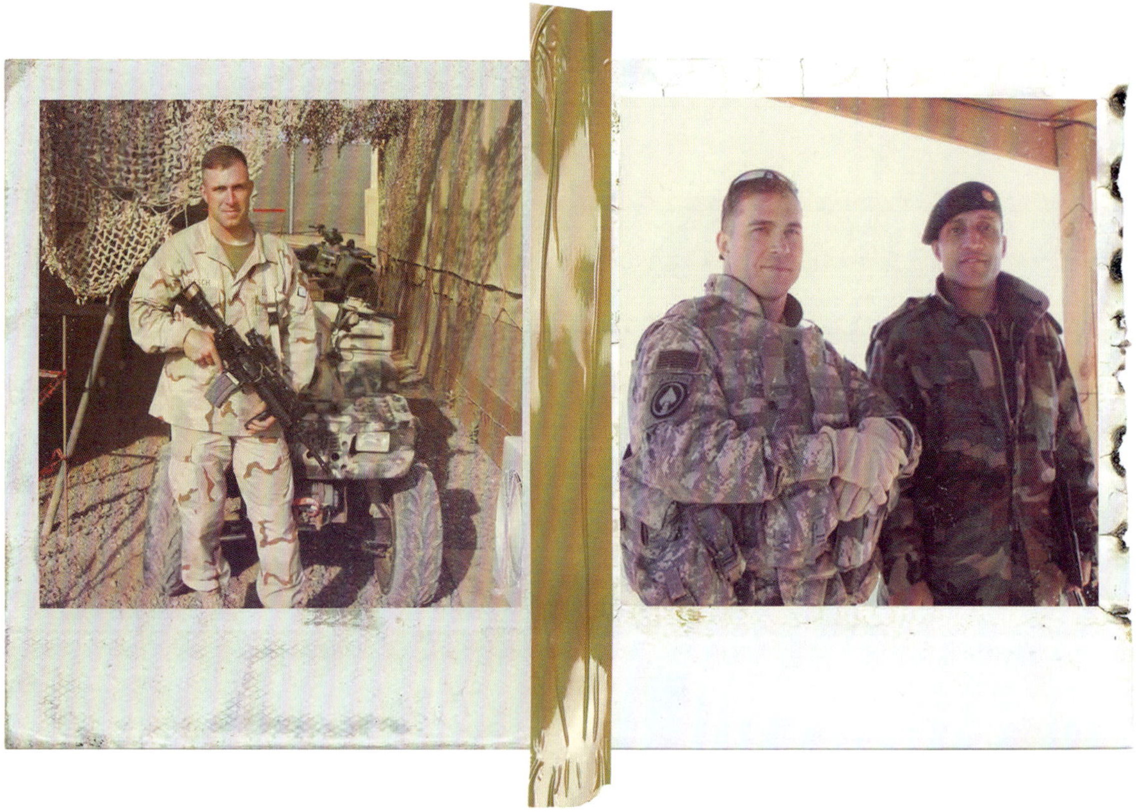

My name is Eric Flesch, and I am a Colonel (O-6) in the United States Army. I enlisted at seventeen for the college money; to leave a future of farm and factory work for something else. I was a part of the peace-enforcement mission in Bosnia, '96-97, was serving in South Korea as the Division LRSD Commander on 9/11, and deployed multiple times as part of both Operation Enduring Freedom and Operation Iraqi Freedom. I was very lucky to do these deployments with the 82nd ABN and multiple joint special operations units. Currently, I'm commanding a training brigade at Fort Jackson, SC, where I attended basic so many years ago. Before that I was the SATMO commander at Fort Bragg, NC, working to build partner capacity missions globally. The spectrum of missions was phenomenal: from training helicopter pilots to teaching counterfire radar systems to training counter-narcotics missions. It was truly a worldwide assignment helping our allies. I would travel to different countries every month. Definitely saw the world. Usually, you only command once at the O-6 level but I was fortunate that the powers that be decided to "re-green me" and allowed me command again here at Fort Jackson.

If I could change anything, I would have tried to adjust my career-family balance. A lot of the deployments were hard on the family. It led to divorce and serving as a single parent. Wisdom I give the young soldiers coming through my command is: you can achieve anything you set your mind to. I'm not in a spot most people thought I'd be. I grew up a pretty poor kid and my first plane ride took me to basic. Not many thought I'd go this far when I enlisted. You can go far if you put your mind to it.

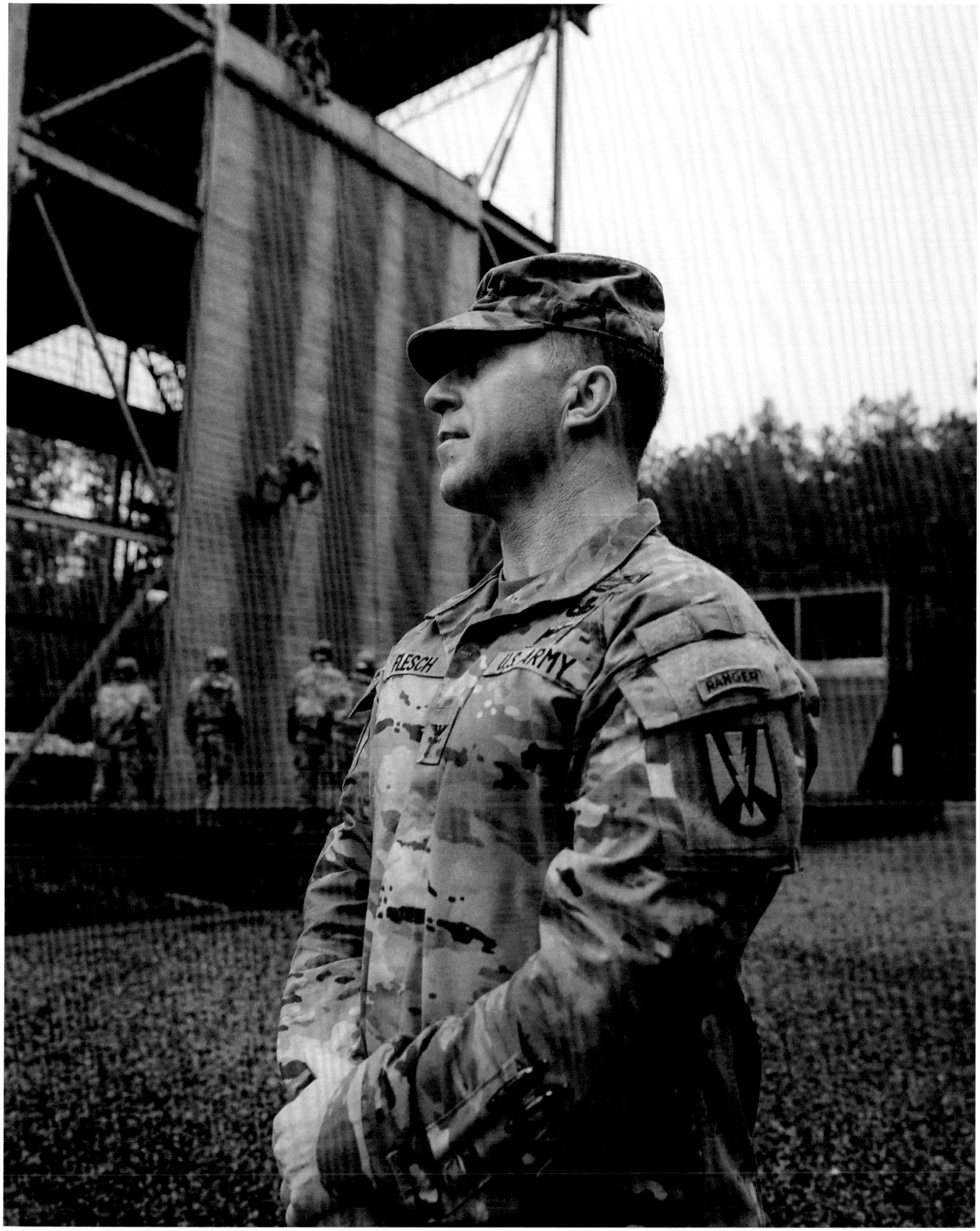

Eric Flesch, Colonel
Army - 11A / Infantry Officer
Years in service - Active
Deployments - 6 [Operation Iraqi Freedom & Operation Enduring Freedom]

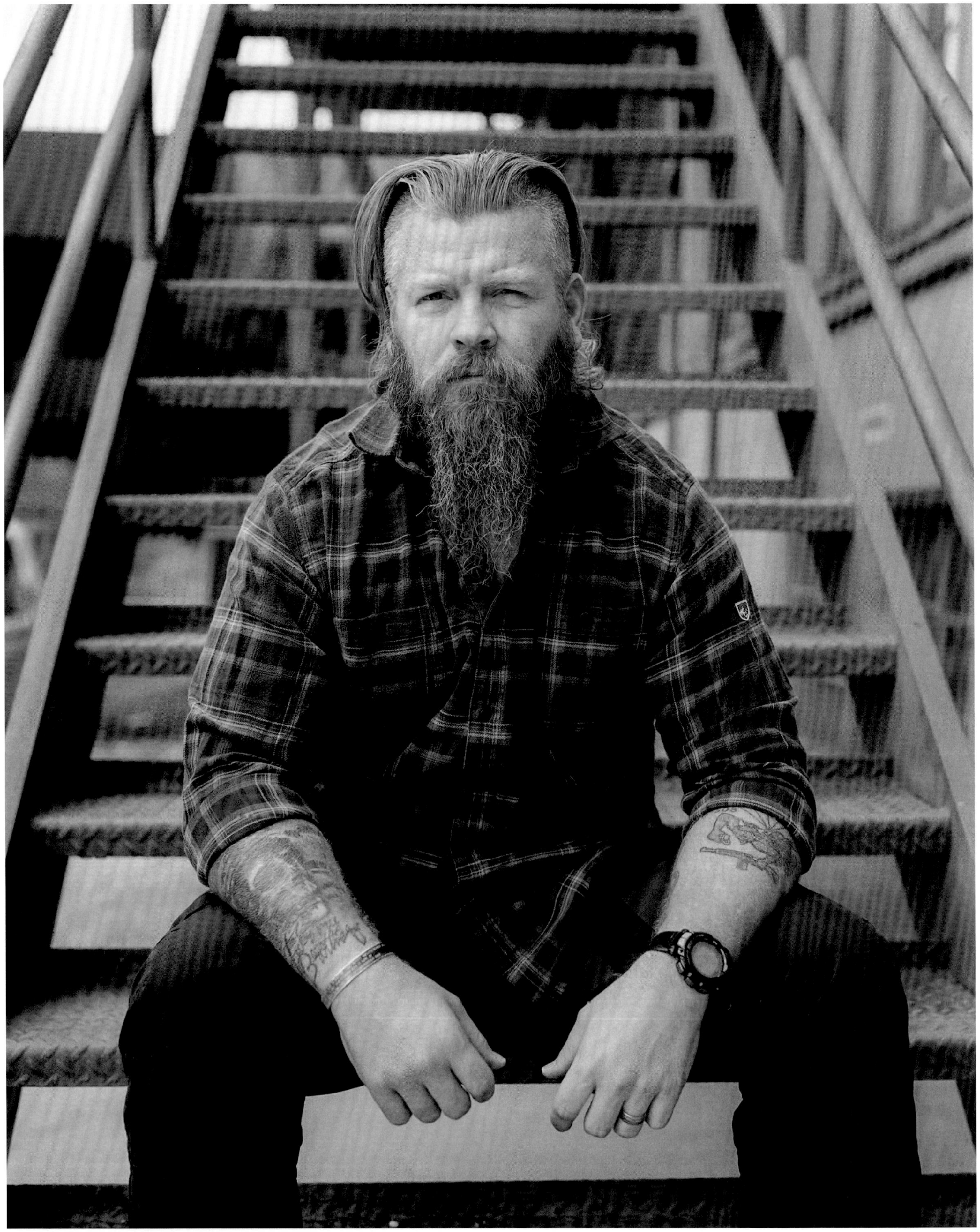

Donald McAllister, Command Sergeant Major
Army - 11B / Infantryman
Years of service - 24 years
Deployments - 5 [Operation Iraqi Freedom & Operation Enduring Freedom]

I enlisted in July of 1992, after watching the movie The Longest Day. I thought the army was cool and wanted to be a paratrooper. I became an infantryman and paratrooper and retired out of the 82nd Airborne Division at Fort Bragg, North Carolina as a Command Sergeant Major (E-9). I deployed to Operation Iraqi Freedom in 2003 and was severely wounded in an IED-initiated ambush. I've had five plastic surgeries on my face and still have shrapnel in my eye. I redeployed in 2004 back to Iraq. I also deployed as part of Operation Enduring Freedom in 2007, 2009, and my final deployment in 2013 as the Operations Sergeant Major. After retirement, I didn't want to do the typical SGM thing and work on-post at range control.

It hit me really hard when I left the service. I was an angry son of a bitch. Finally, my best friend and my wife had a come-to-Jesus moment with me. When I took a step back, and did the work, I realized there was twenty-four years of guilt and pain from all the guys who didn't make it back or are still killing themselves. Since retiring, I continue to stay involved in the veteran community in every way I can, working with numerous non-profit organizations that serve our nation's wounded warriors. I now have the purpose of using my experiences to show how to get through things as a civilian. To be open and transparent, because there isn't a lot of transparency in the veteran community. I still struggle, but I talk with the guys and we do the work. We all handle separation differently, but we are all still going through similar issues and struggles. It ain't about what we did; it's about what we're doing tomorrow.

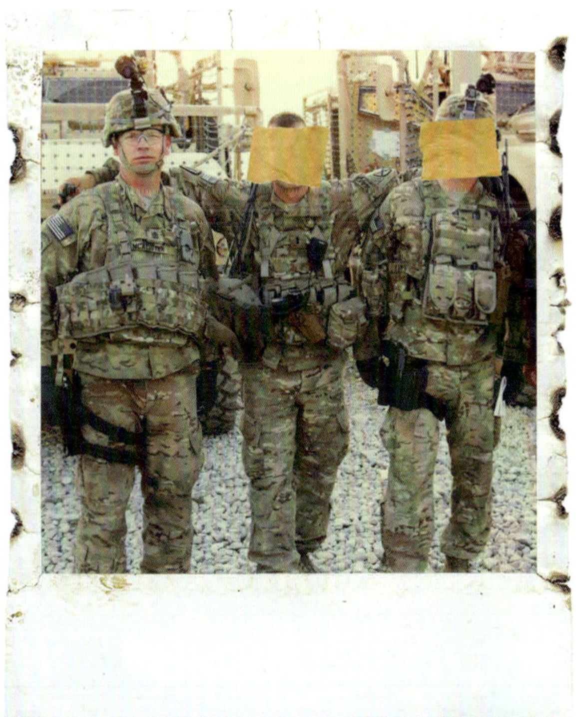

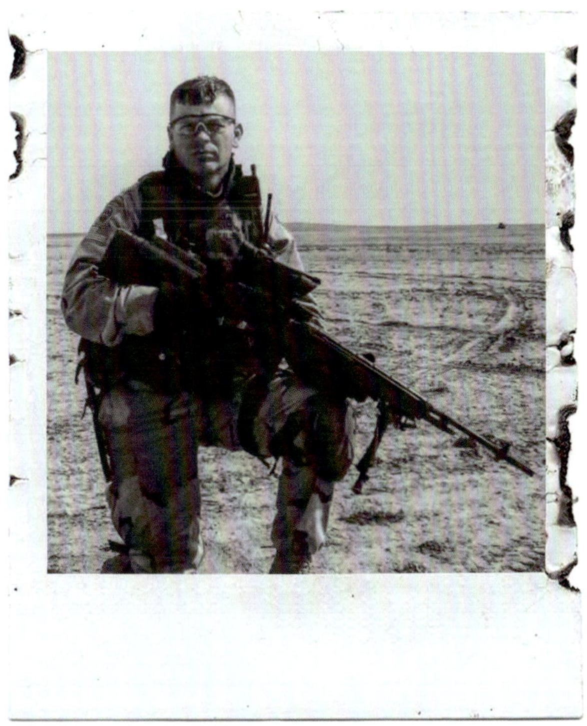

My name is Jose Luis Sanchez and I was a communications specialist, 0621, in the United States Marine Corps. I joined the Marines to do something with my life. I was in and out of trouble as a kid. During a ground patrol with the Brits in 2011, two weeks from coming home, I stepped on an IED and lost my left leg in support of Operation Enduring Freedom. I was hospitalized in San Antonio, TX and medically retired out of the Marines as a Staff Sergeant (E-6). I had also deployed in support of Operation Iraqi Freedom in 2006. Readjustment to civilian life was really hard. I was angry at the world, angry at the military, and was blaming others at the time. It opened up a flood of negative feelings and that mental aspect has been the hardest to deal with. Learning to trust people — to keep relationships with friends and family and kids — and opening up has been a struggle.

Fitness has always been a way of life that I have turned into my non-profit, Rise Above Hardship. Trying to invoke purpose in the human spirit is my passion! Connection and purpose form the core of who we are. We're beasts in the military, and we need an outlet for it and to keep using that drive. To be bonded through strength. I started Rise Above Hardship because the gym has always been a part of my life. I use fitness as a medium to help others rise above their own hardships. I can control my fitness; it's an outlet and a healthy, therapeutic practice to let out my anger and anxiety. I'm also an ambassador for Howitzer Clothing. They have been really good to me and help support the gym and non-profit. No matter what, continue to develop yourself. You have to keep working and growing. There is always something else; it's a continuous process. Every day is a fight, so continue to fight!

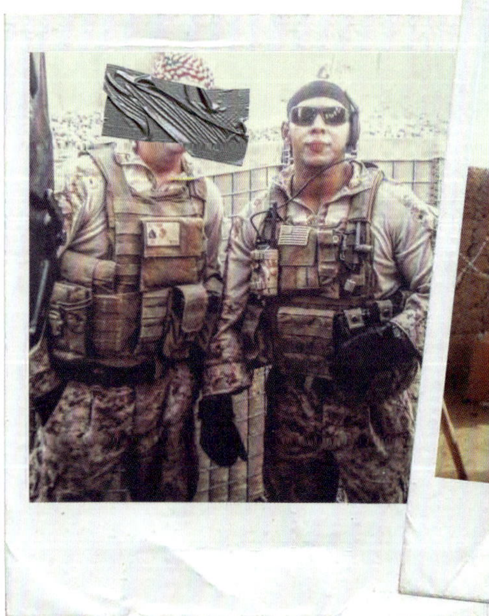

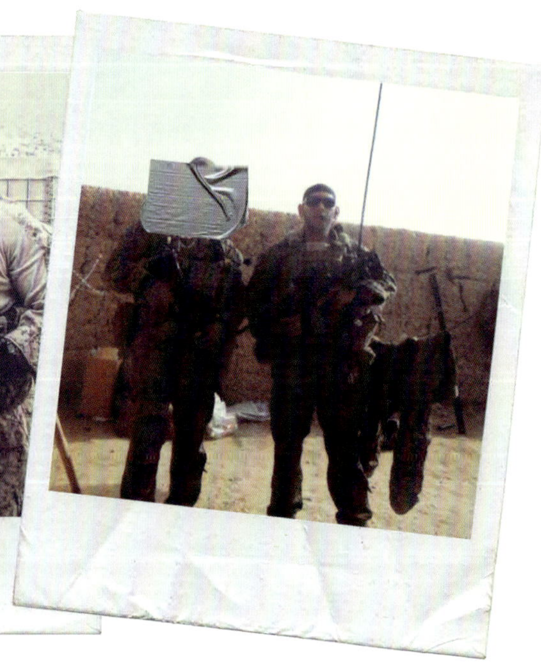

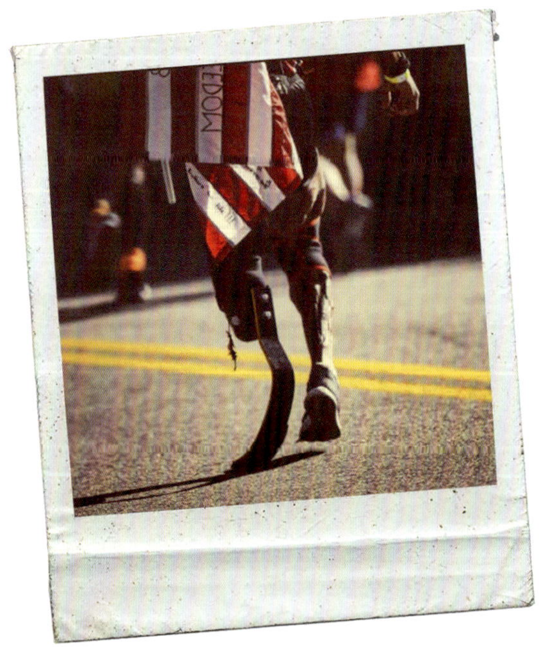

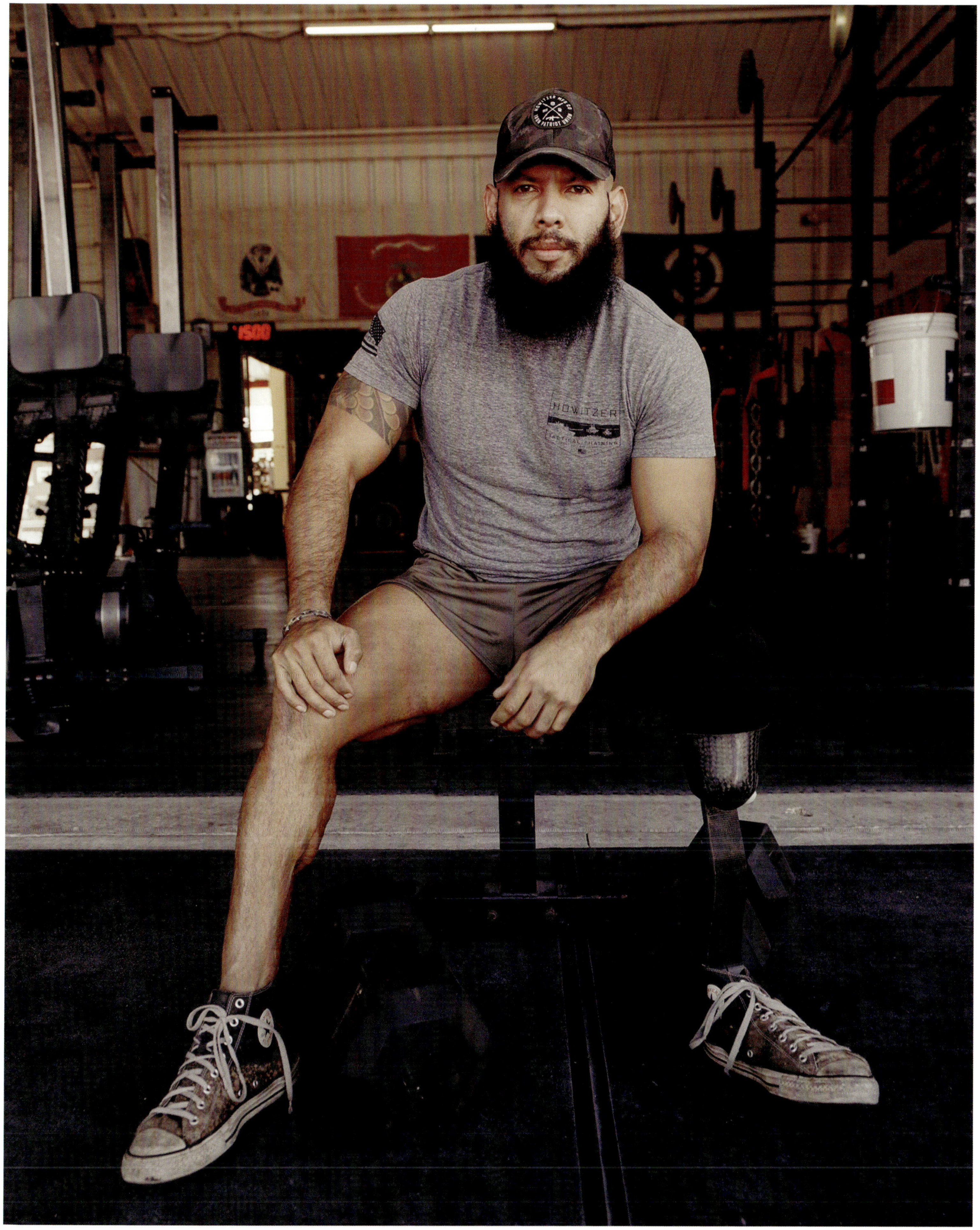

Jose Luis Sanchez, Staff Sergeant
Marine Corps - 0621 / Field Radio Operator
Years of service - 10 years
Deployments - 2 [Operation Iraqi Freedom & Operation Enduring Freedom]

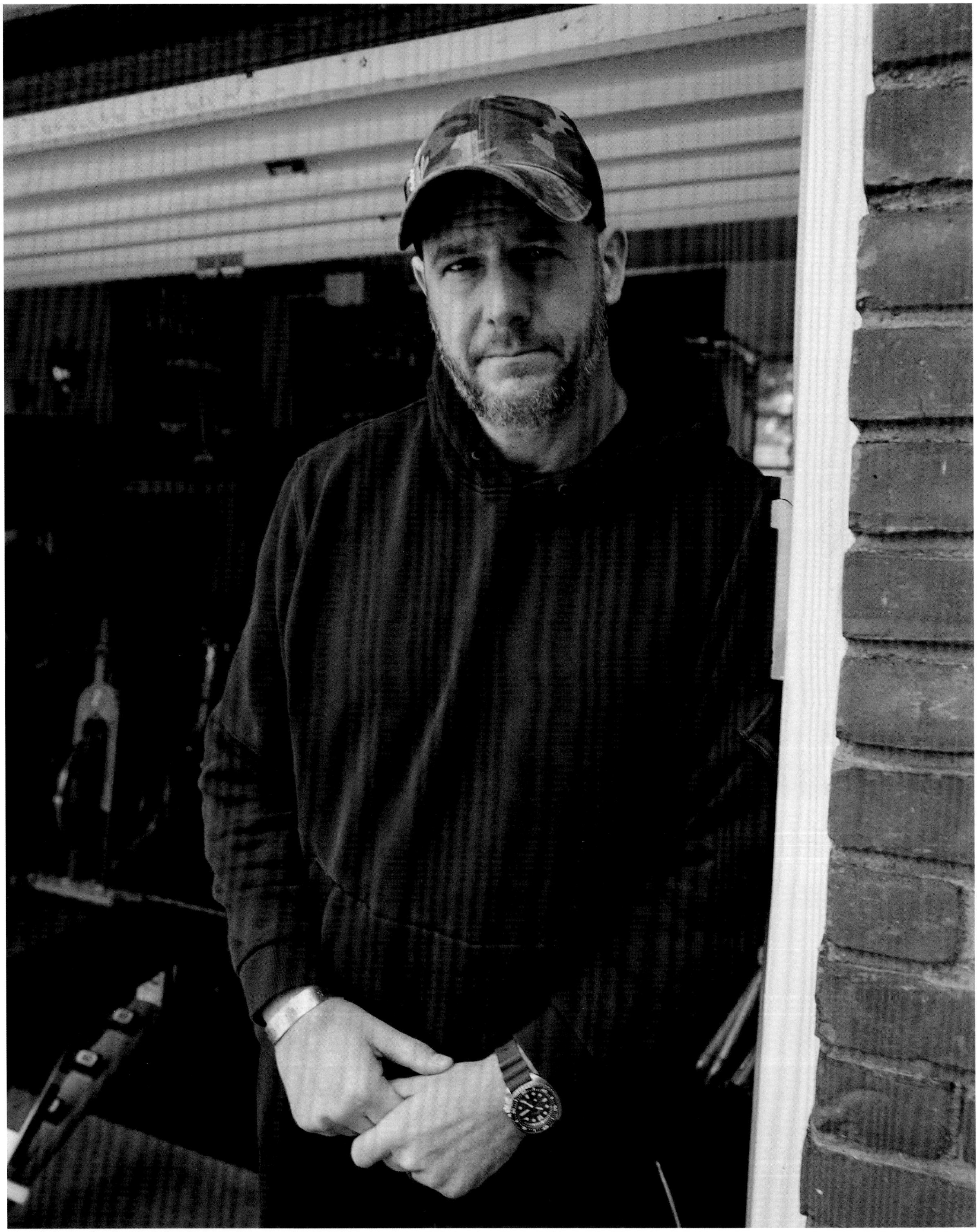

JC Glick, Lieutenant Colonel
Army - 11A / Infantry Officer
Years of service - 20 years
Deployments - 11 [Operation Iraqi Freedom & Operation Enduring Freedom]

My name is JC Glick, and I retired as a Lieutenant Colonel (O-5) infantry officer in the United States Army. I was given an opportunity to go to college and have a career and I felt like that could only happen here in the United States. I am truly grateful. I first deployed as the Ranger Regimental Air Officer in '02, I then did another two deployments with them before taking command of C Company in the First Ranger Battalion, 75th Ranger Regiment and did two more deployments in that position. I then went to the Asymmetric Warfare Group and continuously deployed through 2011. In total I had eleven deployments, spanning both Operation Iraqi Freedom and Operation Enduring Freedom. I was part of the initial invasion of Iraq, among many other missions. Every rotation we were doing something that mattered.

Now that I've separated, I consult for organizations and sports teams on how to get more out of their people and establish deliberate cultures. I've worked extensively with the NFL, including being a keynote speaker at the owners' conference, the NBA, and NCAA (I've worked with student athletes in football, basketball, tennis, and equestrian.) I'm also currently doing some management work for a virtual reality company and am also a writer. In my free time I love to surf and lift weights, and I love to read.

There were three transitions I went through; I don't feel like it was just one. The first was when I stopped deploying to combat. That was the most difficult one for me to adjust to. There were some very low points, and I tried to take my own life four separate times. My second separation, my separation from military service, was actually pretty seamless. I had an offer in hand to be a partner at a large consulting firm. I'd also already written my first book. Finally, this past year, I really went through what I call my third transition of "I'm out but who is my tribe?" I don't want to say it's been difficult, but I have grown and felt the shift. What you do when you leave the service is way less important than who you do it with, in my eyes. You need to find your new tribe. You need like-minded people who share your values. I need people around me who are driven, thoughtful, and curious. Not who are the same, but who are similar to me and hold similar ideas. Which has led me to explore these ideas more thoroughly as a writer.

I am currently working on a paper that argues Maslow's "Hierarchy of Needs" is antiquated and wrong. It's a theory that people have really latched onto, but the data doesn't really support it. We've learned more about the human brain in the past five years than we had in the previous thirty. I'm putting the data together to prove that the real base of human need is connection. That we need our tribe. Anecdotally, the vast majority of homeless in NYC have cell phones. They don't have a home, but they have Facebook, Instagram, and most other social media platforms. They need to be connected. There are charging stations all over New York to assist this program. They aren't off the street, but they are more connected to the people and world around them.

Continued on next page →

The way our brains work is also fascinating. When you study the human brain and see where our brains light up and what happens when we connect, it's near our lizard brain. It's in our baser instincts. It's amazing! Our desire and need to connect is rooted in our evolution. I really believe connection is the building block of human interaction, all the way to believing our purpose is to connect. That is our core purpose: to connect with other human beings.

I am also working on two books. One is on common sense for 2021. Another I am working on with fellow veteran Steve Forti and it centers around the concept of the "unhero." Everyone around us in the civilian world wants to be heroic. When you break down the guys who are "heroes" in the military, it's those that are elevated above others for things they did. Well, they don't think they're heroes. When they did these great things, they weren't trying to be heroes. They did what they were supposed to do. They just did the right thing because it's the right thing to do. Working on being the "unhero" is powerful. Calling 911 doesn't make you a hero, it just means you did the right thing. But there is so much power in that! When I speak with separating service members, I say that I believe there are four things veterans and transitioning service members need:

1. You need a civilian mentor who has not been in. We need to break out of that echo chamber of the military and the veteran community. We need to get to know the other 99 percent of people. We need to speak that language.
2. You need a mentor who is younger than you. There's always a new guy when you're in. They challenge you and they push you and the unit to be better. They help you stay up on what is new and what is changing. They bring in new ideas. It's important to know what is going on. Again, it helps keep you out of that echo chamber.
3. I really can't say this enough: Find your tribe. You need to find your people, so you aren't alone or feeling lonely.
4. Stay physically fit! There is body brain chemistry that is a real thing. The more physically fit you are, the better you perform. Socrates said when you are sick, you don't think as well. Science has now proven it. When you work out and when you are fit, you perform in life better. So, this idea of "I don't need to work out anymore, I'm out." Well, you do. And that's scientifically proven. If you incorporate all four of those in your transition plan you not only feel better, but you usually end up becoming a leader in your community. Americans look to former service members as leaders; it's always been that way. We have a responsibility to step up and lead. We have a responsibility to help the tribe.

"Lead by example" is a myth. To lead is a verb. It requires action. "Lead by example" is passive. It just means you're squared away. If a kid is underage at a party where kids are drinking and he doesn't drink, good for him, but what does that mean for the group? It means nothing. You have to put yourself out there, you have to have the courage of convictions that inform your way to make a decision.

To lead requires action. Live up to the Ranger Creed. Be curious and be kind. There is so much power in being kind and elevating the mood of the people around you!

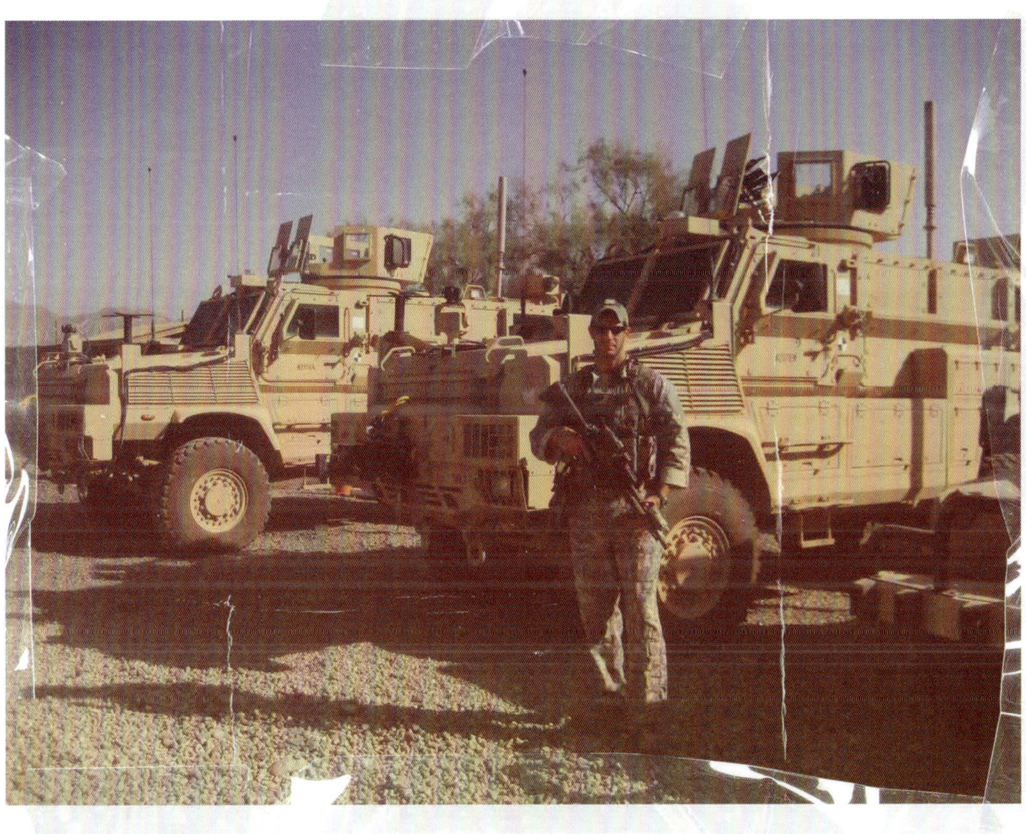

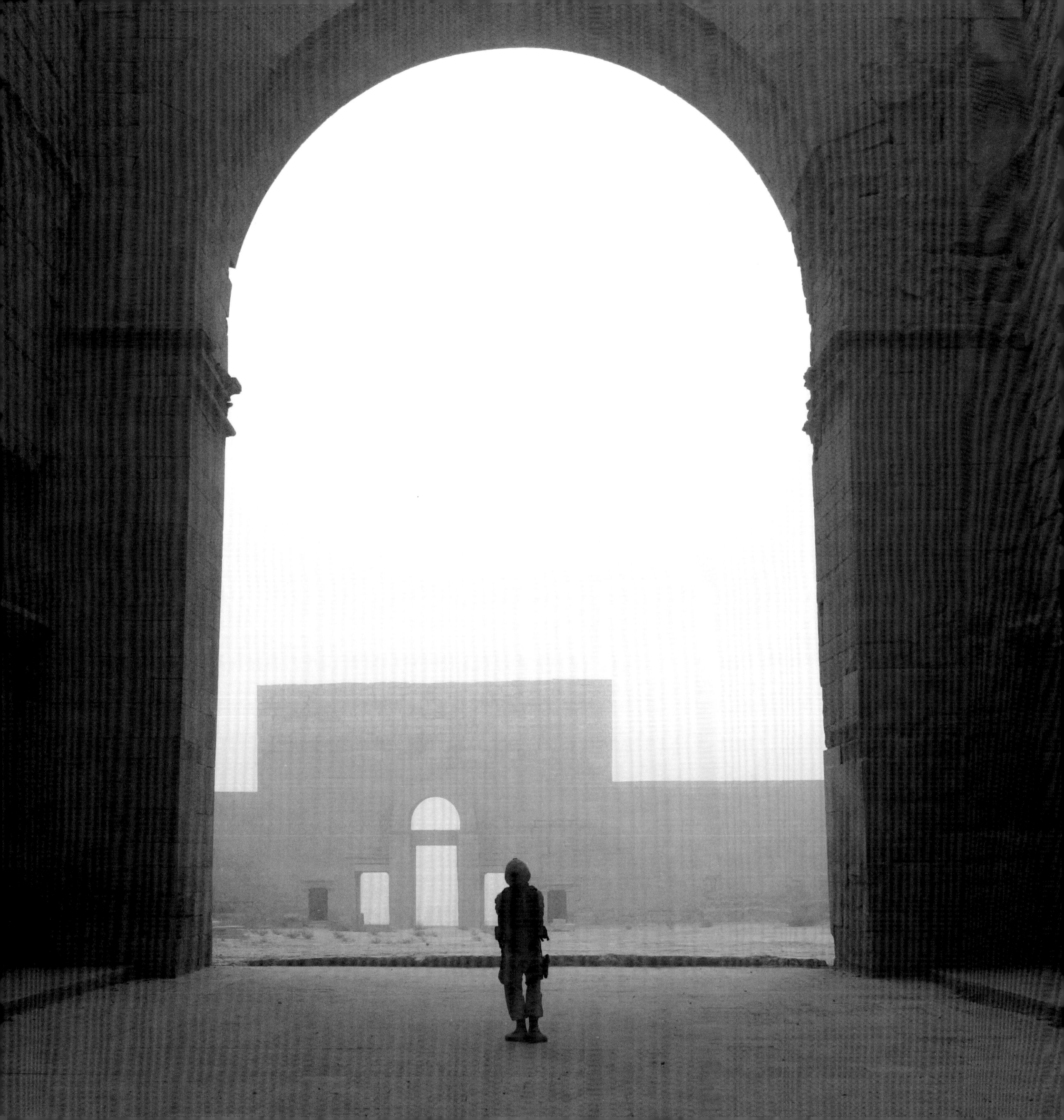

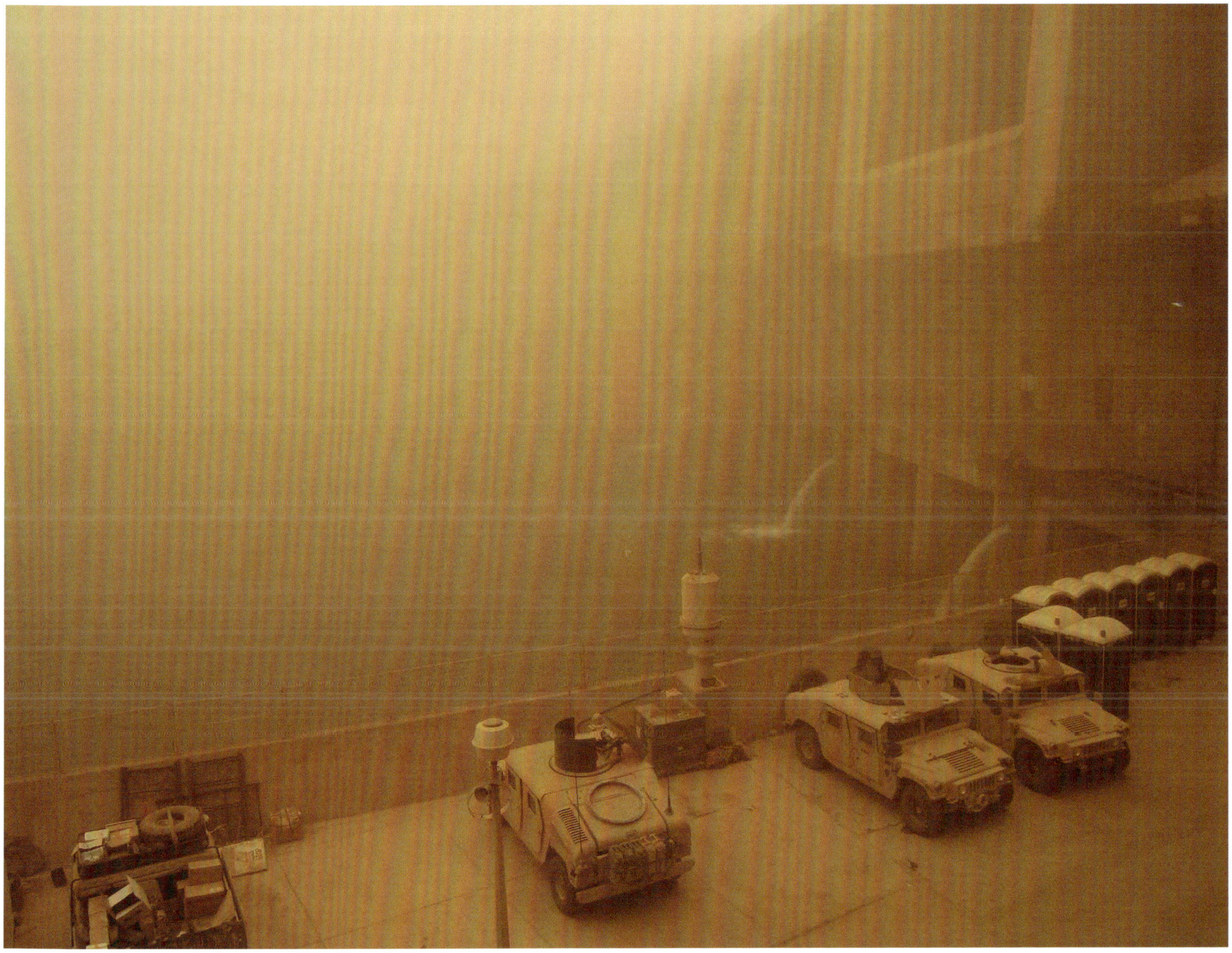

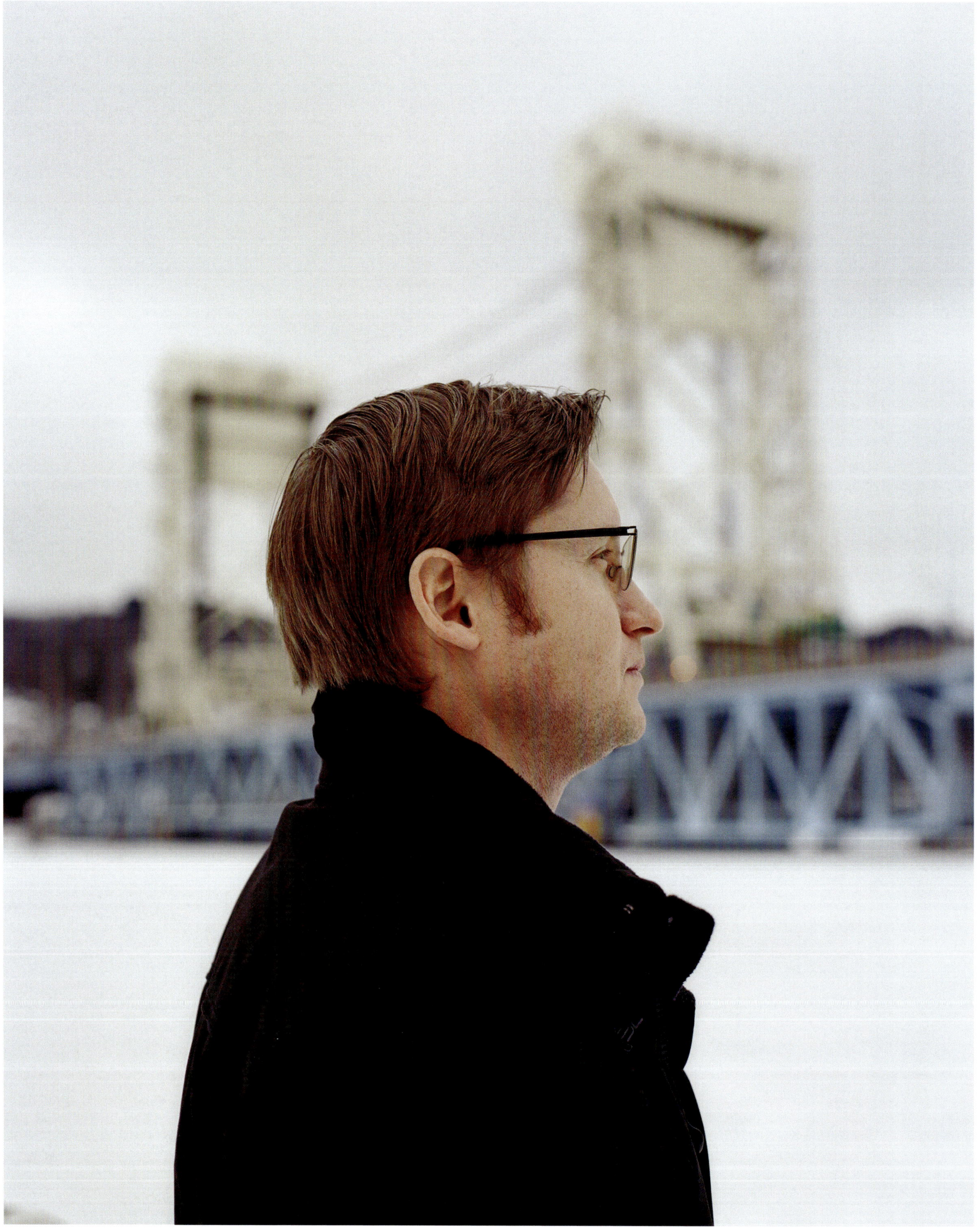

Andrew Hillstrom, Staff Sergeant
Army - 11B / Infantryman
Years of service - 5 years
Deployments - 1 [Operation Iraqi Freedom]

My name is Andrew Hillstrom and I served in the US Army as an airborne infantryman and was medically retired after an IED [improvised explosive device] went off while I was walking a patrol as a dismounted scout team leader in support of Operation Iraqi Freedom. I was roughly two or three meters away from the IED when it went off. I have had roughly twenty-five surgeries during my final two years of enlistment because of the blast.

Initially, I was enrolled in college, but I wasn't ready to finish, so I answered the call to serve. After my medical retirement, I went back to school and completed both undergraduate and graduate studies. Then bounced around as a construction manager in the Midwest and in Washington, D.C. for major building projects like luxury apartment complexes and shopping malls. While in D.C., my wife met the wife of a diplomat, and convinced me to take the Foreign Service Officer entrance exam.

My transition was tough. To me it felt like you lose your sense of purpose. It felt like it was just gone. I also was stuck with certain memories from my experiences in the military. The military gives you a sense of purpose and you have that daily drive to accomplish the mission, but when you get out, that's just gone. When you are a veteran transitioning, the best advice I have is: Leverage the experience you gained in management during your time in service but understand you're in a new place and you need to re-prove yourself. You can't rest on your laurels. You need to prove yourself and continue to push and grow. My personal mantra is "Lead by example." If you lead by example, people follow you and are far more likely to work with you. When you have those things going for you, good things happen.

"Lead by Example."

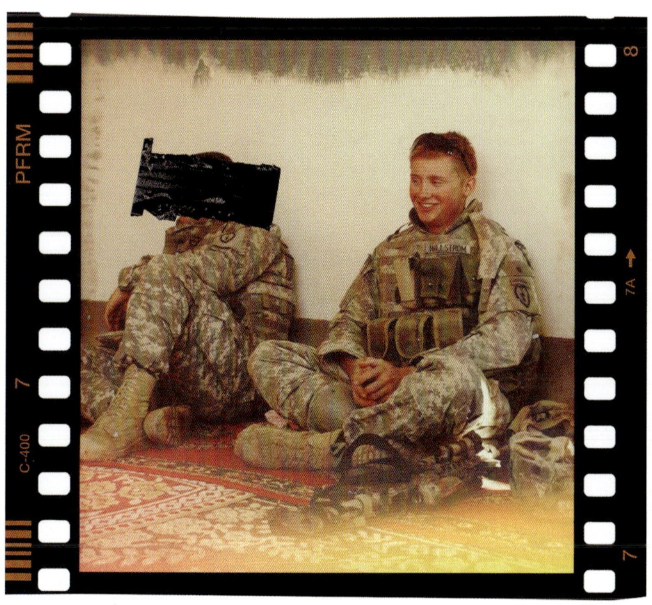

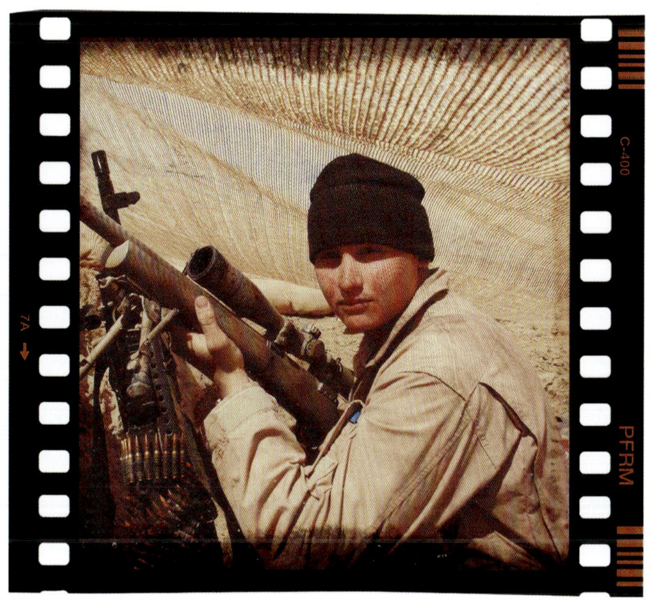

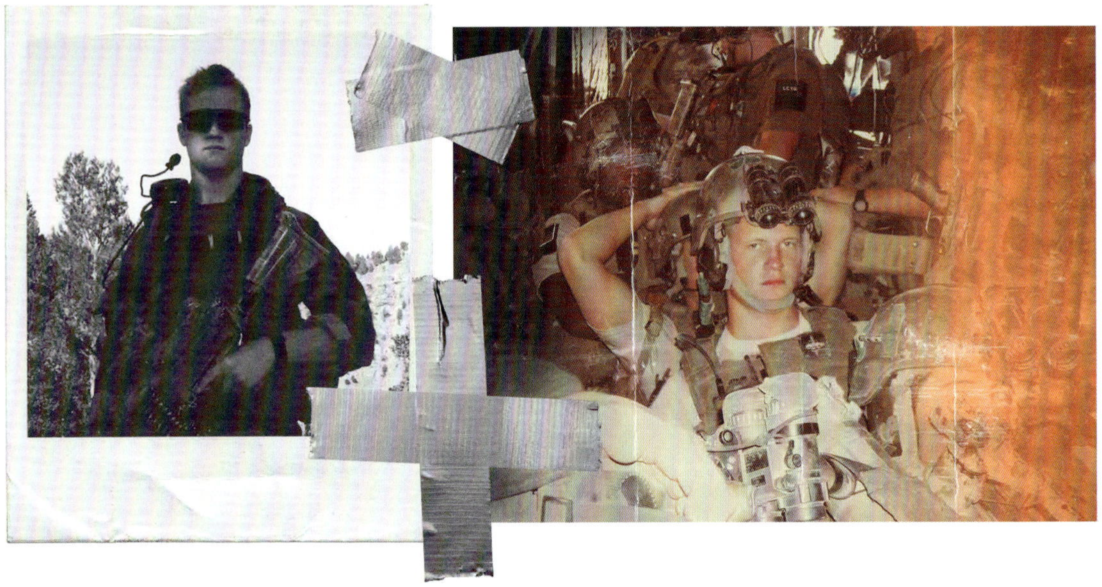

My name is Jordan Hillstrom. I served in the infantry as a member of the Second Ranger Battalion, 75th Ranger Regiment. I served on three deployments, two in support of Operation Enduring Freedom and one in support of Operation Iraqi Freedom. I separated from the service as a Sergeant (E-5). One of the biggest influences that got me interested in joining the military was Mr. Dean, my fifth grade teacher. He often spoke highly of the military and encouraged us to respect those who fought for our country. A few years later, I was in the eighth grade. One fall day, our teacher dropped an armload of papers as she saw an airplane crash into the World Trade Center. The horrible experiences of 9/11 really made an impact on me and ultimately led to joining the military a few years later.

I'm now the Vice President of Business Development for my software company, Xeratec Software Development. Our mission is to provide hope for a better day tomorrow through the technology we create. This new mission gets me excited about life daily. I also work with fellow veteran entrepreneurs on their ideas. One of the coolest projects I'm working on is Apex Target Systems. It's a reactionary modular target system designed to help train the military and law enforcement more effectively. It's amazing to see a vet take an idea, work with other vets, and realize their dreams.

These days, things are going really well. But it hasn't always been this way, particularly when I first got out of service. In the military you rely on the brother and sister to your left and right. After the military, I completely cut ties with those I had worked so closely with, or even the idea of relying on others. I struggled to transition to civilian life. For nearly five years, I felt like I was on an island. So alone. Eventually, I saw a counselor and realized what I'd done and started to plug back into the community. Now I work with some great organizations, in addition to helping veterans' small business ideas.

Live everyday like it's your last and treat people with respect and empathy. I've done enough to see that everyone is going through something. Kindness costs us nothing.

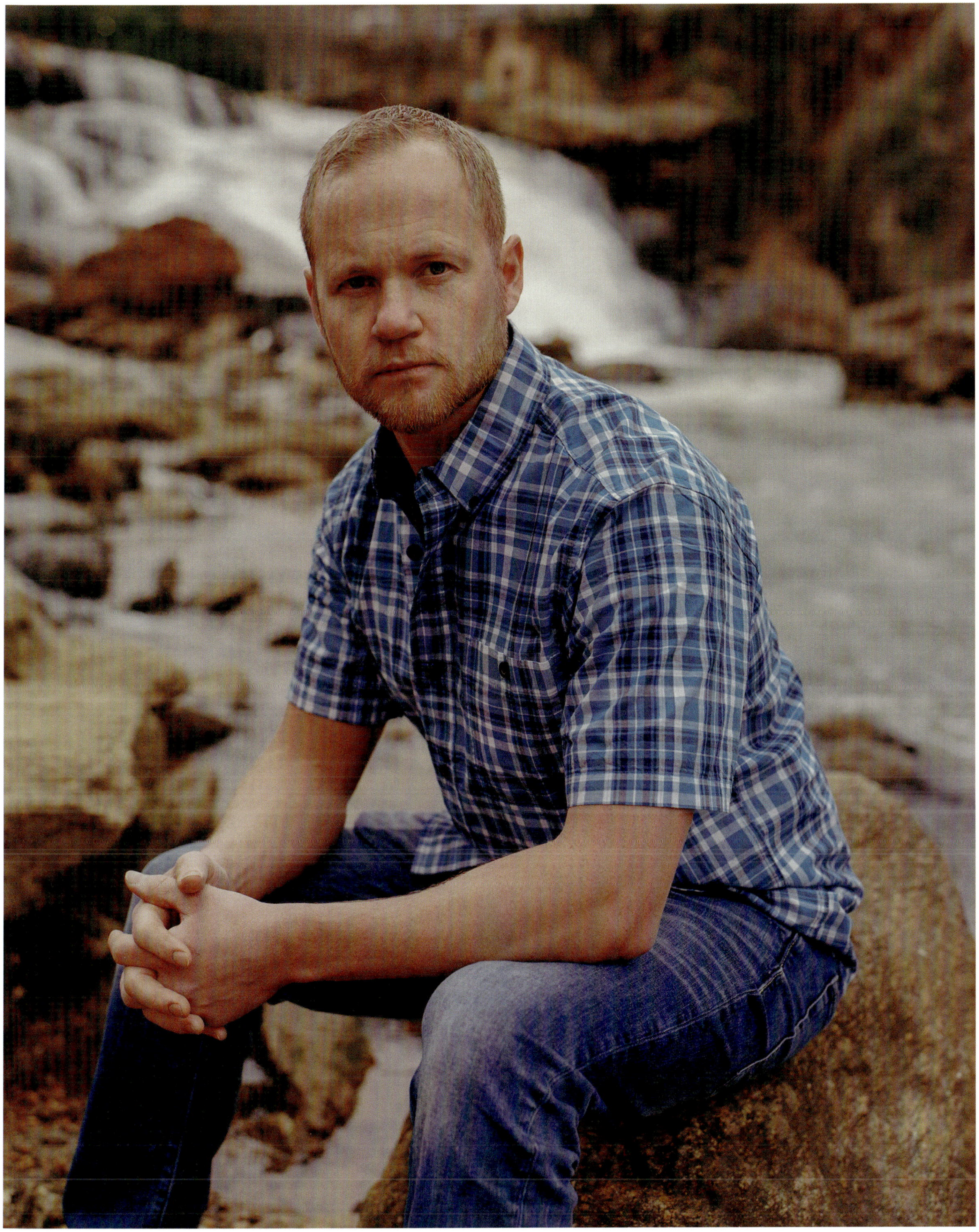

Jordan Hillstrom, Sergeant
Army - 11B / Infantryman / 2/75 Ranger
Years of service - 5 years
Deployments - 3 [Operation Iraqi Freedom & Operation Enduring Freedom]

Drew: "I am the second-oldest of sixteen siblings, and Jordan is only fifteen months younger. We were super close in age growing up. I joined in 2004 and was the first person in the family to enlist. That helped pave the way for Jordan, I think."

Jordan: "My brother was my best friend growing up and my idol. It was natural for me to also serve."

Drew: "Growing up, we were a hockey family and since we were so close in age we were usually on the same teams. We had the same circle of friends. I boxed in high school, so we never really fought; I think he knew what was up [smiles]. We also worked on the same farm as kids and we were super close compared to some of our other siblings."

Jordan: "We ended up in separate units and honestly, we didn't really keep tabs on each other a ton. There were a lot of military schools and deployments going on."

Drew: "There was a lot of our parents passing messages back and forth to us when we called home."

Jordan: "I was in Afghanistan when Drew got hurt in 2007. We had just gotten to Bagram Air Base and I saw an AOL instant message about Drew.

"Instantly my stomach was gone and I went right to the worst-case scenario. At first my team leader thought I was joking when I told him, so I actually went out on mission that night, which was a really bad idea. I notified my squad leader after and once that happened, the ride began."

Drew: "I, of course, don't remember any of this [chuckles]."

Jordan: "As more information came, we were told Drew's death was imminent, and the Army network just kicked in. They quickly got me on a flight to Iraq to meet up with Drew. As I was in the air, headed to Iraq, they told me he wasn't there anymore. So we landed, I waited a couple hours, and took off again to Qatar.

"Once we got there I was told I was gonna be stuck for eight hours to clear customs and paperwork and stuff like that. I was deflated. I was still being told my brother was going to die. I started to break down a little bit, my brother, my hero, was hurt and as it's happening, someone knocks on the door of the airplane.

"'We're looking for Private Hillstom,' the man said. He's in a full suit, driving a blacked out suburban. No idea who this dude is. 'Come with us,' he said.

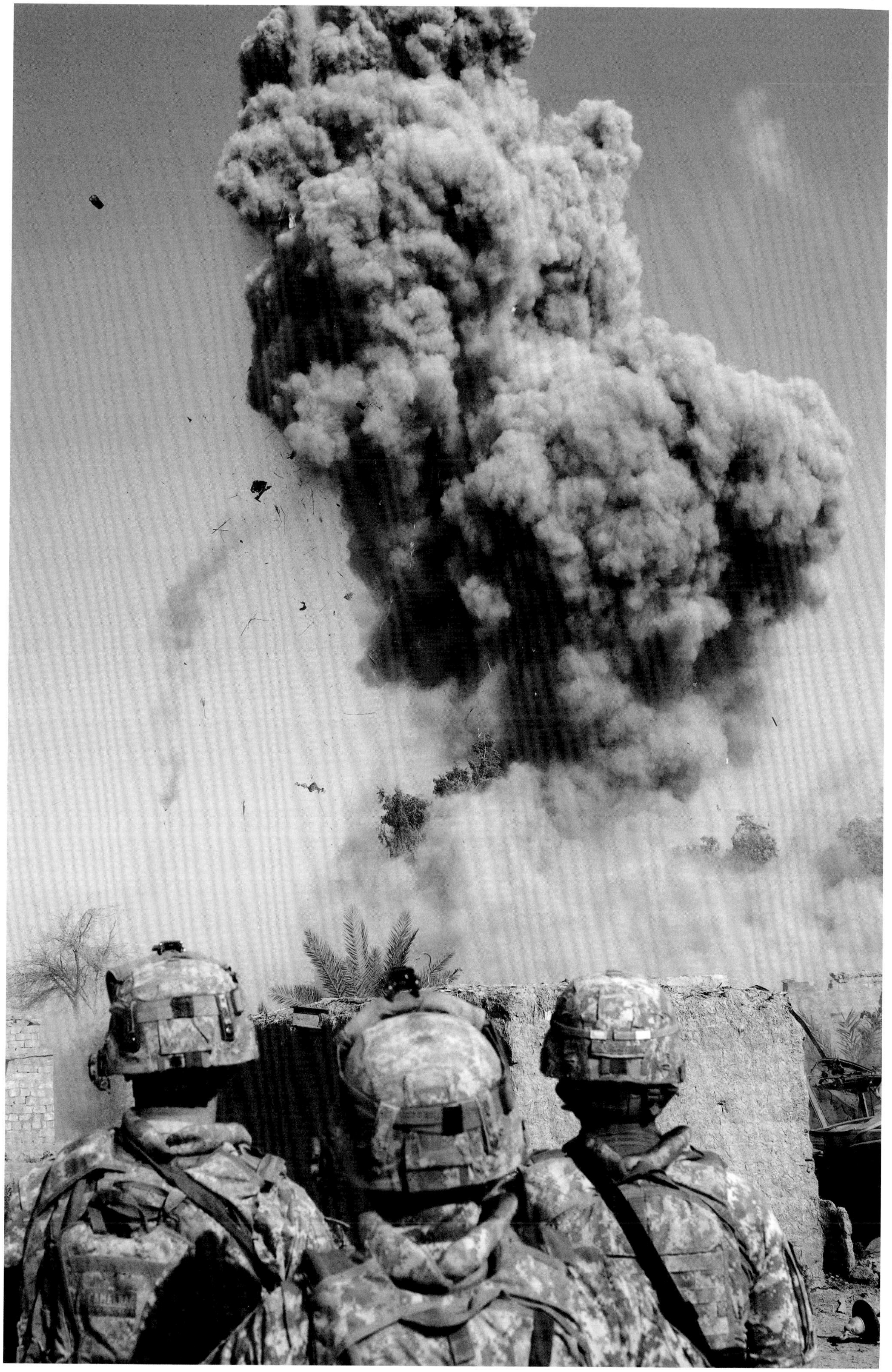

He grabbed my stuff, loaded me in the back of his SUV, told me to lay down, threw a blanket over me, and off we went.

"To this day I have no idea who he was or where he took me, but I got a shower and some food. It was so crazy. They gave me civilian clothes and some paperwork magically appeared to get me on the civilian flight out of Doha. The next thing I know I'm waiting for my flight to Germany.

"I got there, met my contact, and they told me I had just missed Drew and he's already on his way to Washington, D.C. but he's alive.

"It was so confusing, and I'm again in another airport not really knowing what's going on, so I spent forty-five bucks for a payphone call. Again, whatever magic was going on behind the scenes for me, just happened. In like an hour I had a plane ticket, and eventually landed in D.C.

"You need to understand, this all happened really fast. From the time Drew got blown up until I got to see him was about a day and a half. I hadn't even talked to my parents yet. The logistics and the support that everyone in the Army had to support me getting to my brother was insane.

"As I walked into the room, his head and shoulder and leg were all in metal halo stuff and he was on some really heavy pain killers. First thing he said to me when he finally woke up was, 'You can have my golf clubs.'"

Drew: "We grew up playing golf together and I'd spent a ton of money to get a custom-fitted set. Apparently, that was really important to me at the time. It was equally chaotic for our parents."

Jordan: "When I saw my mom, she looked really pale, you could just tell she was going through a ton. No kid should ever have to experience it, but that hug she gave me when we saw each other was like every Christmas bottled up into one moment. It was very bittersweet."

Drew: "For the first month, month and a half I was either in surgery or on heavy pain meds. I remember Jordan coming, but I had a pretty traumatic brain injury. When you are medevaced, it's so abrupt and so fast, they are working solely to save your life. It took years to fully process it."

Jordan: "He was really angry for a while, and rightly so. I remember there was this random commercial that came on for Dry Eyes, the eye drops. We're all sitting there watching TV and Drew was like, 'Shut it off.'"

Drew: "The commercial was just annoying me for some reason."

Jordan: "And then, no shit, Ben Stein, the actor in the commercial actually walked in right at that moment! It was completely unreal."

Drew: "It was '07 and at that time famous people were coming in every couple of days. I met the president; he actually painted a portrait of me for his book. My wife also painted him and his wife when I was part of Team 43.

"We went to the Army-Navy Game and did a whole bunch of other things too. It's a good thing because it helps you focus on your physical recovery, and it's a needed distraction on some levels. I'm really grateful my time at Walter Reed allowed me to process everything in phases. In total I was at Walter Reed for over a year. Eventually, I was active duty but working for a Wisconsin National Guard unit."

Jordan: "I went back and forth a few times to visit him while I was on leave. We actually had the whole family for Thanksgiving one year. We hadn't all been together as a family for years and it was a really classy and gracious gesture. First-class flight, black car for the whole weekend."

Drew: "A lot of events like that are, very quietly, put on by Vietnam Veterans. They would sometimes come in and share stories of their experiences. I'm so grateful they did that. Their experiences are so different from what I experienced.

"Flash forward to today and Jordan and I are working together. It's so weird, if it wasn't for my son's pneumonia, I wouldn't have been evacuated from Brazil as a part of the diplomatic corps. Our timelines wouldn't have lined up. Now as we're older and we know how to work together, it's really fun and exciting, being able to do this together. It's great to work around people you want to work with. And that that person is my brother is ... just great."

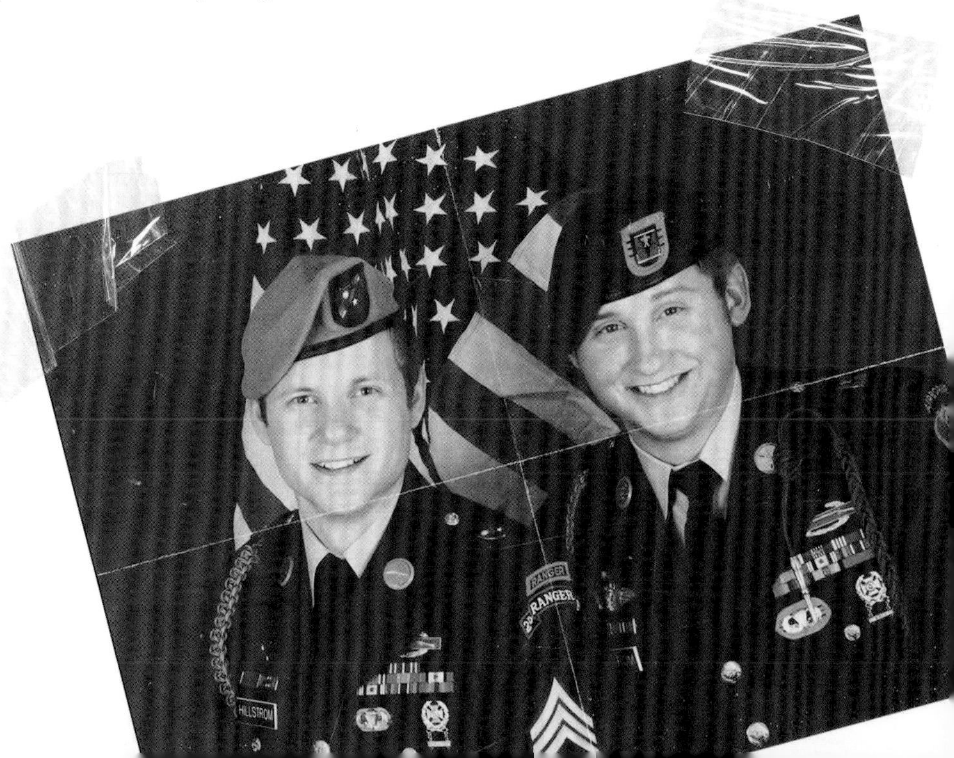

My name is Mike Schlitz and I was an 11B, infantry, in the United States Army. I got in a lot of trouble in high school. I didn't want to work dead-end jobs, and those were my options, so I joined to figure out what I wanted to be. I fell in love with it. I medically retired as an E-7, Sergeant First Class, in 2010. I deployed once, in support of Operation Iraqi Freedom.

In Feb. of 2007 my unit took over the road clearance mission from a sister platoon. One of the vehicles we had wasn't fully up-armored, so I made that one mine since I was in the back and those vehicles are rarely targeted. But that day: boom. Before I could even get a four-letter word out, I was hitting the ground. I remember looking back at my vehicle and I couldn't see my guys. I got up and I could feel the flames on my face and torso. I dumped my body armor and started to roll 'til my body locked up. I thought I was going to die and accepted it. Then I got hit with that fire extinguisher and two things hit me at once:

1. This massive, indescribable pain that just didn't let up
2. An ice blanket that was just instantly cool

I've had ninety-nine total surgeries to date.

I'd get invited to events, usually because of my disfigurement, and it didn't feel right. I didn't like being on the receiving end like that. I also knew I wasn't going back in after losing my hands and because I now have such bad vision. I got really depressed. I lost my purpose. One organization I did something with was Operation Comfort. I asked if I could volunteer with them, rather than being supported, and that was the start of me becoming a veteran advocate. I've also worked with the Gary Sinise foundation and Gallant Few, among many great organizations. One of the first questions you should ask any non-profit you are thinking of supporting is, "What's your follow up?"

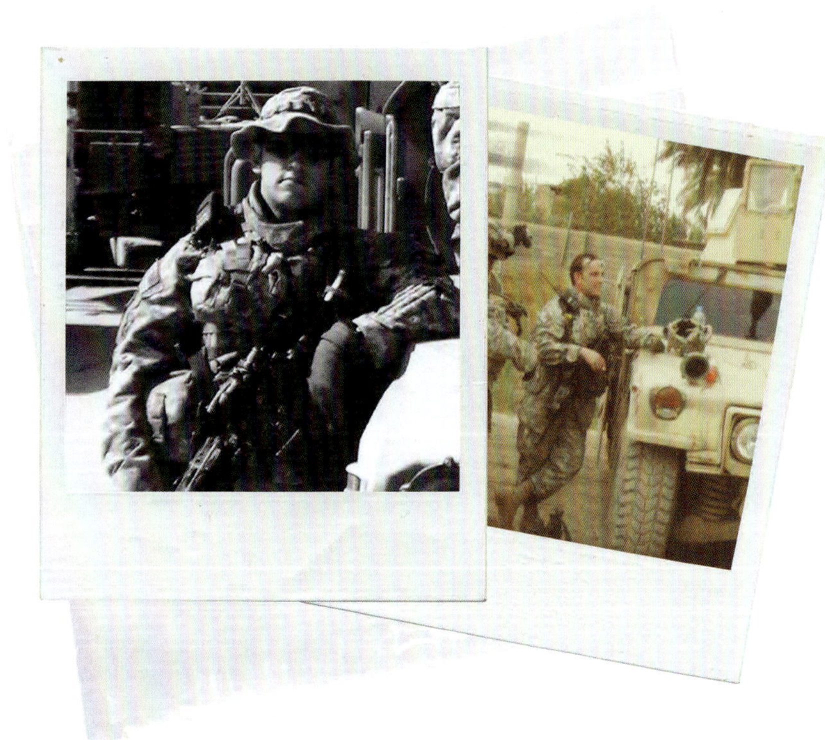

Doing an event is great, but if veterans just go back to where they were — still isolated, self-medicating or even medicated on prescriptions they temporarily went off, we aren't giving these men and women the help we think we are. At events, they've just been given a huge emotional high of the experience and then they're going to hit a low when they get back. Following up is absolutely critical. Find your purpose. Every day is a struggle in some aspect or another for everyone. Find what gets you out of bed in the morning. If you don't have that right now, FIND IT.

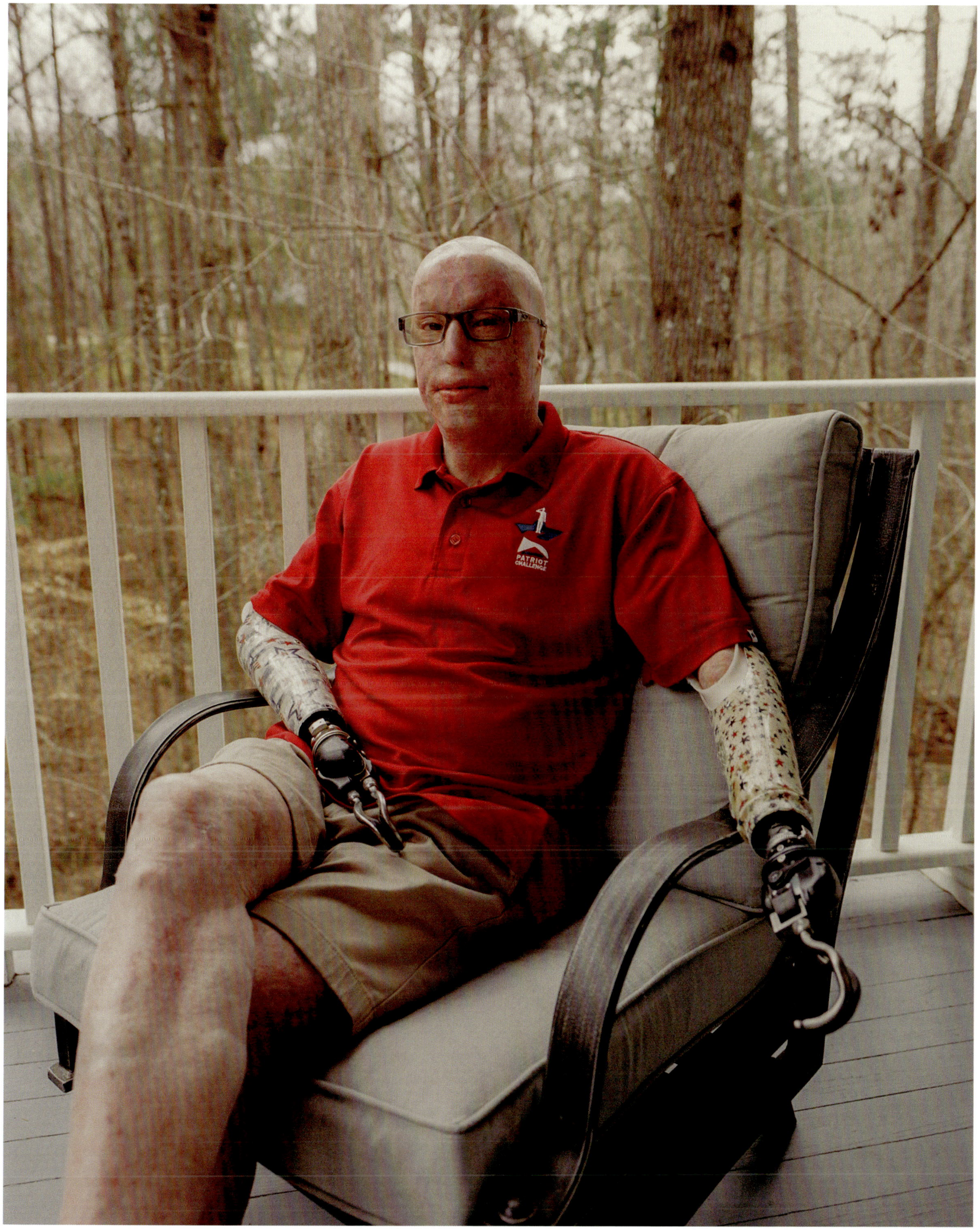

Mike Schlitz, Sergeant First Class
Army - 11B / Infantryman
Years of service - 14 years
Deployments - 1 [Operation Iraqi Freedom]

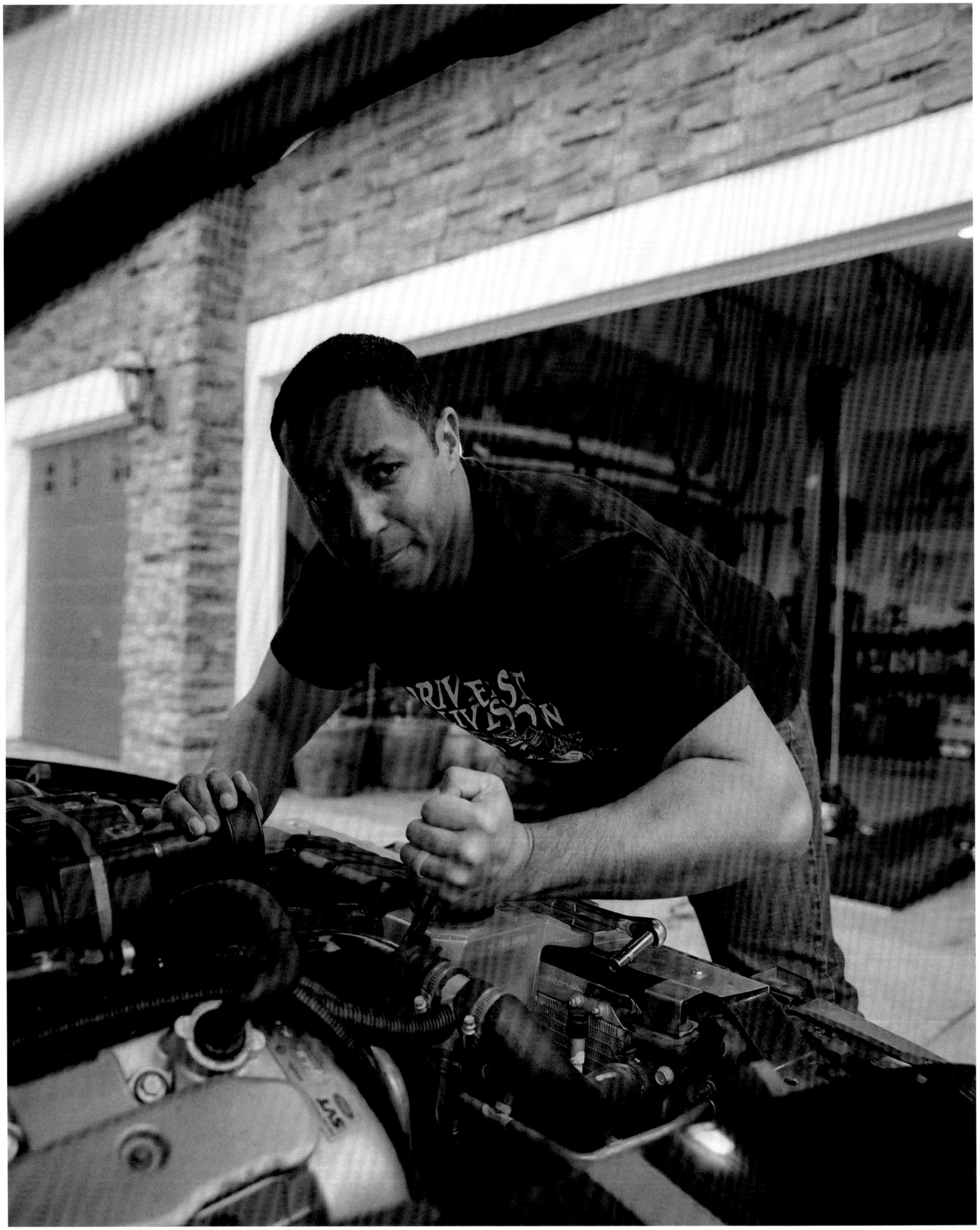

Alex Gruber, Lieutenant
Navy - Cryptologic Warfare Officer (CWO)
Years in service - Active
Deployments - 4 [Operation Iraqi Freedom & Operation Enduring Freedom]

My name is Alex Gruber and I currently serve as a Cryptologic Warfare Officer in the United States Navy. As a senior in high school, I watched the events of 9/11 unfold in real time. The events that day drove me to enlist in the armed forces, to serve my country and do my part to protect this great nation. I joined as an Electronic Warfare Technician in 2002 and was promoted to the rank of Chief Petty Officer in 2012. In 2014, I was commissioned as a Limited Duty Officer in the Cryptologic Warfare designator. My last deployment in 2017 was aboard the USS Oscar Austin (DDG 79) where we built and developed relations with allied nations and defended freedom on the high seas.

My team of Cryptologic Technicians was among the best in the fleet. Day in and day out they would work tirelessly to intercept and decipher electronic signals and stand ready to engage any threat that stood in our way. Other deployments included the USS Rushmore (LSD 47), the USS Harry S Truman (CVN 75), and the USS Trenton (LPD 14), supporting various exercises and real-world operations, including Enduring Freedom and Iraqi Freedom.

Deployments have taken a toll on my family, but I thank God that my wife of over eighteen years has the resolve and the strength to raise a family when the Navy took me into harm's way. I have no regrets but if I could change one thing, it would certainly be finding more time to spend with my family. When I am not in uniform, I enjoy Brazilian Jiu Jitsu, working on my Mustang, travelling, and spending as much time with my family as possible. If I could give you one piece of advice it would be: “Don’t sweat the small stuff.”

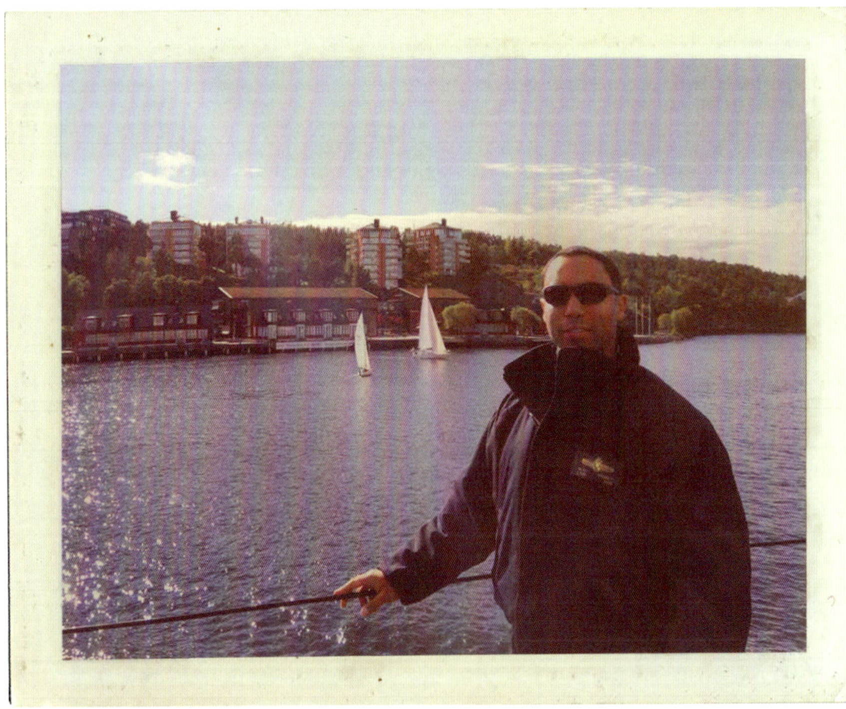

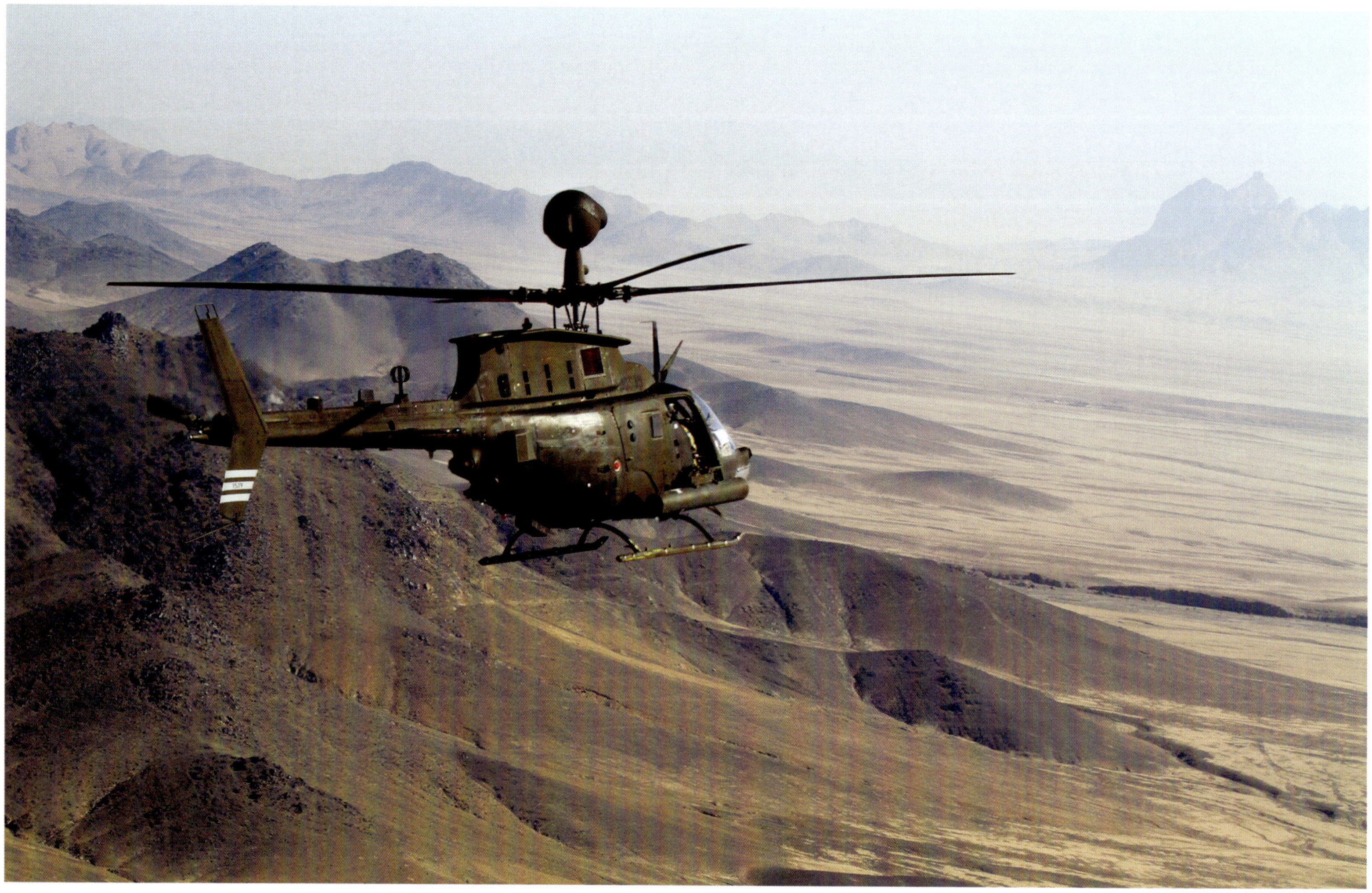

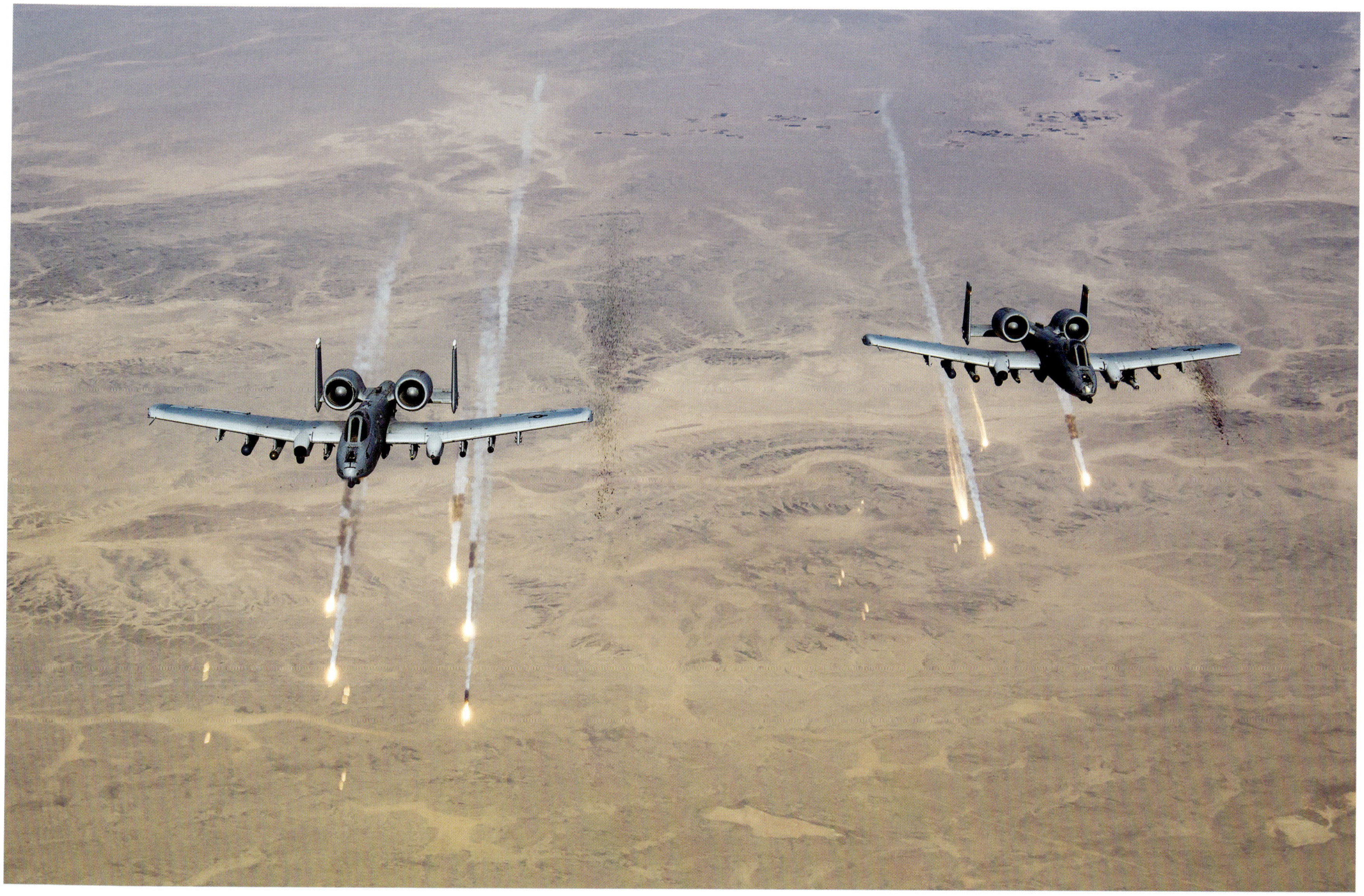

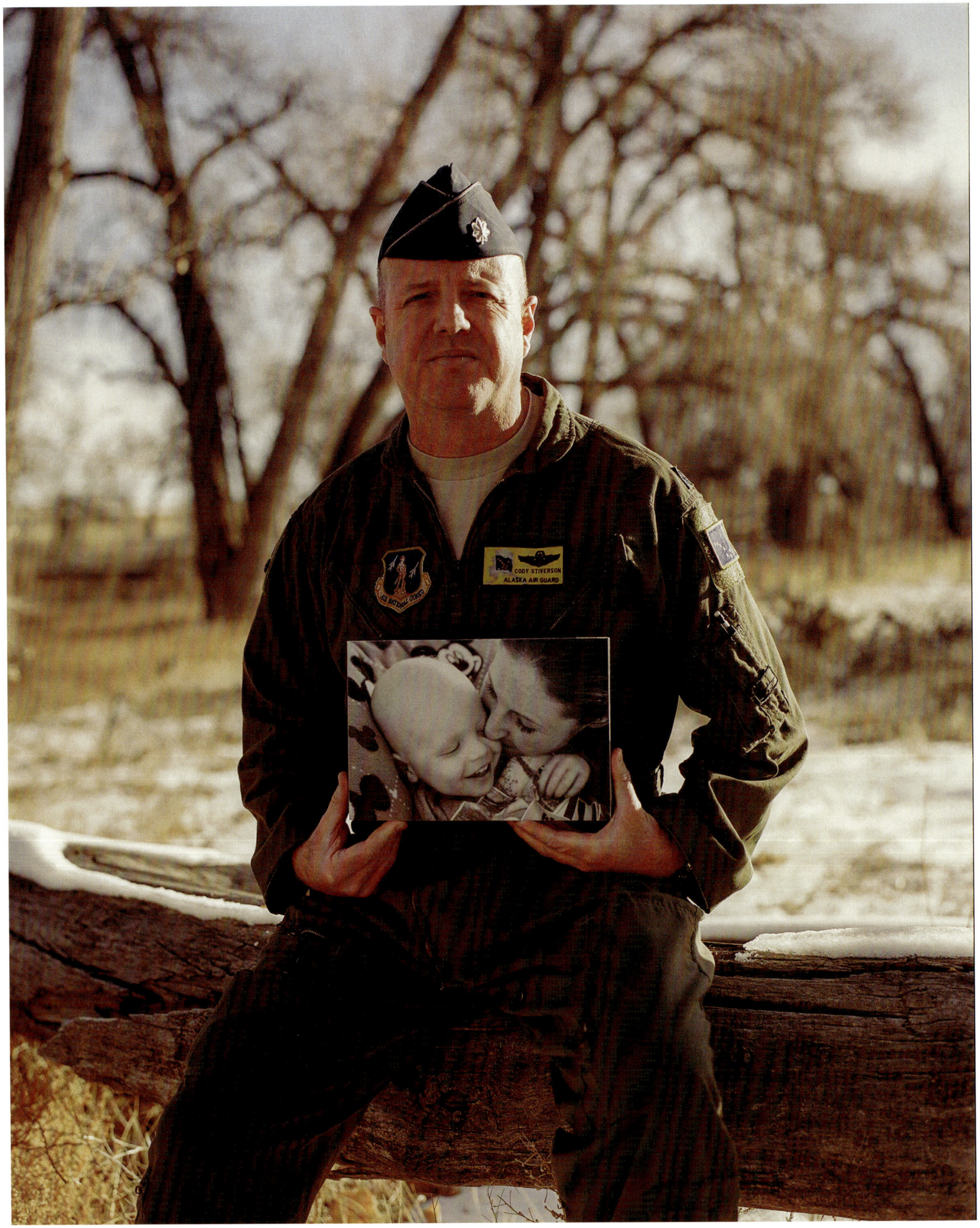

Cody Stiverson, Lieutenant Colonel
Air Force / Air National Guard - 11F3F / F-15C / MC-12 / KC-135
Years in service - Active
Deployments - 3 [Operation Noble Eagle, Operation Inherent Resolve & Operation Enduring Freedom]

My name is Cody Stiverson, and I am currently a Lieutenant Colonel (O-5) in the Air National Guard and have previously been on active duty. I started off as an F-15 pilot and moved to the MC-12 while on active duty, before transferring to flying the KC-135 refueling tanker in the Air National Guard. My family has done a lot of service. Both of my grandfathers served in WWII, my uncle was in Vietnam, and my cousins served as well. I've deployed as part of Operation Noble Eagle (Presidential Support), Operation Inherent Resolve in 2015, and finally Operation Enduring Freedom 2009. When I completely separate from service, I'm planning to transfer into pediatric cancer research. I will be working with a Biopharma called Artisan to find more options for kids with cancer.

Right after my first trip to Operation Inherent Resolve, my daughter, Adelaide, was diagnosed with a rare pediatric liver cancer. It started a call to action for me and my family. Currently, we've raised over half a million dollars to support research in Adelaide's name. Our daughter died in 2016. Working to get better treatments for rare cancers, I am involved with hospitals all over the world. The leadership and organizational skills I've learned in the military have been a real asset to moving the ball forward and getting things done. Being a vet has also been a great way to open doors. I'm also big into woodworking. I really like making things with my hands; it's really satisfying. I have been selling growler butlers (a way to dry your beer) as a hobby and also a way to raise money to continue the fight against pediatric cancer. What I've learned along the way is a d-eep sense of serving others. I grew up as an only child but have found it's key to work more outside myself and serve others more.

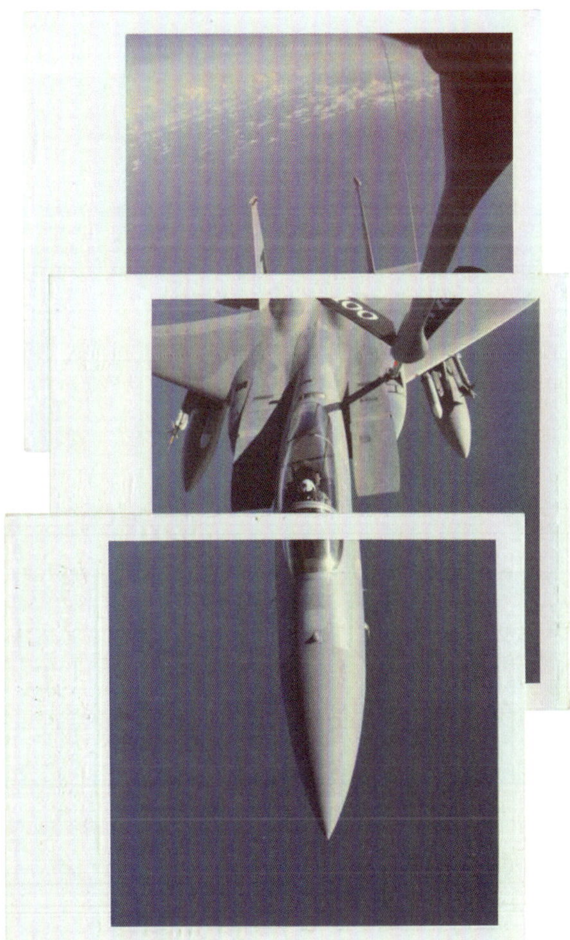

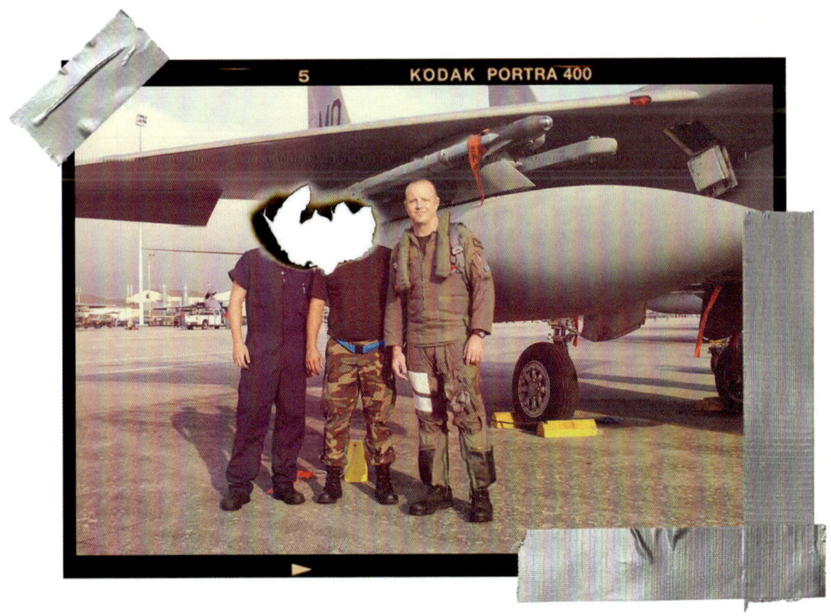

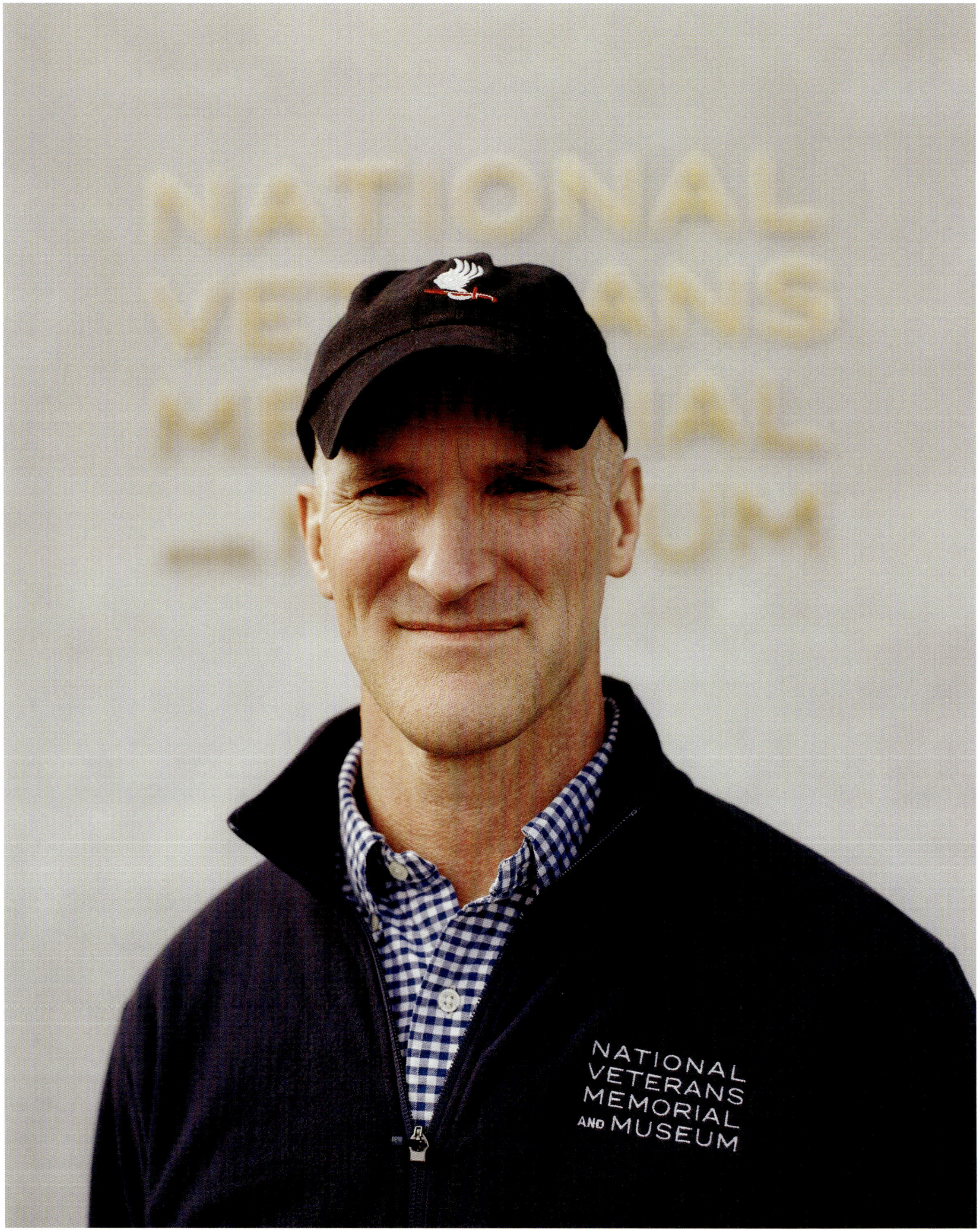

William “Bill” Butler, Colonel
Army - 11A / Infantry Officer
Years of service - 27 years
Deployments - 3 [Operation Iraqi Freedom & Operation Enduring Freedom]

My name is Bill Butler and I served as an infantry officer in the United States Army, retiring as a Colonel (O-6). My dad served in Vietnam and then in the 11th SF Group in the Reserves. He had an old Ranger handbook in this footlocker in our basement and he said it was the one school he regretted not going to. That piqued my interest in Army ROTC and the Ranger Challenge team in college. In 1989, I was in the last class to let cadets go through Ranger School. I earned my Ranger Tab and branched infantry after I graduated. Eventually I was assigned to C Company, 3/75 and was there when B-Co deployed to Somalia. I deployed to Iraq in 2003 in support of Operation Iraqi Freedom after I became an executive officer with the 173rd Airborne Brigade. We turned around almost immediately to Afghanistan for Operation Enduring Freedom VI. Finally, I was a battalion commander with the 173rd Airborne and deployed to Afghanistan again for OEF X.

I retired in 2017 and struggled with the question, "What do I do?" I didn't want to go into the business world and chase a paycheck. I wanted a sense of purpose, so I joined the Department of the Army as the Deputy G7, for US Army Africa for two years. I was really happy in that job. I went to Boston for a course. Right before leaving for the course, one of my former soldiers, an incredibly successful one, committed suicide out of the blue. It rocked me. In the middle of this course, they were talking about finding your purpose, finding your true north, and it really got me thinking. As I was pondering these things, I saw that General Ferriter had posted for a chief of staff for the National Veterans Museum and Memorial. I flew in for the interview and in the museum, there are fourteen thematic alcoves talking about the various steps to service. Big spaces, massive pictures, and stories. One was about combat, showing wounded troops from the 173rd. It turns out that one of the soldiers in this picture was part of my personal security detachment on a different tour of Afghanistan. I knew the story of that picture and told it all. From that moment I knew that's where I needed to be. I've been working here for the past two years.

COVID-19 has been a challenge for us here at the museum. Our bread and butter had been in-person attendance, corporate and private events, and dinners. The shutdown forced us to aggressively pivot to digital events and content curation. We keep improving, we keep working to curate veteran relationships and create content to honor veterans. We'd get fifty to 350 people through the door for an in-person ceremony, depending on the event or day. We had almost 6,000 on our digital channels when we broadcast a VE-Day ceremony. People from France and the Netherlands were watching the speaker tell his story about his experience serving in WWII. The pivot has gotten our name out there even more. We're a half-day drive from almost half the country. We will now be able to fully do hybrid events while partnering with organizations around the US. Working together helps everyone.

Continued on next page →

We also made the decision, right away, that we weren't laying people off. Our staff was surprised because so many people in the museum space were furloughing and laying staff off. In the Army, it's mission first, people always. We are always trying to accomplish the mission and take care of our people. People are our most important resource. If we lay someone off, we need to replace them eventually. In life, just like the Army, you need to get everyone to the objective. That's what we did, we prioritized our people and kept everyone gainfully employed helping us pivot to virtual programs.

I always remind people that relationships matter. The people you meet today, at some point, good or bad, are going to come back to you. A gentleman by the name of Mike Ferrusi was a platoon sergeant when I was company commander. We knew of each other, but we didn't really know each other until we had to briefly work together on staff assignments for about six months. Fast forward to 2009, our battalion just completed mission rehearsals and outload to go to Afghanistan. I lost my command sergeant major (CSM). He could no longer deploy. I put in a call to Fort Benning and told the Maneuver Center of Excellence CSM that I'm in a bind and I need a CSM. He's like, "I got the perfect guy for you, it's Mike Ferrusi." Mike was at Benning and was about to move up to a more senior position in the enlisted ranks. But when he heard that I needed a guy he said, "I know Bill, I'm going to deploy with him." Relationships matter.

I am also blessed to be married to my wife of twenty-eight years. We have two children who both selected professions serving others. Our son went to West Point and is now an infantry officer and our daughter is a nurse at a children's hospital. My mother-in-law has always said, "You need to bloom where you are planted," and my wife always took that approach as an Army brat. We always tried to go out and do and see things wherever we were stationed, so our kids could see that this is a great big world. They really embraced that. I am a very lucky man.

"Relationships matter."

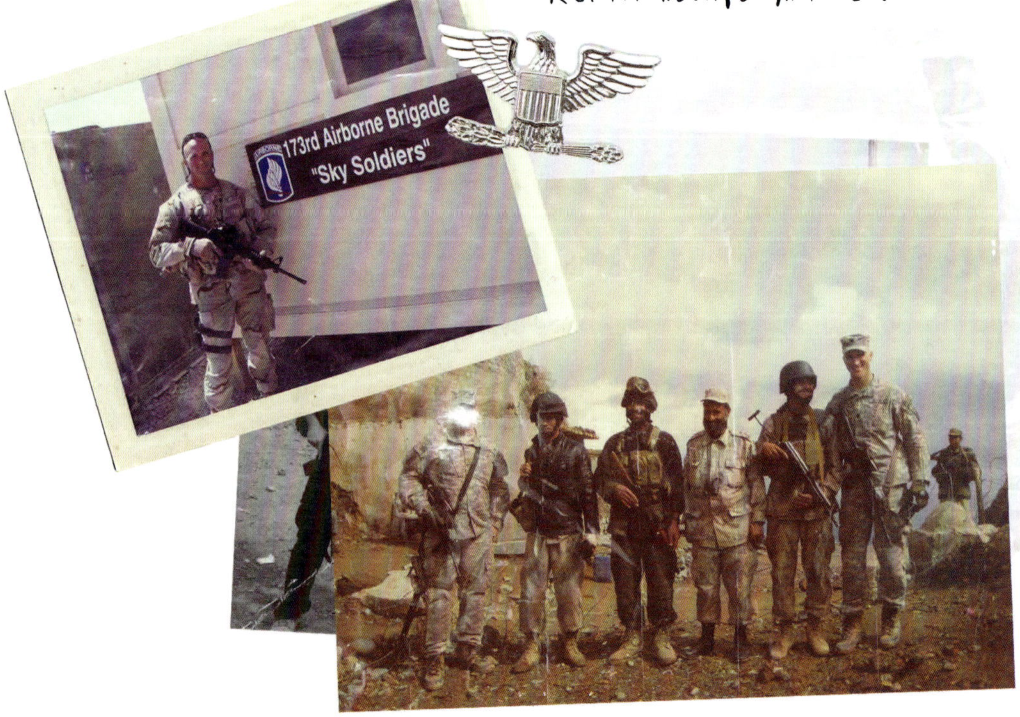

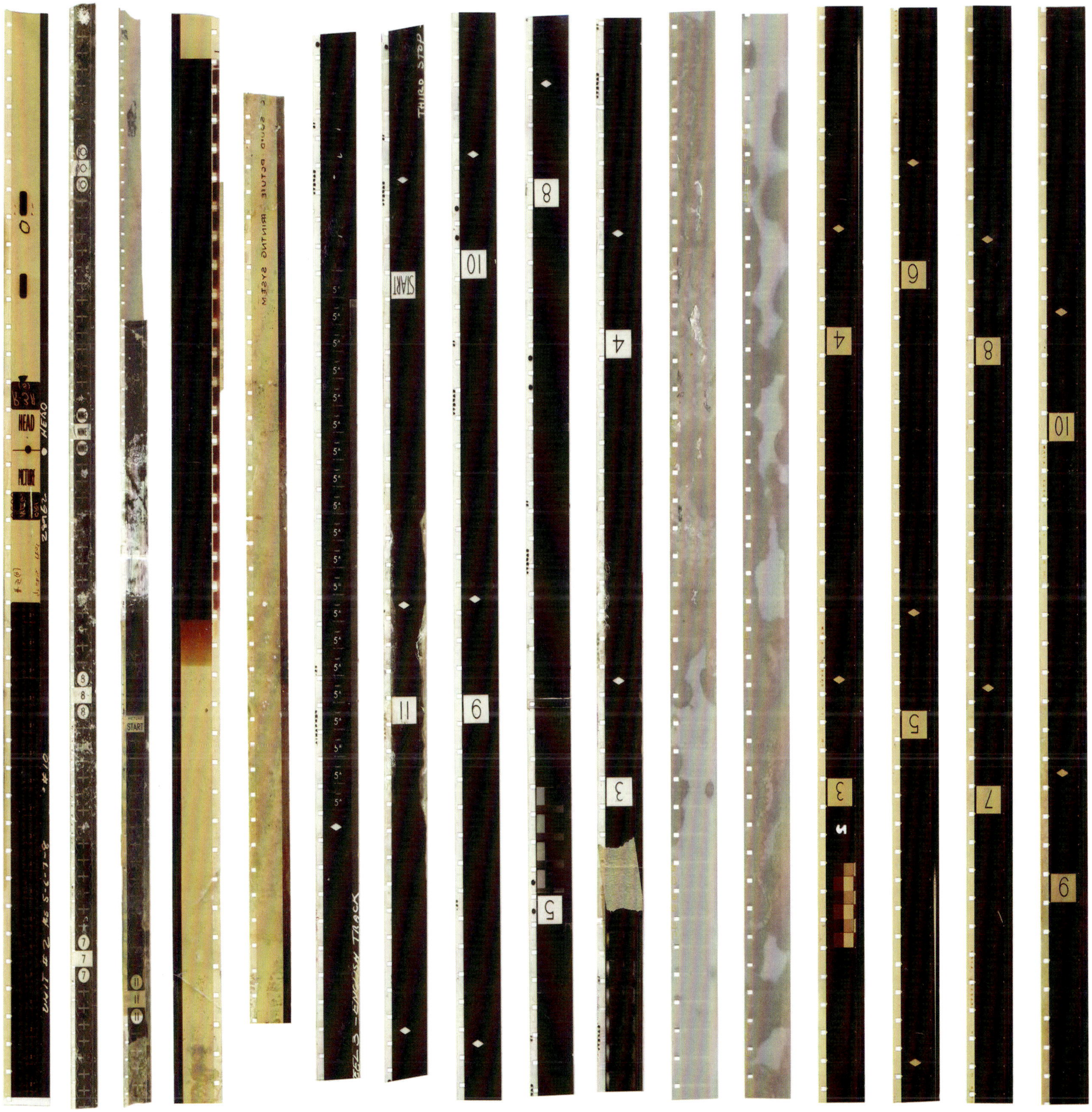

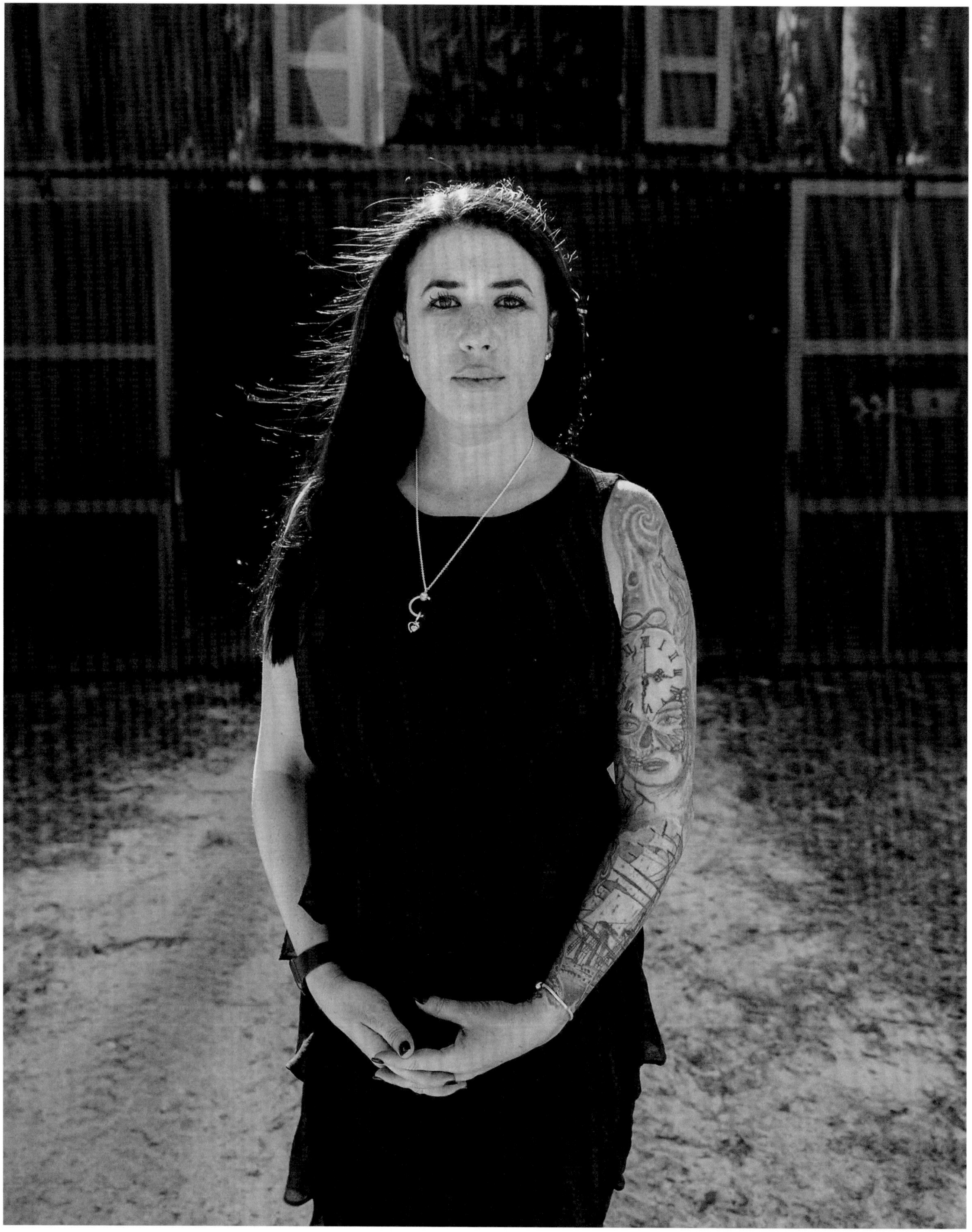

Jacqueline Griffin, Sergeant
Army - 35N / Signals Intelligence Analyst
Years of service - 4 years
Deployments - 1 [Operation Enduring Freedom]

My name is Jacqueline Griffin and I served in the US Army for four years as a 35N, Signals Intelligence Analyst. I was Varsity cheerleading captain at my high school and had an opportunity at a full ride scholarship, but I turned it down for the military. I enlisted in the military because people told me I couldn't, and I was bound and determined to prove them wrong.

When I separated from service in 2010, I was a Sergeant (E-5) coming off a year-long deployment to Afghanistan in support of Operation Enduring Freedom. When I came back from Afghanistan I had less than 90 days before leaving the Army. I was told not to rush the process and later found out why, but at the time I was committed to moving on to the next chapter in my life. The female veteran community is much smaller than it is for men. So many civilians don't really realize that we deploy alongside our male counterparts and are always astonished when I tell them about my service. We are just as much a part of the fight! After separating I started going back to school again. I got married and moved around the country (even to Hawaii) with my husband. We had two beautiful boys. While we were in Hawaii, I became obsessed with scuba diving and used the GI Bill to become an instructor. I love being in the water and scuba diving really helped me heal. Everything seems to fade away when you are under water. Now I work for Crown Defense, where we help warfighters work on armoring vehicles to help keep our troops safe.

Take everything with a grain of salt. This is something I struggle with but I'm getting better at applying it in my own life. I separated from the military so quickly that I bottled everything up. I had children and a marriage to worry about and the stress just piles up. I turned to drinking as a solution to my worries, and have recently started dealing with my alcoholism. As of now, I am a month sober thanks to the program I am in.

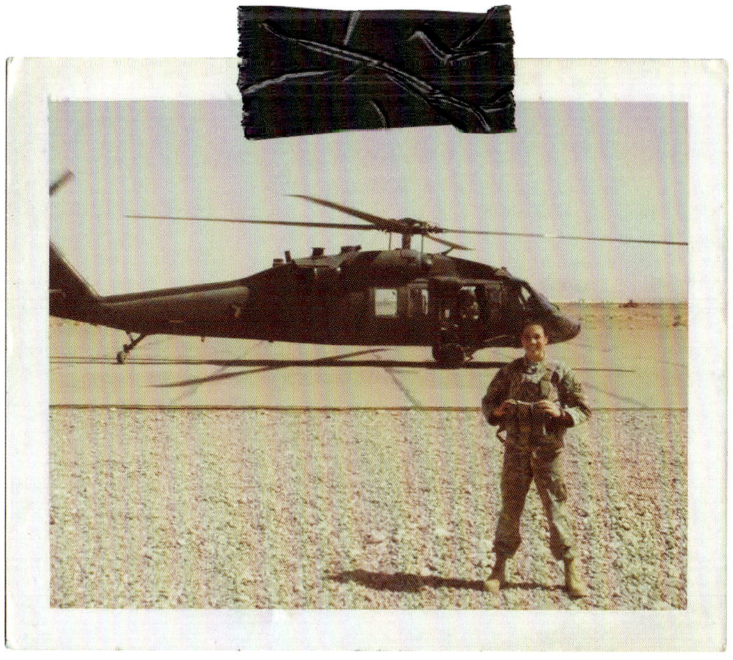

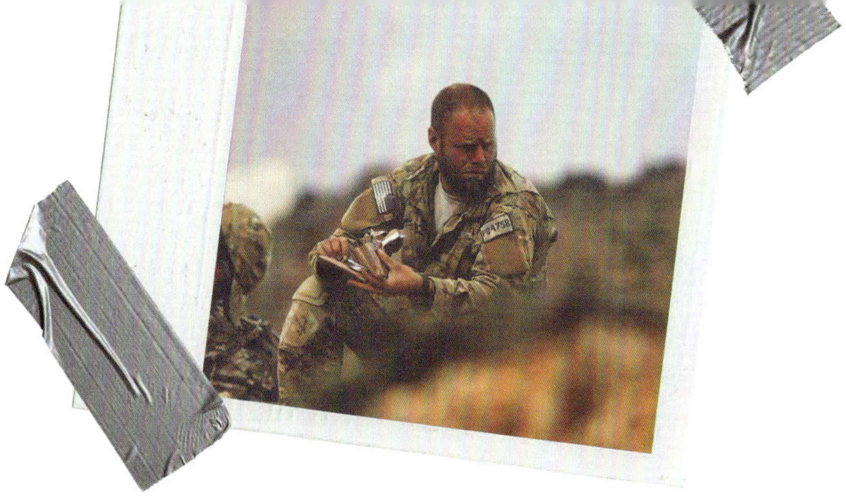

My name is Steve Forti, and I am currently an 18F Special Forces Intel Sergeant in the United States Army National Guard. My last deployment in support of the Global War on Terror was to Afghanistan in 2009 . My father passed away when I was in college, from cancer related to Agent Orange exposure in Vietnam, where he served in '65 and '66. It affected me in many ways, obviously, but the ability to afford college was at the top of the list. Frankly, I was a bit lost, and I just wasn't getting the job done in college. I enlisted in 1991 and completed Special Forces training in 1994. The Guard is interesting, because it allows you to live multiple lives. After my '01 tour, we were informed we wouldn't be deploying anywhere, so I went to nursing school and became an ICU nurse. Nursing felt like a good profession of service. In '07 I re-enlisted. Then I worked for a trauma consultant but eventually decided I wanted to do my own venture and started a fitness competition app: Fit Fight.

In March of 2020, I became a consultant for crisis management at the Hospital for Special Surgery (HSS.) I was immediately thrown onto an amazing team and assisted with converting an orthopedic surgical hospital into a COVID treatment facility. Being a combat vet and a medical professional, I was able to say, "Yes it's bad but it's going to be okay." In June 2020, HSS asked me to be the Chief Wellness and Resilience Officer. I'm deeply grateful. In the military we used to high-five ourselves that we could outperform, outshoot, out ... whatever, with a hangover. But even now, this far into the wars, we're seeing so many problems with people not being able to cope. The culture is different, but there are similar problems across the country in the medical community with burnout. So, my charge is to figure out what we can do to effectively support our staff in these challenging times. There's such a lack of education around PTSD. People used to ask me, "Are you going to freak out if you see blood?" I'm like, "Are you kidding me?" It's taken until my role here at HHS for me to feel like I was in a place that understood my background and experiences. War fighter culture shouldn't be high-fiving about overconsumption and crazy behavior. Life isn't about "Sleep when you're dead," that's bullshit. We should be figuring out how to promote sleep hygiene; confronting burnout and insomnia; looking to decrease divorce and substance abuse rates, all the unsexy things. We aren't celebrating the positive choices enough, and it's catching up with us.

The gentler you are with yourself, the kinder you are to others. The idea of empathetic leadership is so powerful. Think of how impactful it'd be if you said, "This is the thing I'm weakest at, and for the next month I'm going to work on it" and then show your team the incremental improvement.

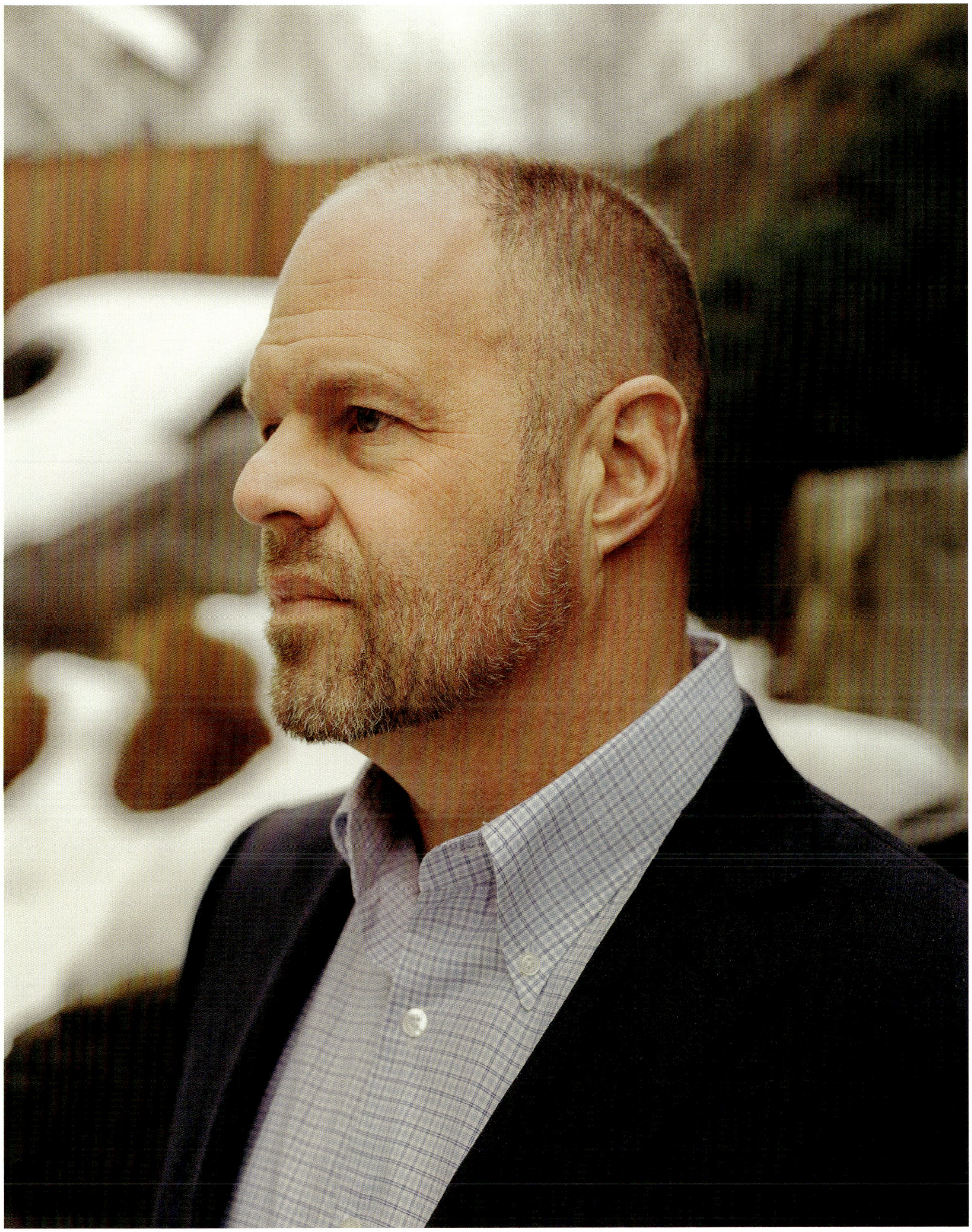

Steve Forti, Master Sergeant
Army - 18B, 18F / Special Forces
Years in service - Active
Deployments - 2 [Operation Enduring Freedom]

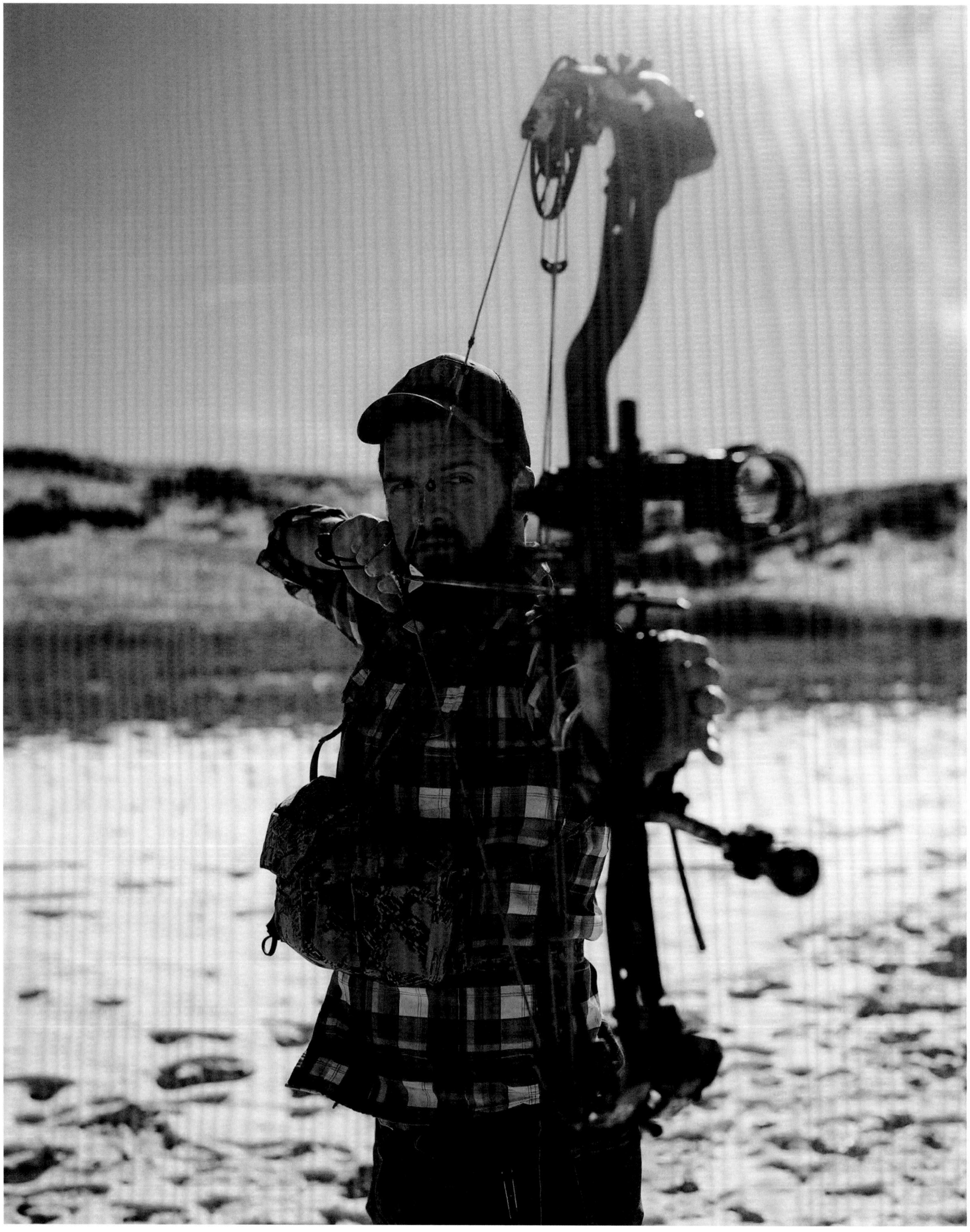

Nick Orzechowski, Major
Army - 11A / Infantry Officer
Years of service - 12 years
Deployments - 3 [Operation Enduring Freedom]

I'm Nick Orzechowski. I am a former infantry officer in the United States Army. I deployed in support of Operation Enduring Freedom three times: from 2010-2011, again in 2012 and finally in 2013 as a member of the 75th Ranger Regiment. I recently separated from service as a Major (O-4) and now I manage a business development team at a technology firm, which was the goal of my transition. I joined the service because of 9/11. Watching the Twin Towers fall was a galvanizing moment for me, and I answered the call of service. As I was separating, Breakline, a veteran networking group, really helped me gain my new career. The COMMIT Foundation and Breakline really helped me and are organizations I have actively supported post-service. My transition was unique. On one level, it was easier for my family to adjust, because of COVID-19 considerations and how that affected my terminal leave. I was able to be at home more than most people have. It also made it much harder to find a job and to network my way into a role because of limited opportunities and how I had to do so much of it through a screen and not face-to-face. Leveraging the veteran network made it so much easier.

Since separating, I've taken up bow hunting. I enjoy how technically focused you need to be to shoot properly and how humbling hunting with a bow is. You need to be focused and careful; I love it. My mantra while I was in the service was "Vanquish mediocrity." As I've started my civilian career, it's morphed into simply "Slow down," but it's still hard to shake some of the stink of the military off. We were always going, 110 percent of the time, especially in special operations. But we aren't training for or deploying to life-or-death situations anymore. It's okay to slow down.

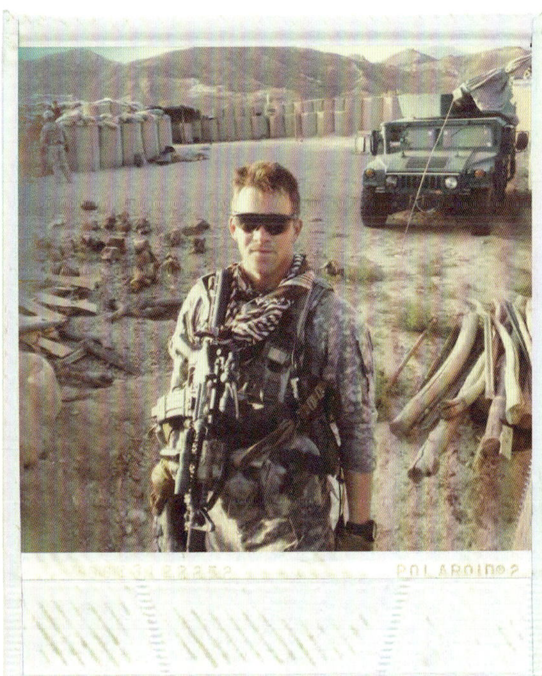

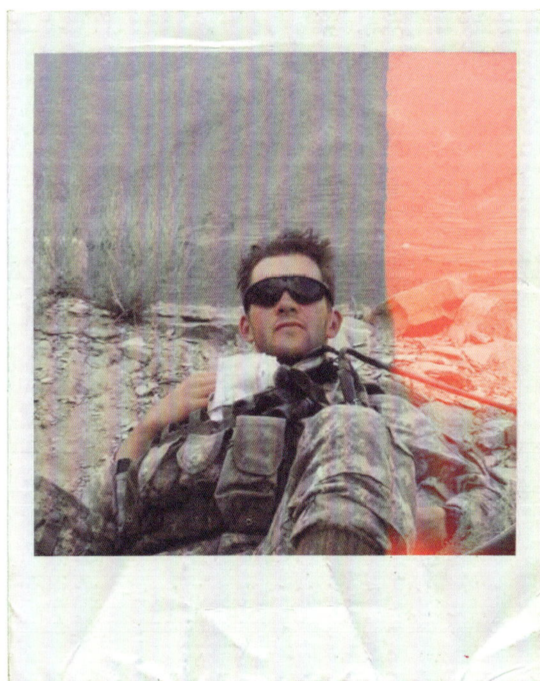

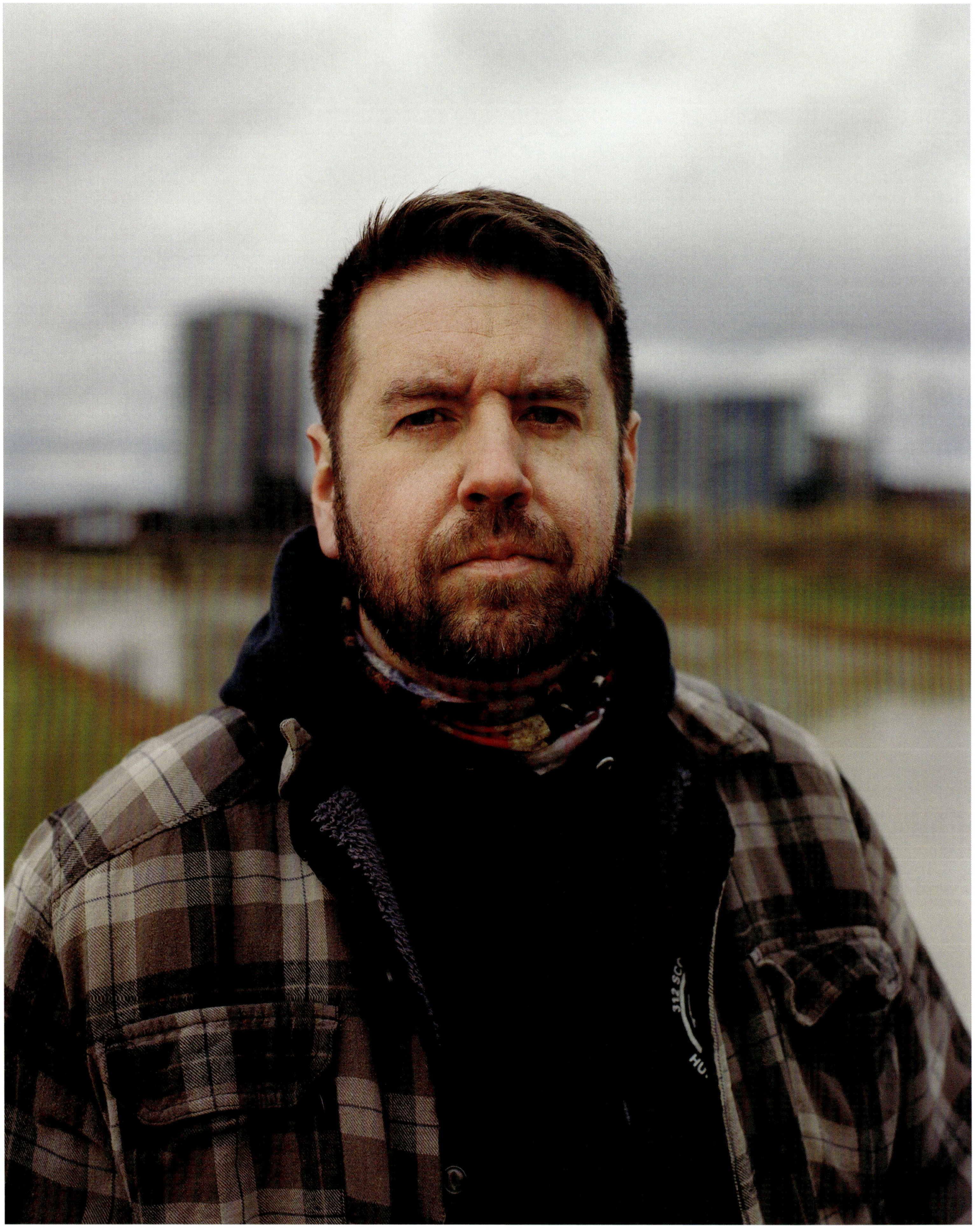

Jake Lowe, Sergeant
Marine Corps - 0311 Rifleman / 8541 Scout Sniper
Years of service - 4 years
Deployments - 4 [Operation Iraqi Freedom]

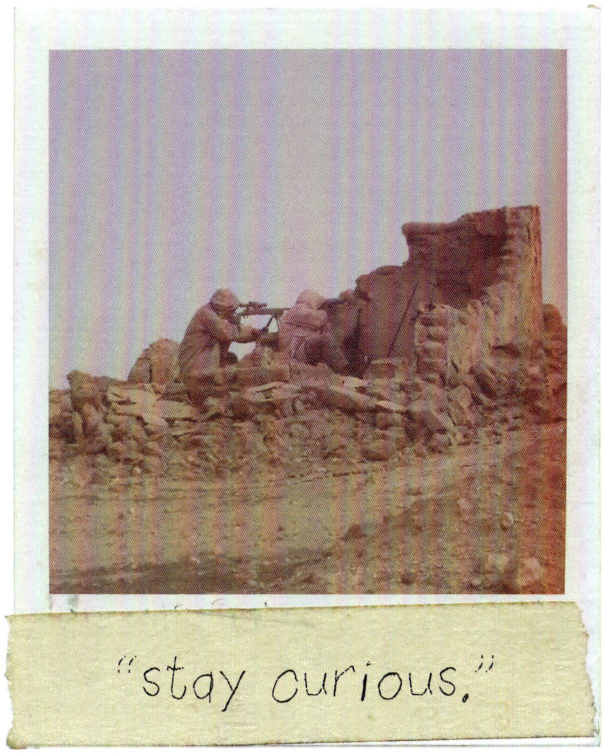

My name is Jake Lowe, and I was a rifleman and eventually a scout sniper in the United States Marine Corps. I joined the Marines because it was similar to the way I grew up; a lot of service members are in my family. It also gave me a chance to travel the world on the government's dime while serving my country. I deployed in 2003 with the initial invasion of Iraq, was in Djibouti, Africa in 2004 and finally went back in support of Operation Iraqi Freedom in 2005. I separated from the service as a Sergeant (E-5). When I enlisted, I scored high enough on the ASVAB to do any job I wanted to do, but I chose to be a grunt: something my recruiter scoffed at, but as someone who enlisted post-9/11, I really wanted to make an impact.

I signed up for the VFW and Disabled American Veterans, really got all the boxes checked for a disability rating with the VA. Got out in June of 2006. It was rough. We had our first kid, so I was working full-time, then went back to school full-time, and we had another baby in the process. I landed my job off the USJobs.com website. It's a good job. I started in inventory management, moved to business process analysis, and now manage a team to supply US military service members.

As you transition, social networking is huge. It's all who you know, and if you aren't actively searching for opportunities or talking with friends, no one else is going to do it for you. Also, stay curious! Always keep learning.

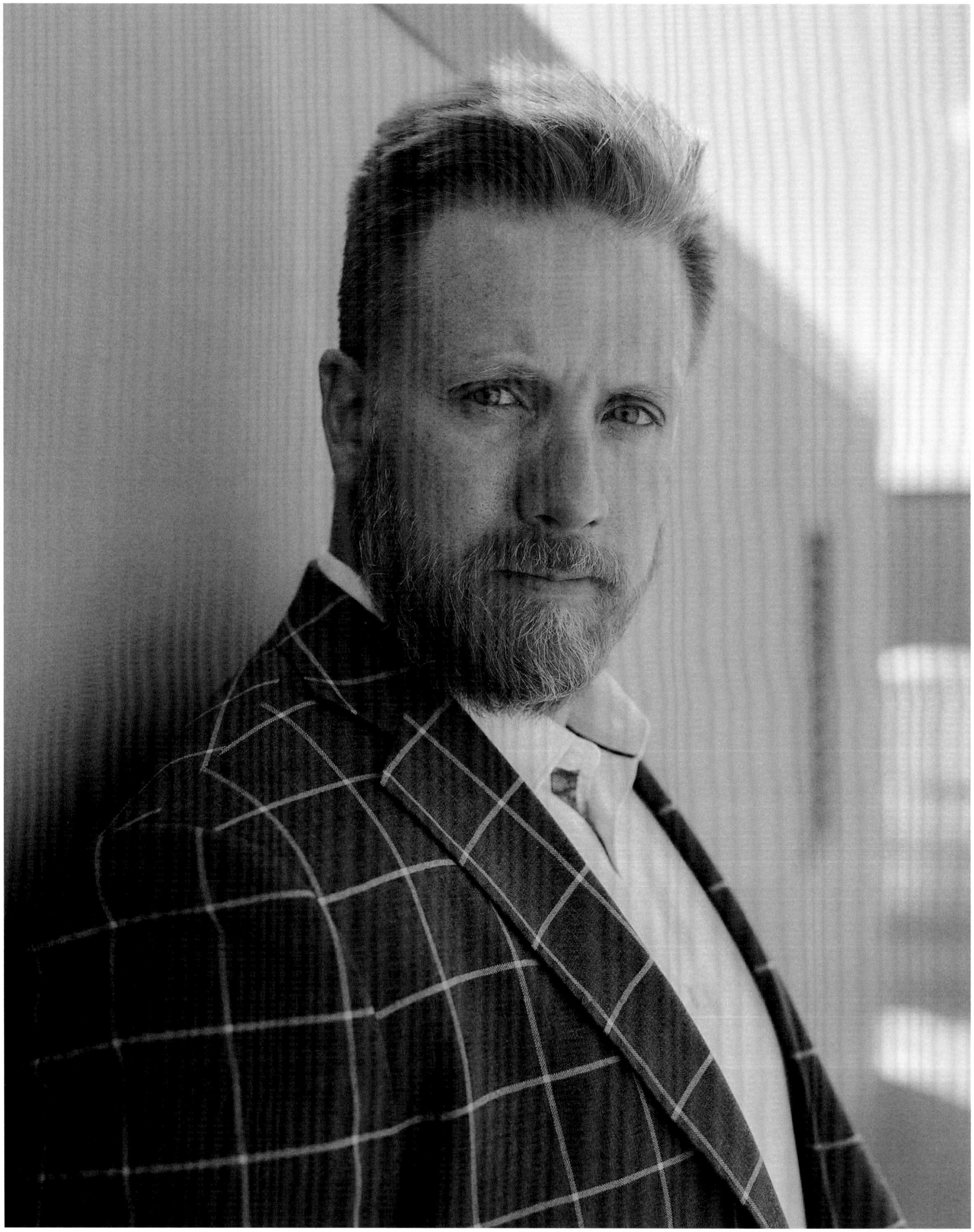

Jason Van Camp, Major
Army - 18A / Special Forces
Years of service - 14 years
Deployments - 3 [Operation Iraqi Freedom]

My name is Jason Van Camp, and I was an 18A Special Forces Officer. I was medically retired after fourteen years as a Major (O-4). I deployed three times in support of Operation Iraqi Freedom. I started my military career at West Point. I had never aspired to go there. I was a good athlete, but I was undersized for major football programs. West Point fell in love with me and recruited me hard. My parents supported it and I said, "I'm in." Other than West Point being in New York and the Army-Navy game, I knew nothing about it. It was quite a culture shock.

When 9/11 happened, I was in the Officer Basic Course at Fort Sill, Oklahoma. There were like two hundred people crowded around this little coffee shop TV as we watched the second plane hit the World Trade Center. I remember thinking, "We're under attack." Knowing that we were going to war, I wanted to really push myself. Just like West Point, I wasn't initially set on becoming a Ranger School graduate or member of Special Forces. I was looking for the next story or the next adventure. Courses like that are tough. You really have to be dedicated. For Ranger School there were about a hundred guys who started our Pre-Ranger course. In the end, only three of us graduated from Ranger School. Then I made it through Special Forces Assessment and Selection and was accepted into the Q-course where I worked my way up to Special Forces Officer.

In 2010 I got really sick on a deployment to Africa. I finished our deployment and went out with friends right when I got home. I woke up that night and there were cops and paramedics all around my bed and my girlfriend was crying. They took me to the hospital and ran some tests. At first, they didn't have any idea what was going on, but I started having more and more seizures and eventually I was medically retired for it. So, I had to ask, what's next? My mom used to say, "Show me your friends and I'll show you your future." A lot of my friends who had gotten out had started businesses. So, I started my for-profit firm, Mission Six Zero.

Our company delivers keynote speaking engagements and workshops. I was doing an event with the Oakland Raiders through Mission Six Zero. I was talking to the veterans at the event. The wounded guys were talking about things non-profits had done for them and to me it sounded awesome, but they were telling me it was weird. It was uncomfortable for people to just give them stuff or take them on trips because after that high wore off, it all felt even worse. They started talking about how they wanted to own businesses and how they had ideas. The conversation kept playing over and over in my head: there was a real opportunity to help. So, I started Warrior Rising, a non-profit that empowers veterans who want to start their own businesses.

Continued on next page

Embrace discomfort! You're only going to grow if you get comfortable being uncomfortable. You can choose to do the hard things, or the hard things are going to choose you. Serving in the military is just something you did. It was a job. Now, you need a new purpose, you need a new job. Fellow veterans, open your minds! Be willing to learn new things and apply your time in service where appropriate.

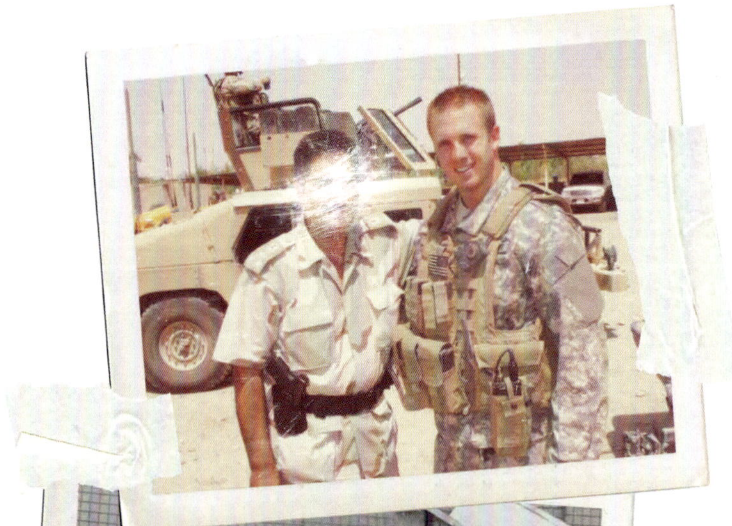

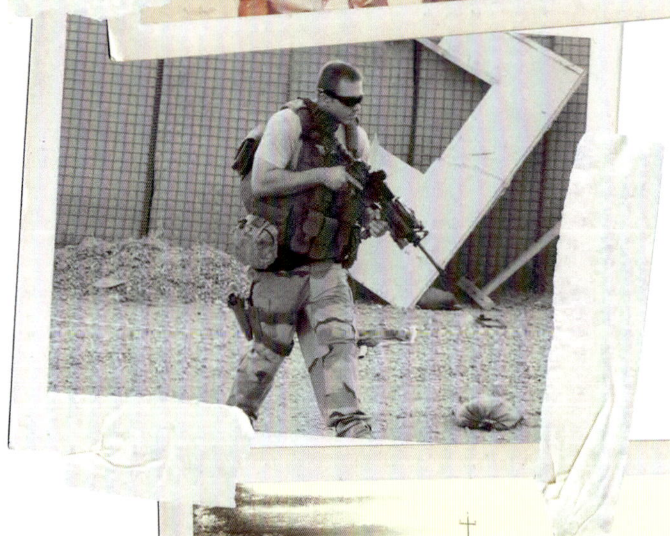

"Get comfortable being uncomfortable."

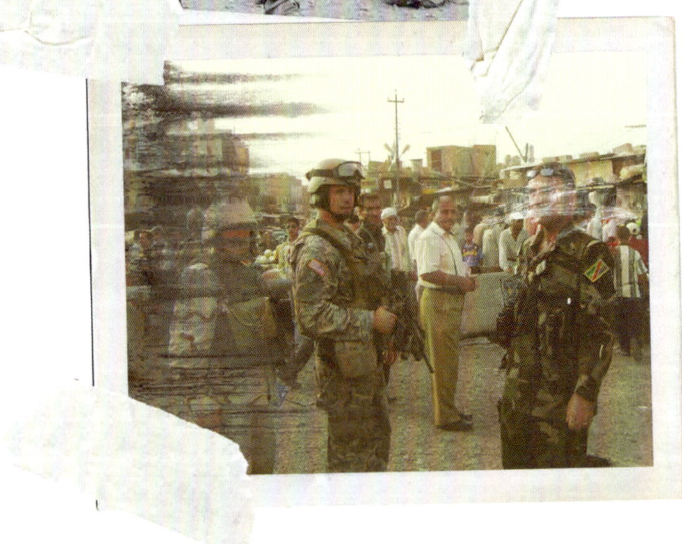

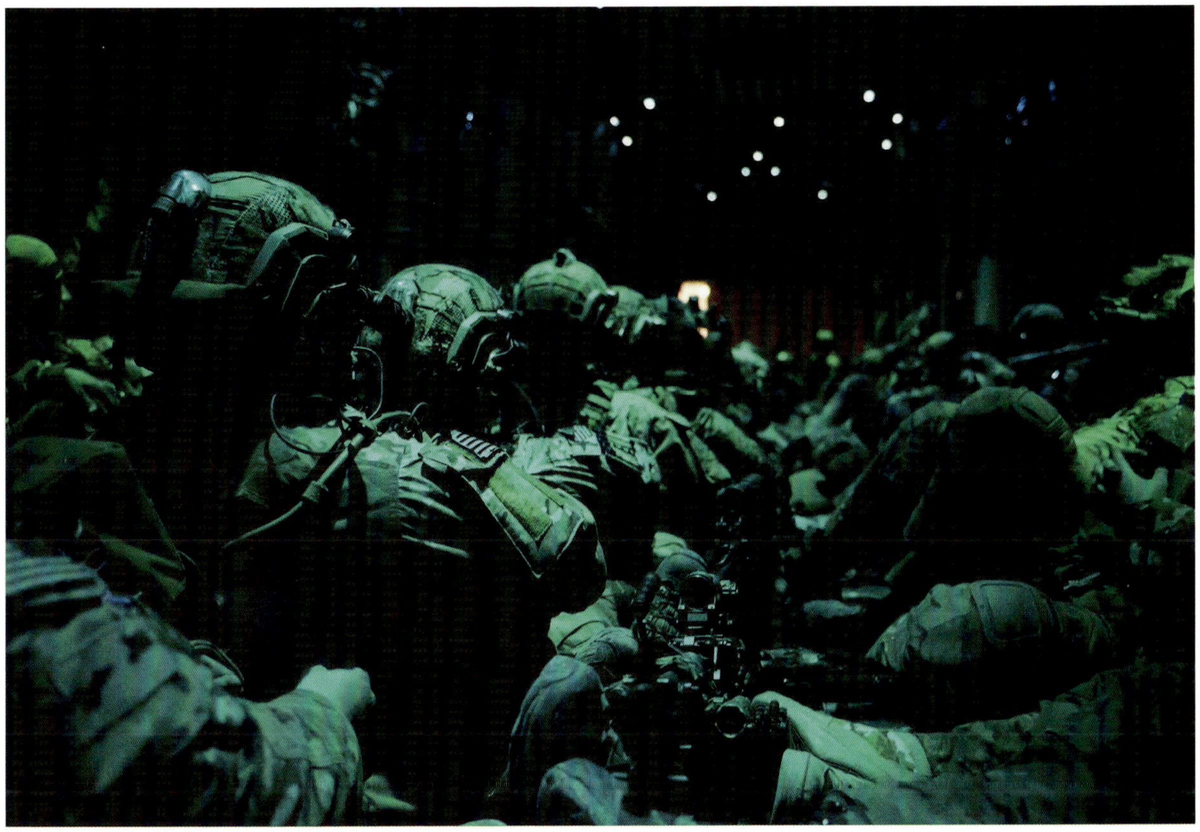

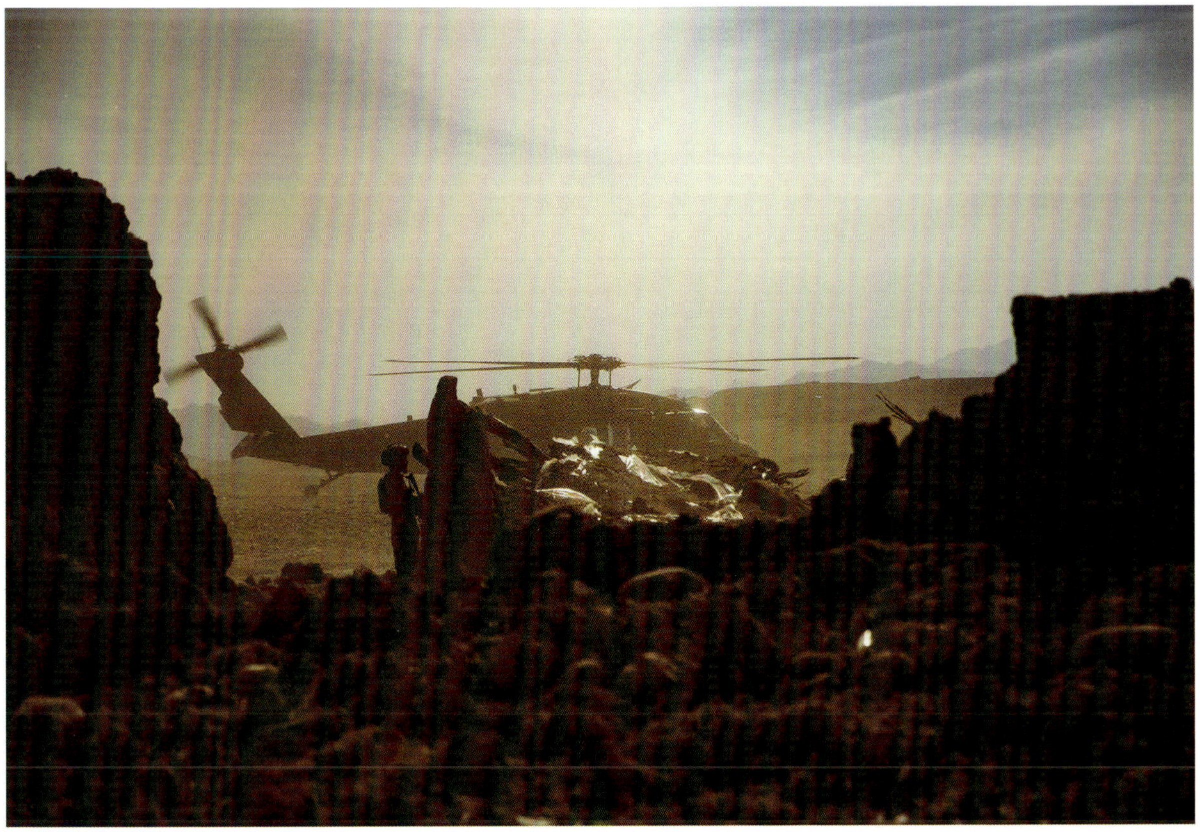

My name is Brent "Fancie" Looby. I was an EA-6 Prowler pilot in the United States Marine Corps. I started off as a peacetime service member. I grew up around aviation as a kid and have always been enthralled with the romance of landing high-performance aircraft on the deck of big ships. I retired as a Lieutenant Colonel (O-5). I deployed first as a part of Operation Northern Watch, supporting the no-fly zones immediately after 9/11. I also did a tour of the Western Pacific in 2003, supported Operation Enduring Freedom in 2005, and supported United States Special Operations Command in 2007 and again between 2011 and 2012. After I retired, I went to business school at the University of Texas-Austin, as an alternate path to being an airline pilot. I met my two business partners in our distilling venture, Desert Door. We all bring completely different attributes, experiences, and skill sets to the table, but since we're all veterans, there is tremendous shared trust and understanding. This shared drive to complete the mission and our collective sense of service and hard work has created a form of shorthand as we've worked together. It's allowed us to rapidly grow Desert Door. We are the first sotol distillery here in Texas. Our goal is to make sotol here what bourbon is in Kentucky, and we feel like we are well on our way! One of the biggest lessons I've learned in life is to dream big and then stay true to your dreams. If you don't believe it, and you don't go out and work hard for it, no one else will!

My name is Ryan Campbell and I enlisted into the Army as a 98G crypto linguist. We did a lot of counter-IED work in support of Operation Iraqi Freedom between 2005 and 2007, during the surge. We were figuring out who was setting up and manufacturing the bombs, trying to give the commanders on the ground as much information as possible. I joined right as I was graduating from college. Then 9/11 happened, so I met with a recruiter after graduation. I was an E-5 sergeant when I separated. Working in military intelligence is less rigid than some areas of the military. Free thinking is really important, and I'm a rebel by nature so it wasn't a particularly hard transition to leave the military, it was just different. I was a consultant for the next ten years, working directly to support the warfighters. I'm an entrepreneur by nature, and working for others wasn't scratching that itch. This led me to getting an MBA at the University of Texas-Austin. Desert Door spirit company was my team's class project. Since I worked with fellow vets on it, we all gravitated to each other because of our shared values. After we got an A on the project, we sat around for a few hours, then started building our company. Our shared military experience helped because we had shared values and understood shared purpose. We raised some money, got our facility, and built out the company. It went from nothing to viable in about four months, largely because of our shared collective work ethic. The military teaches that the limitations you put on yourself are all in your head. You can get more done and take things further if you are efficient with your time and willing to accept a certain level of discomfort.

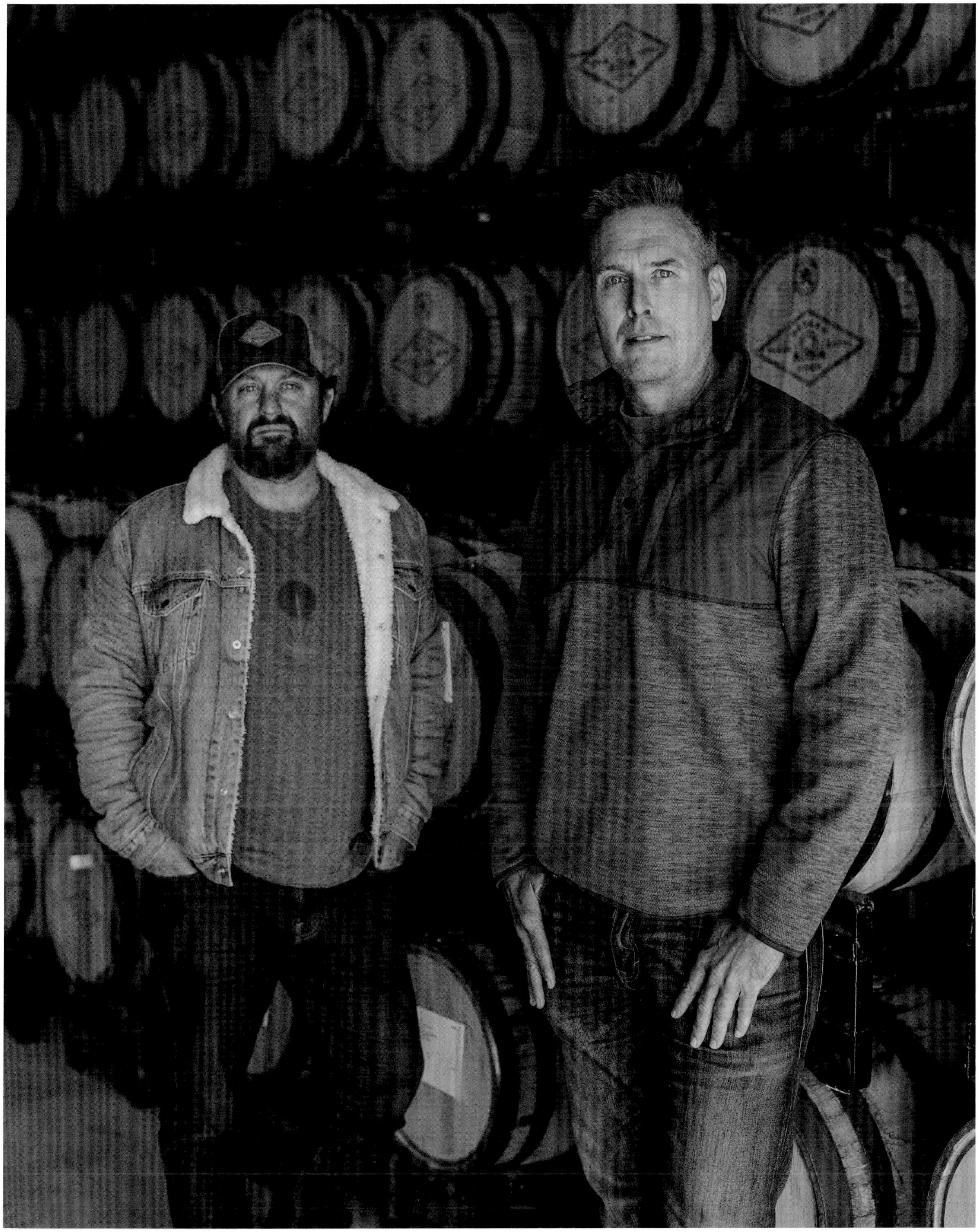

Ryan Campbell, Sergeant
Army - 98 Golf
Years of service - 5 years
[Operation Iraqi Freedom]

Brent "Fancie" Looby, Lieutenant Colonel
Marine Corps - 7543 Prowler Pilot
Years of service - 21 years
Deployments - 5 [Operation Enduring Freedom]

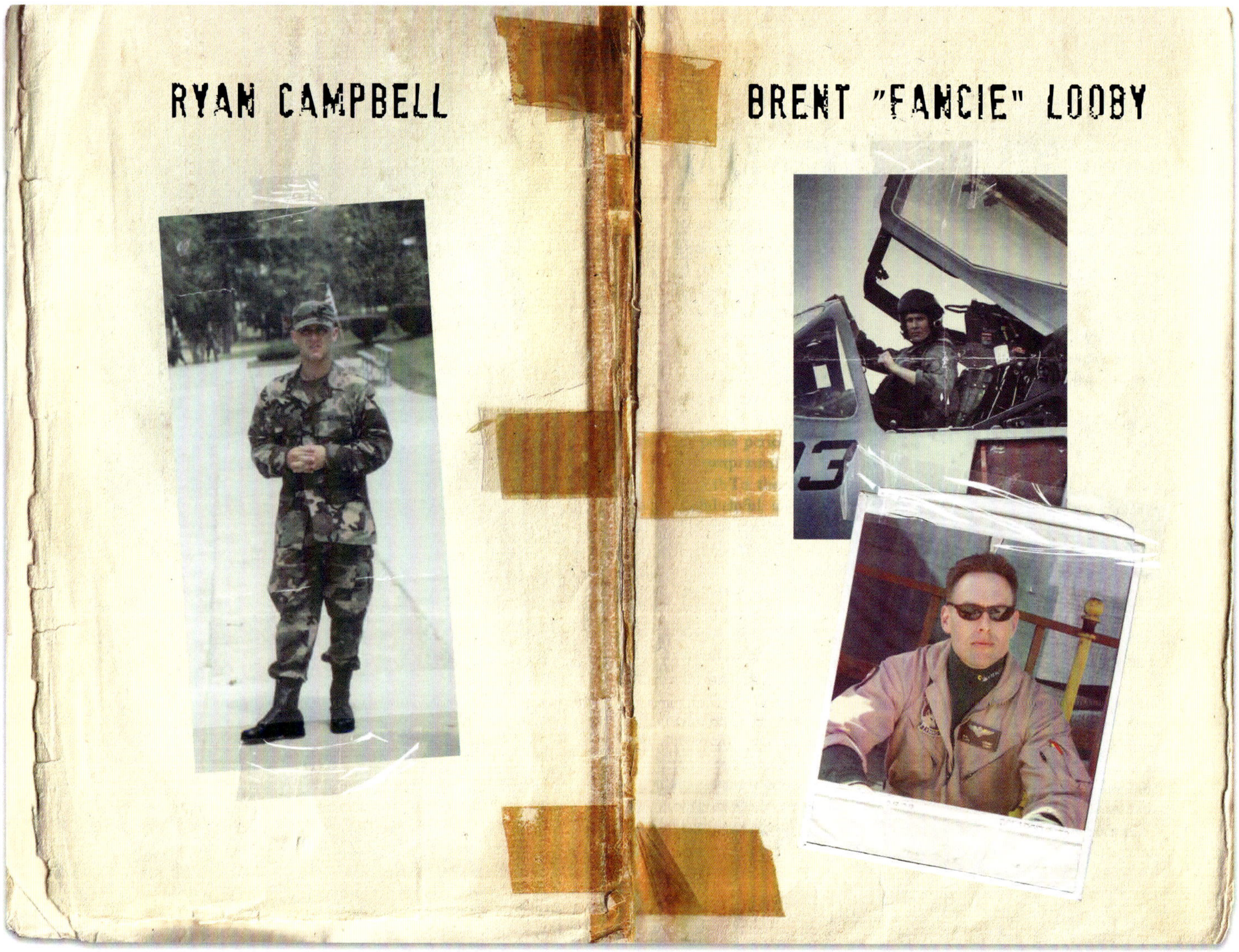
RYAN CAMPBELL
BRENT "FANCIE" LOOBY

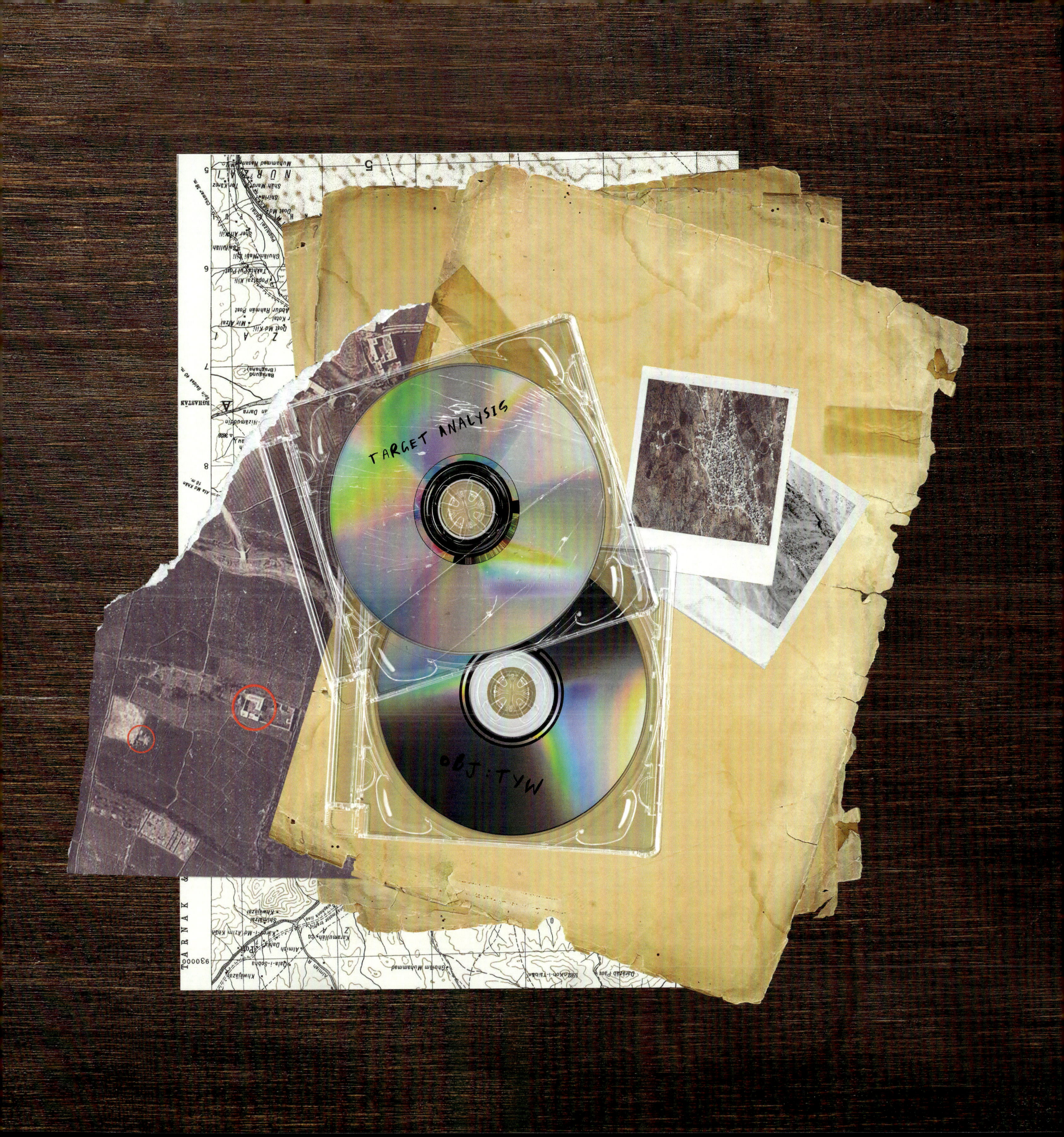
TARGET ANALYSIS
OBJ:TYW

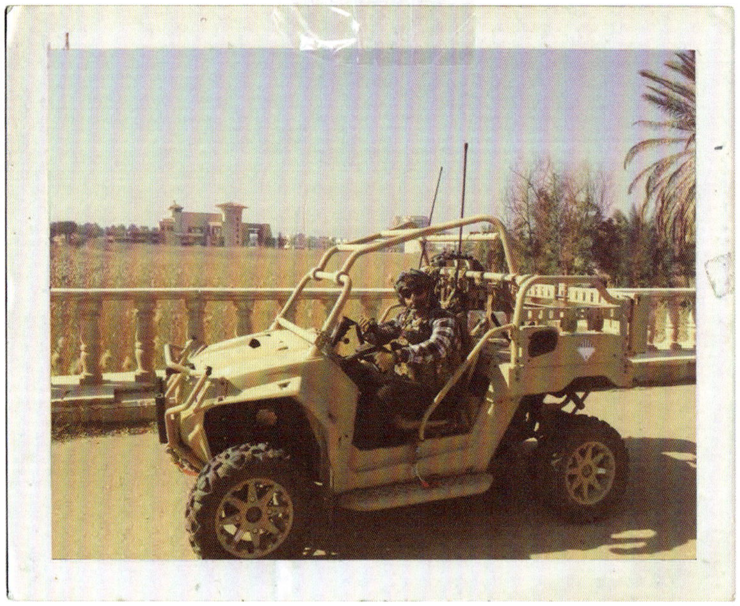

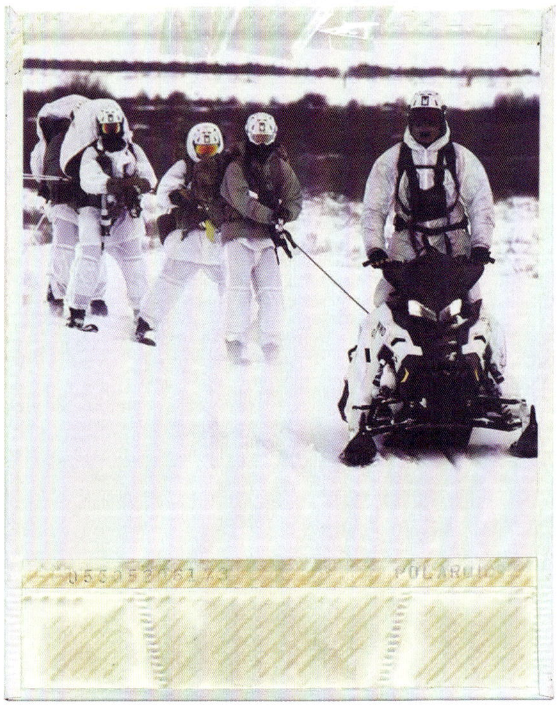

"today is a good day to die."

I'm Jeremiah Wilber, and I retired from the United States Army as a Special Forces Team Sergeant. I'd wanted to serve since I was three or four. My grandfathers and father had all served, and it was kinda my calling. I loved the warrior path they laid out for me. I deployed on teams as part of Operation Iraqi Freedom, Operation Inherent Resolve, Operation Enduring Freedom, and twice to counter Russian aggression North of the Arctic Circle. I had a really good transition, because of the people I'm surrounded by. I am the National Partnership Manager for Uncanna, a CBD company founded by a fellow veteran. I am also a hunting guide and outfitter here in Colorado. I have also worked with Clandestine Media as a model. The veterans at these companies, especially the former Special Operations guys, really took me in and helped me and my family post-retirement.

When I'm not working or spending time with my family, I like to backcountry ski and compete in SkiMo: think ultra-marathons, but on skis. Also, any time I can be on a horse in the wilderness is the best therapy I've ever experienced in my life. I really enjoy spending the day riding. Growing up in a Native American family, my mom used to always say, "Today is a good day to die." Warriors have long said it before battle because it meant they were right with their family and The Great Creator or God. It's about honor and always being prepared for death as you are right with the ones you love. That's how I choose to live my life.

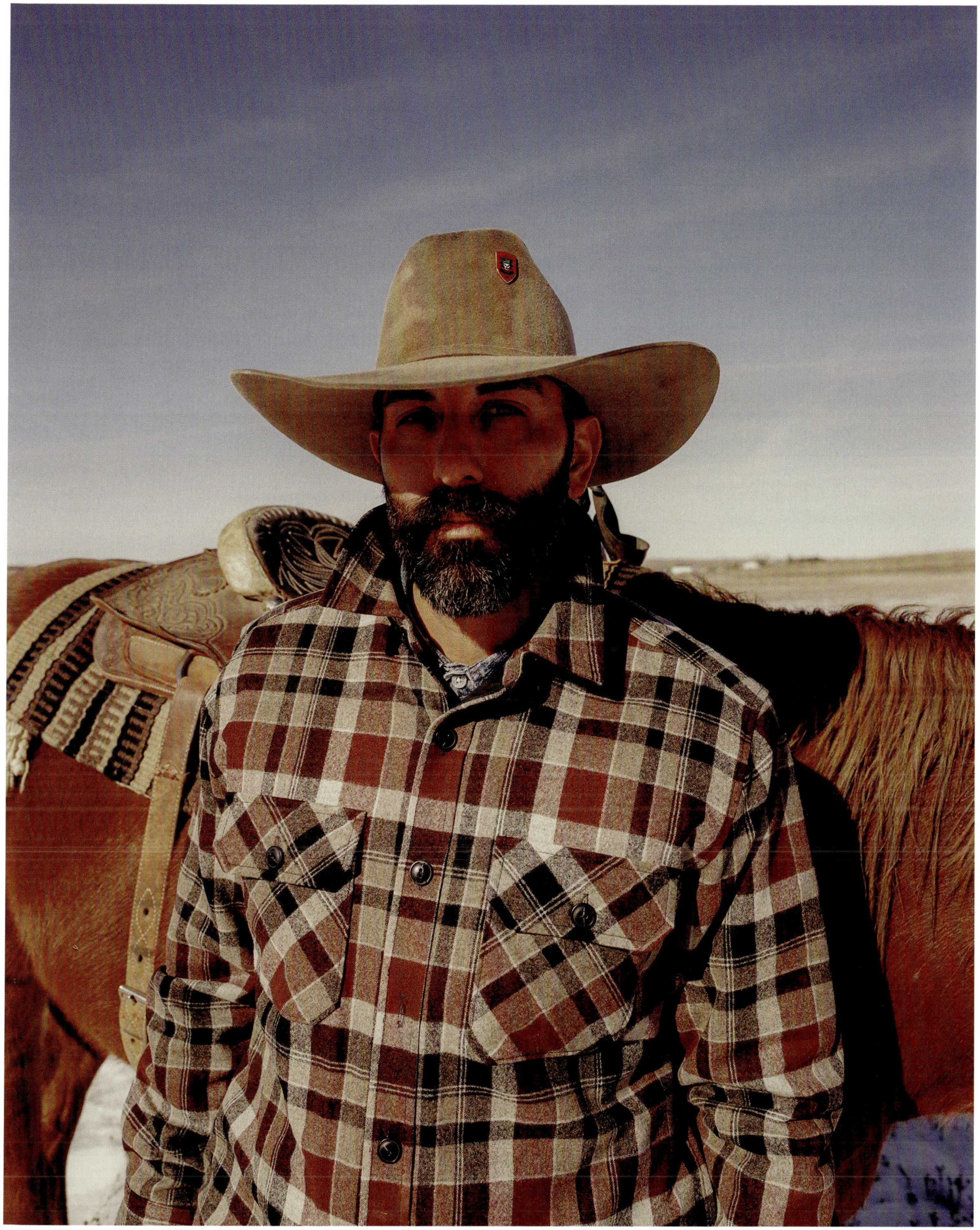

Jeremiah Wilber, Sergeant First Class
Army - 18E, 18Z / Special Forces
Years of service - 20 years
Deployments - 5 [Operation Iraqi Freedom, Operation Enduring Freedom, & Operation Inherent Resolve]

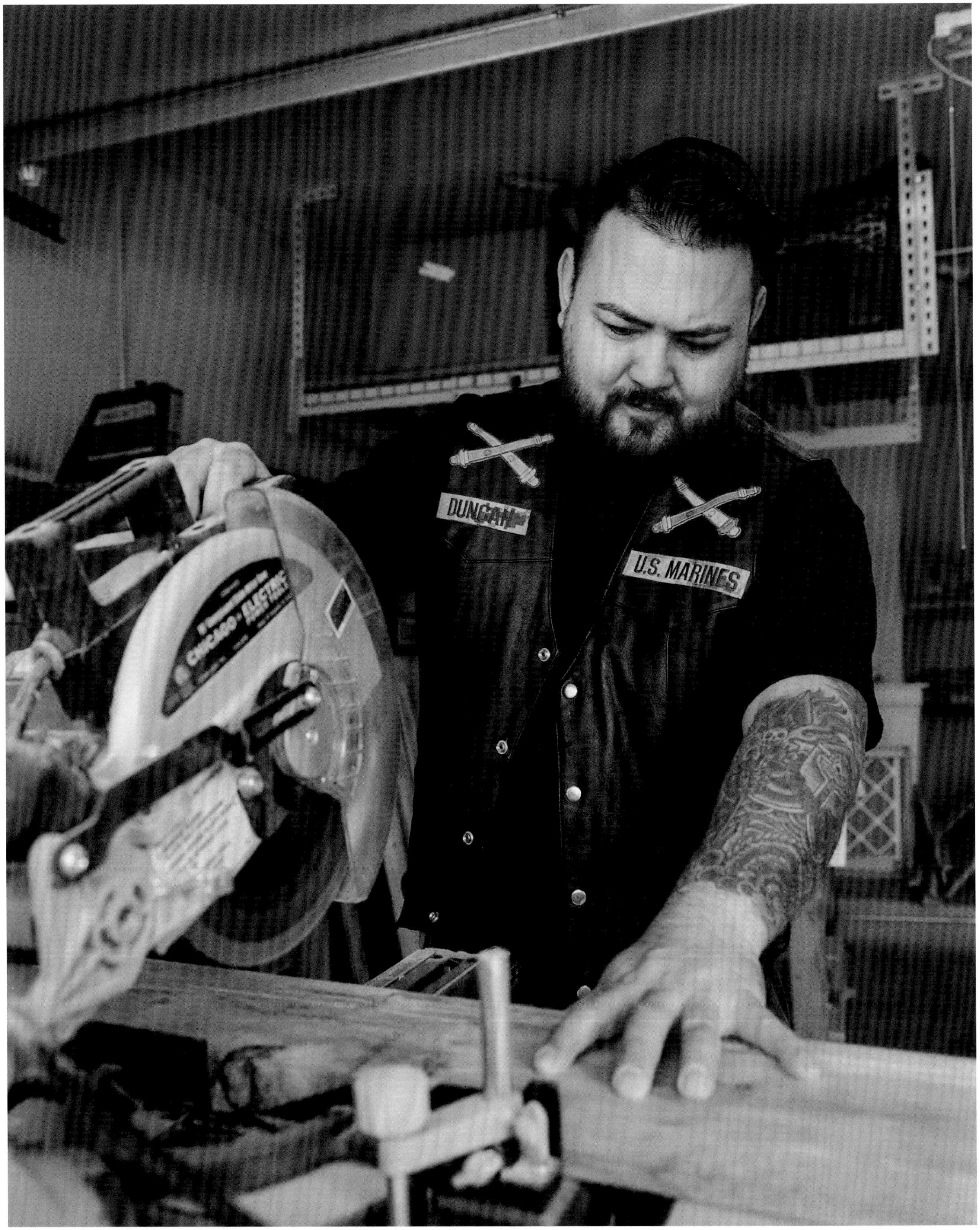

Jeff Duncan, Corporal
Marine Corps - 0844 / Field Artillery Fire Control
Years of service - 4 years
Deployments - 1 [Operation Iraqi Freedom]

I'm Jeff Duncan and I was enlisted in the United States Marine Corps as an 0844 fire direction control member. I joined because of family, and because I wanted to serve and see the world. I also wanted to get money for college. I separated as a corporal (E-4) and deployed in support of Operation Iraqi Freedom in 2006. We got extended due to the surge, and ended up being in-country for over ten months.

I had a rough transition. I started drinking a lot, post-service. Had a hard time sleeping due to drinking and nightmares. I was initially turned away by the VA, but the substance abuse counselor recommended me to the mental component and got me some help. I started to heal but I hated the meds. I'm a big advocate against the meds. I didn't like the side effects. I could tell they're supposed to help but I could also feel them messing with my head. I ultimately decided to do sixteen weeks of exposure therapy and that really helped me.

Since separation, I started by attending school and then moved to San Antonio to own a house and have a better quality of living. I'm currently working in building maintenance right now and am considering getting into HVAC because of my building manager's recommendation. I'm part of a private Facebook group for Marines in San Antonio to help each other and connect, to have a sense of the military brotherhood. Life Mantra? Take care of number one. It's good to have your family and brothers and everyone to help as much as they can, and you can do the same. But if you aren't helping yourself, it's hard to help others. Take care of yourself, but never be afraid to ask for help.

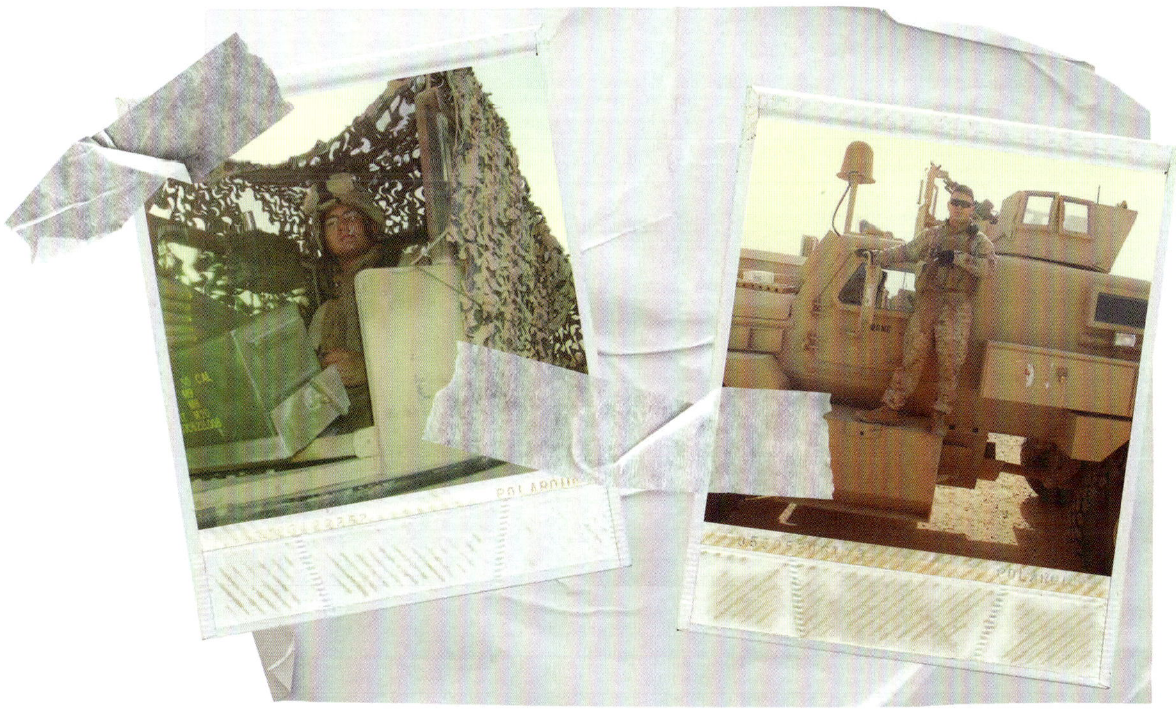

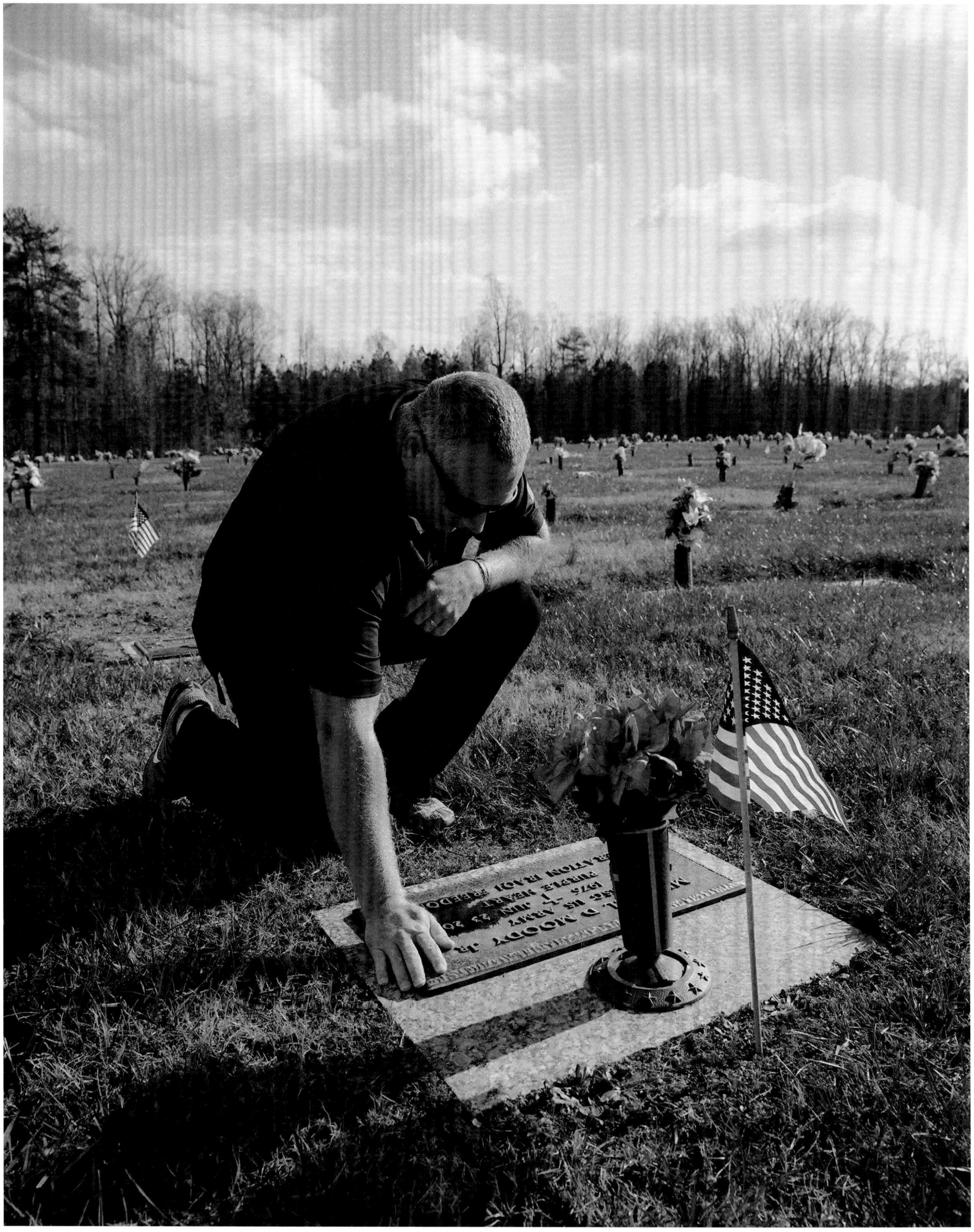

Roy Bell, Sr., Staff Sergeant
Army National Guard - 88M
Years of service - 12 years
Deployments - 3 [Operation Iraqi Freedom]

My name is Roy Bell, Sr. and I was an 88M truck driver in the United States Army National Guard. I started my career in 1997 at seventeen, while still in high school. I was super enthusiastic to join. I'm an Air Force brat and at sixteen had already started talking to the recruiters. The Army Guard had a split option where you could go to basic training the summer between your junior and senior years of high school if you had your parents' permission. I was in such a hurry to get into the military. There's a sense of pride that you need when you're serving. I signed up originally to be a generator mechanic. That's what my dad did in the Air Force, so we were always working on cars. Halfway through my senior year, I got a letter from Spartanburg Methodist College to run cross country on scholarship, but if I wanted to take the scholarship, I couldn't be a generator mechanic. Which is how I became an 88M.

I eventually transferred to active-duty Army and was on Hunter Army Airfield when 9/11 hit. For me, it was really serious; my unit directly supported the First Ranger Battalion and at the time they were on rapid deployment alert, and my platoon supported them. Everything was packed and ready to go, we just needed to draw weapons and we were off. It didn't get to that point, but it was really tense. I eventually deployed three times in support of Operation Iraqi Freedom. I have thirty-nine months in-country. I was medically retired and still live in the Fort Lee area of Virginia.

When we deployed the three different times, we saw three different sides of the country. In 2003, it was new, and the kids were always asking us for candy and waving. In 2005 they weren't as friendly, throwing rocks and sometimes even bottles of piss at us. In 2007, it felt like they were throwing grenades all the time. It just kept escalating.

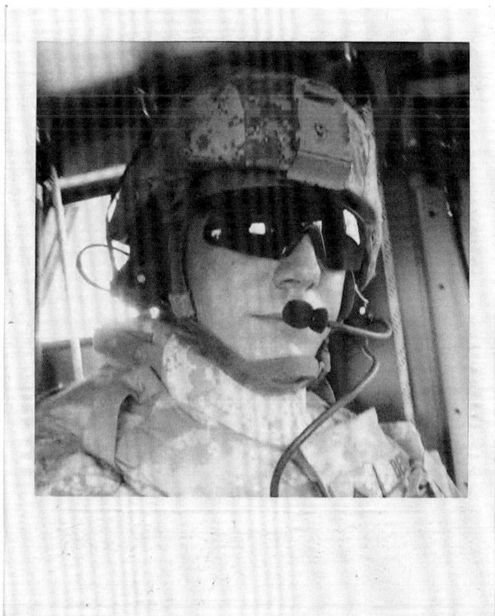

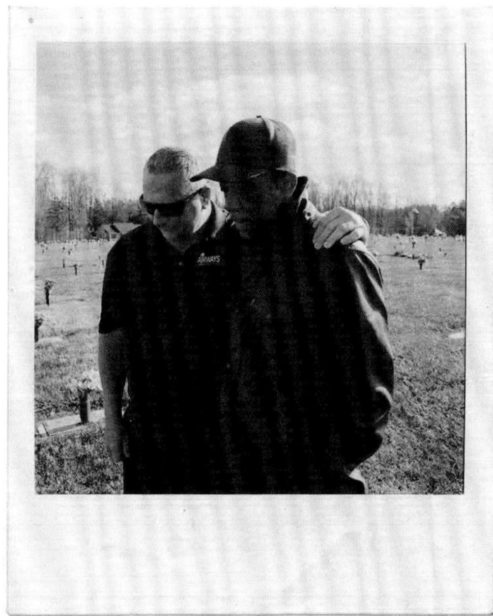

When I joined it was the decision of a cocksure seventeen-year-old kid, I really wanted to fight back. I was young, in shape and 9/11 happened when I was a sophomore in high school. When that happened, I really had no other choice, I enlisted as soon as I could — initially as an aircraft mechanic, because my parents wouldn't sign off on the enlistment papers if I didn't have a "safe" job. But on my eighteenth birthday, before my ship date, I changed my job to infantry, 03xx. I wanted to be in the mix, in "the shit" so to speak. I left active duty as a Lance Corporal, and was promoted to Corporal in the IRR, after two tours in the Iraqi city of Ramadi in support of Operation Iraqi Freedom. Many of the men I served with were phenomenal men, whom I learned a great deal from, but many of those lessons didn't take until after I'd left the service.

I sold cars when I first got out; it being 2008, car sales were suffering. It was hard for a while, it showed me how ill-prepared I was to enter the civilian world. It showed me things I needed to get better at, but it was difficult to put on a smile every day and talk to people, like I hadn't just been to war. Initially with the economy and my lack of job skills, it was very difficult for me. Now I'm the Editor-in-Chief at a digital marketing agency here in San Antonio, Texas. In 2019, I participated in the Talisker Whisky Atlantic Challenge, a trans-Atlantic rowing race, as part of an all-veteran crew with an organization called Fight Oar Die. We worked with the graduate school of psychology at the University of Denver, using this experience as data gathering to provide better treatment for veterans suffering from PTSD. Most teams have over a year to prepare and have known each other their entire lives. We'd known each other for seven months and only spent about seven weeks together, in person before the row. Despite that, we set a new record for our class, the Spindrift, coming in at 50 days, 11 hours and 35 minutes.

My life mantra is: Nobody is coming, it's up to me. No one is coming to do the work for me. You have to go out and earn it.

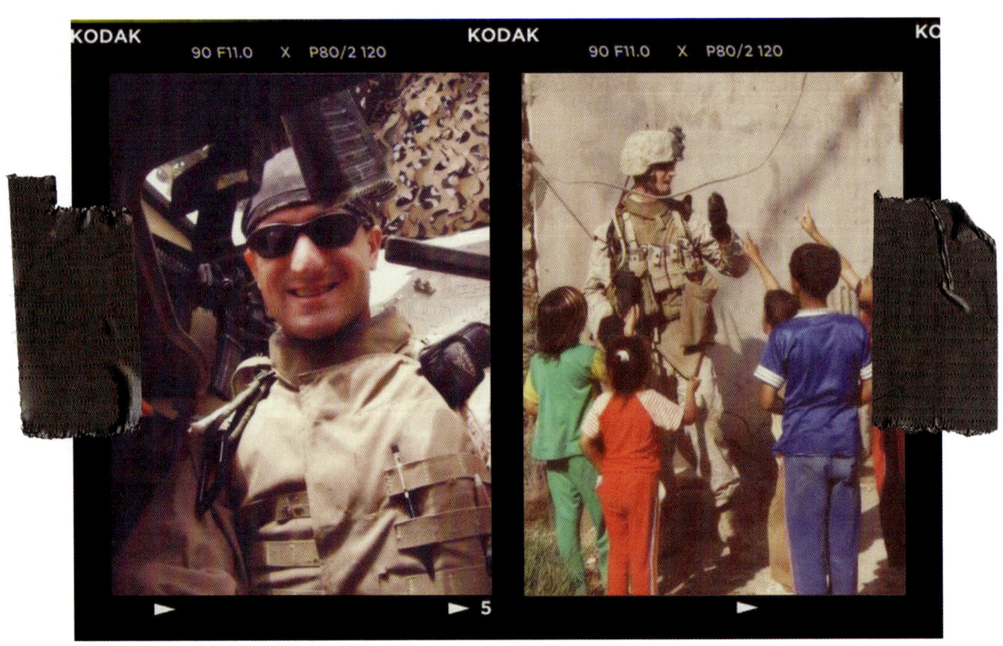

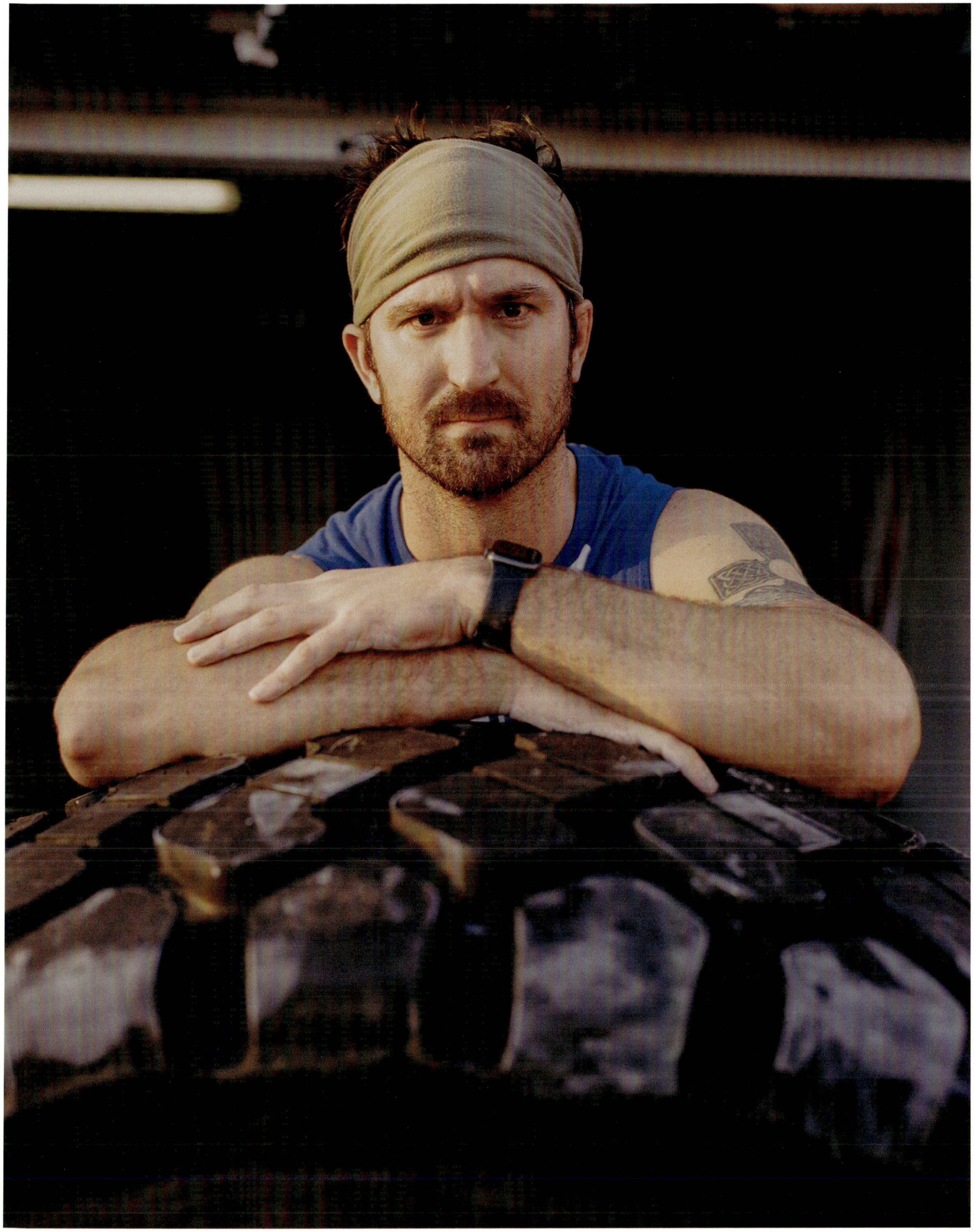

John Fannin, Lance Corporal
Marine Corps - 0351 / Assaultman
Years of service - 4 years
Deployments - 2 [Operation Iraqi Freedom]

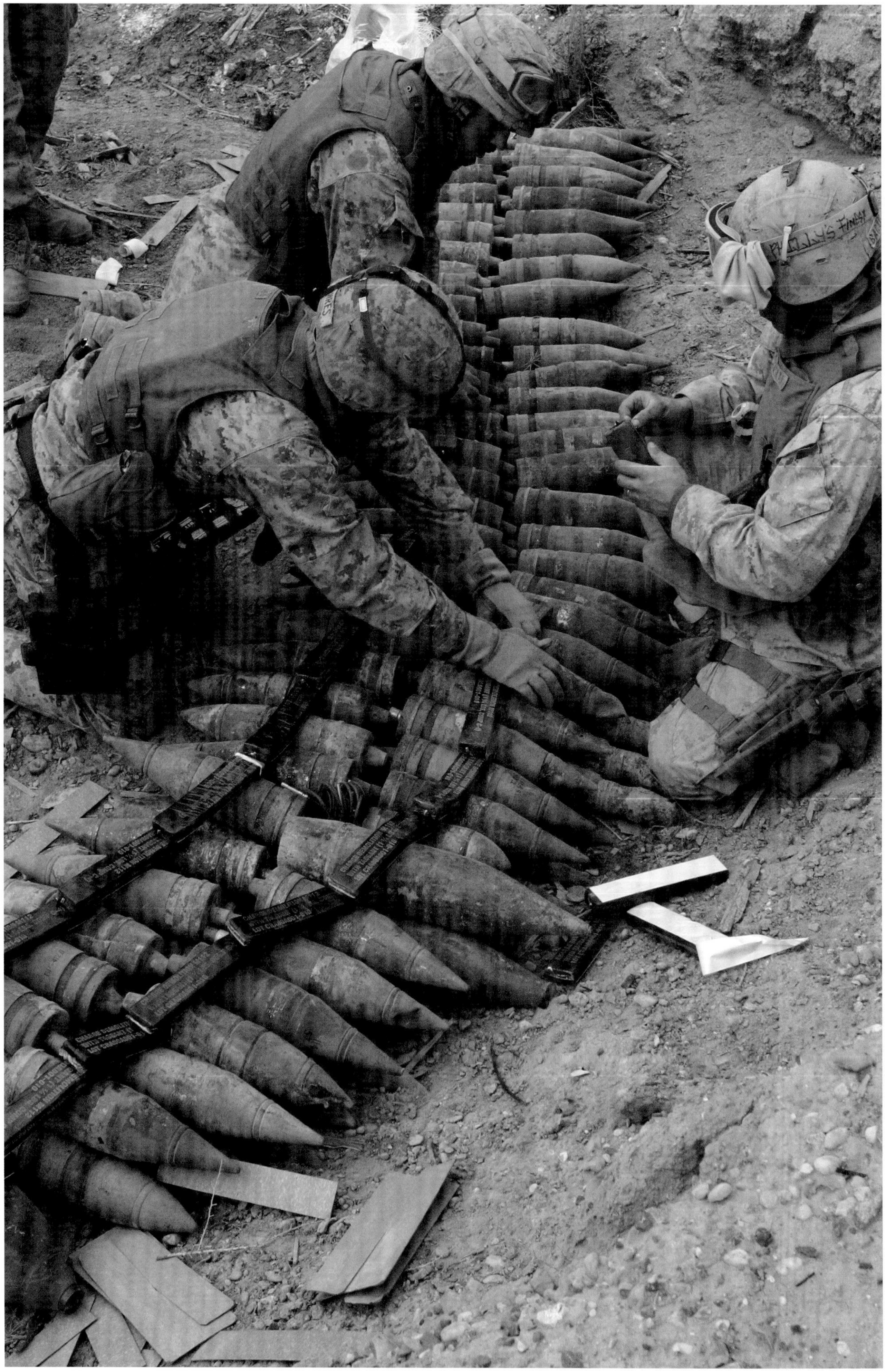

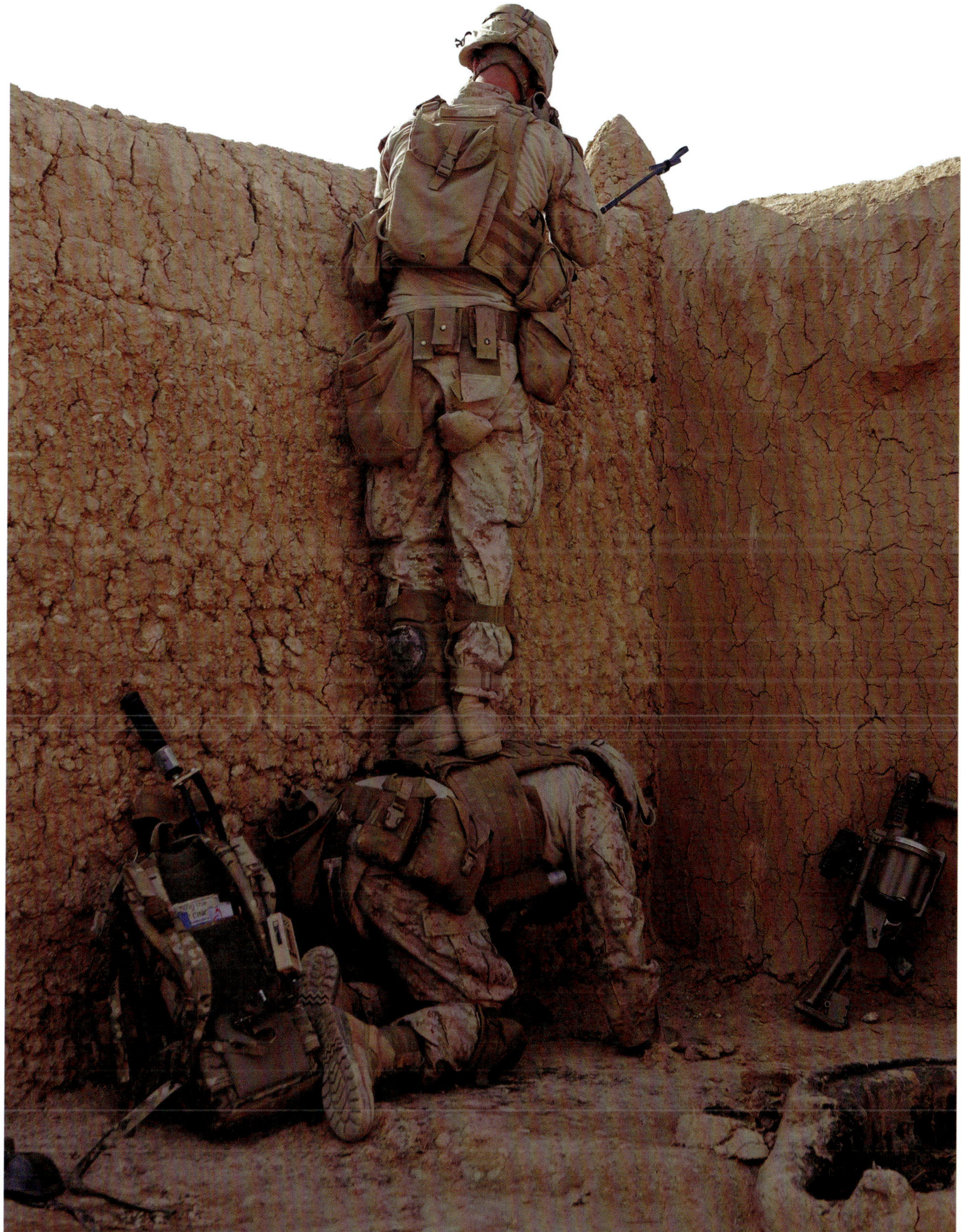

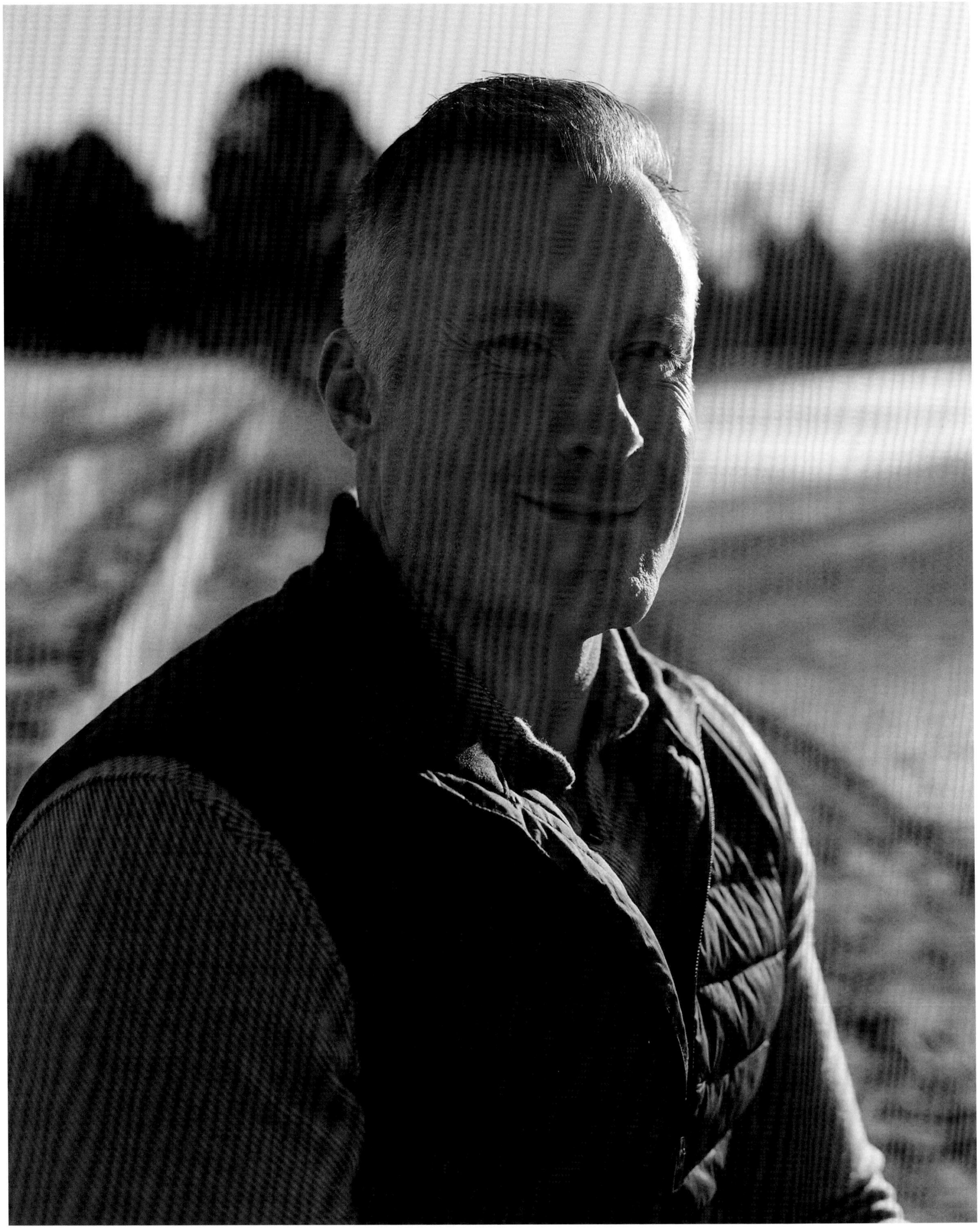

Jack Aalborg, Colonel
Air Force - Special Operations Pilot (11S) MC-130H Talon II
Years of service - 25 years
Deployments - 8 [Operation Iraqi Freedom & Operation Enduring Freedom]

I'm Jack Aalborg, and I was a member of the United States Air Force as a Special Operations pilot. I flew the MC-130 in support of spec-ops missions all over the globe. I joined the Air Force to scratch the adventure itch and see the world. It might sound kinda trite, but I wanted to leave home and set out on my own. I left the Air Force as a full Colonel (O-6) and served in Operation Enduring Freedom and Operation Iraqi Freedom, along with other special operations missions during my career. Post-9/11 missions shaped the majority of my career.

Readjusting to civilian life was hard, much harder than I thought it would be. I thought I had a pretty good game plan, and I am now flying cargo routes for UPS. Most people don't realize just how much of your identity is military, so as I looked beyond my time in service, I ended up having a harder time than I anticipated parlaying that experience and skill set in the civilian sector. If I could go back and change something, I would have definitely done more homework on working for an airline. I also would have thought much harder and asked myself the questions: 'What can I do?' and 'What do I want to do?'

I am now a member of the VFW in Colorado and am looking to work with the Air Force Academy as an alum. I feel like short shrift is paid to mentoring soon-to-be officers and increasing their understanding that in a few short months after graduation, they could be leading a couple hundred enlisted airmen. To transitioning veterans, I'd say: Don't be afraid to fail. Don't be afraid to take a chance. When you served, you were the tip of the spear for a reason, so don't be afraid to take that mentality with you into civilian life.

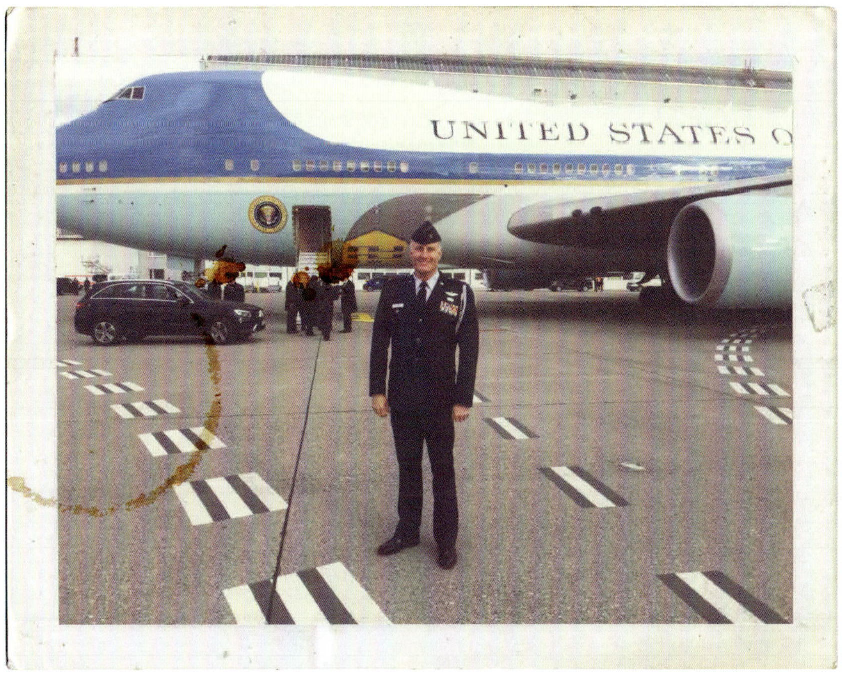

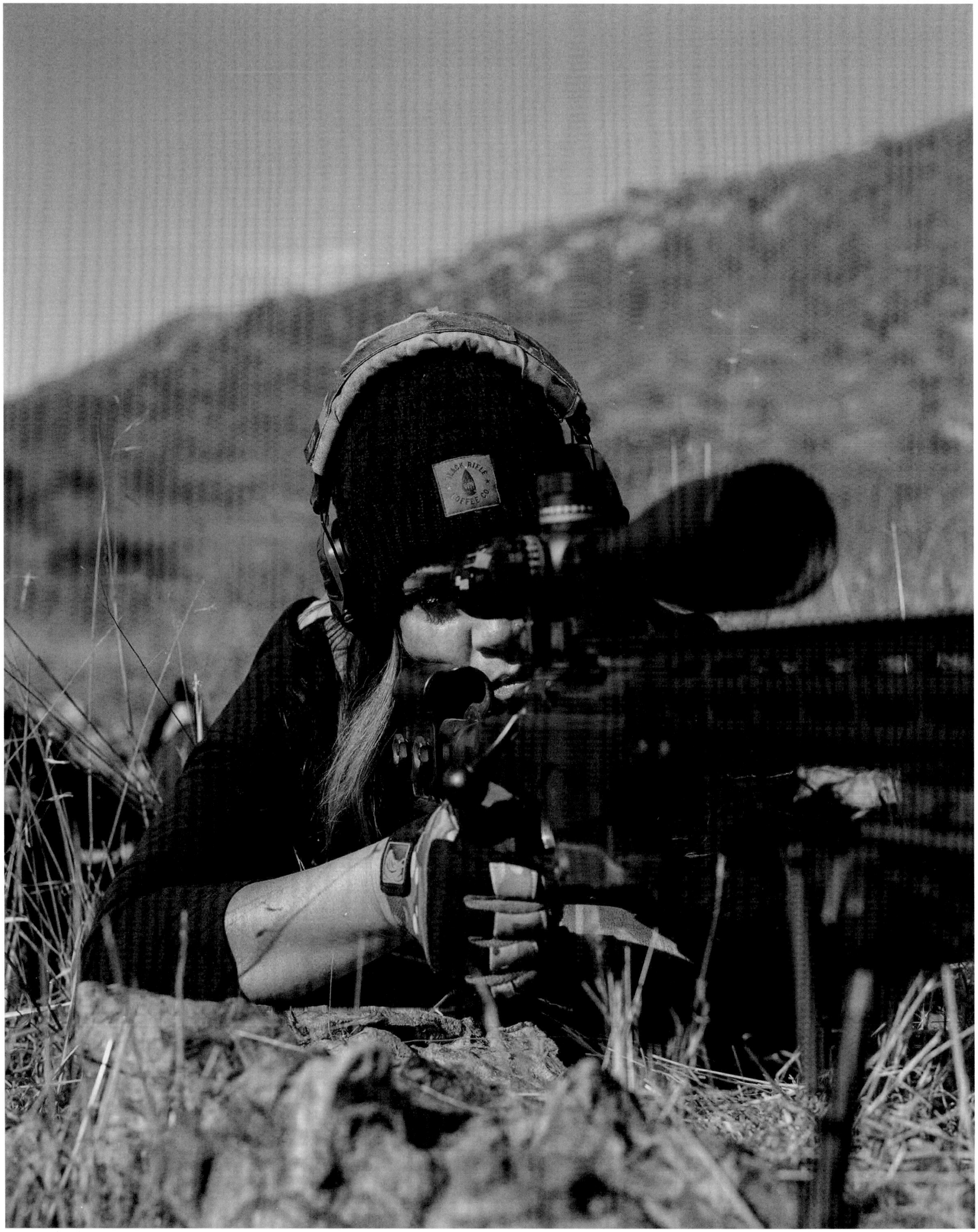

Domenique “Nikki” Selby, Lieutenant Commander
Navy - 8401 Corpsman / Critical Care & Trauma ER Nurse
Years of service - 10 years
Deployments - 3 [Operation Iraqi Freedom & Operation Enduring Freedom]

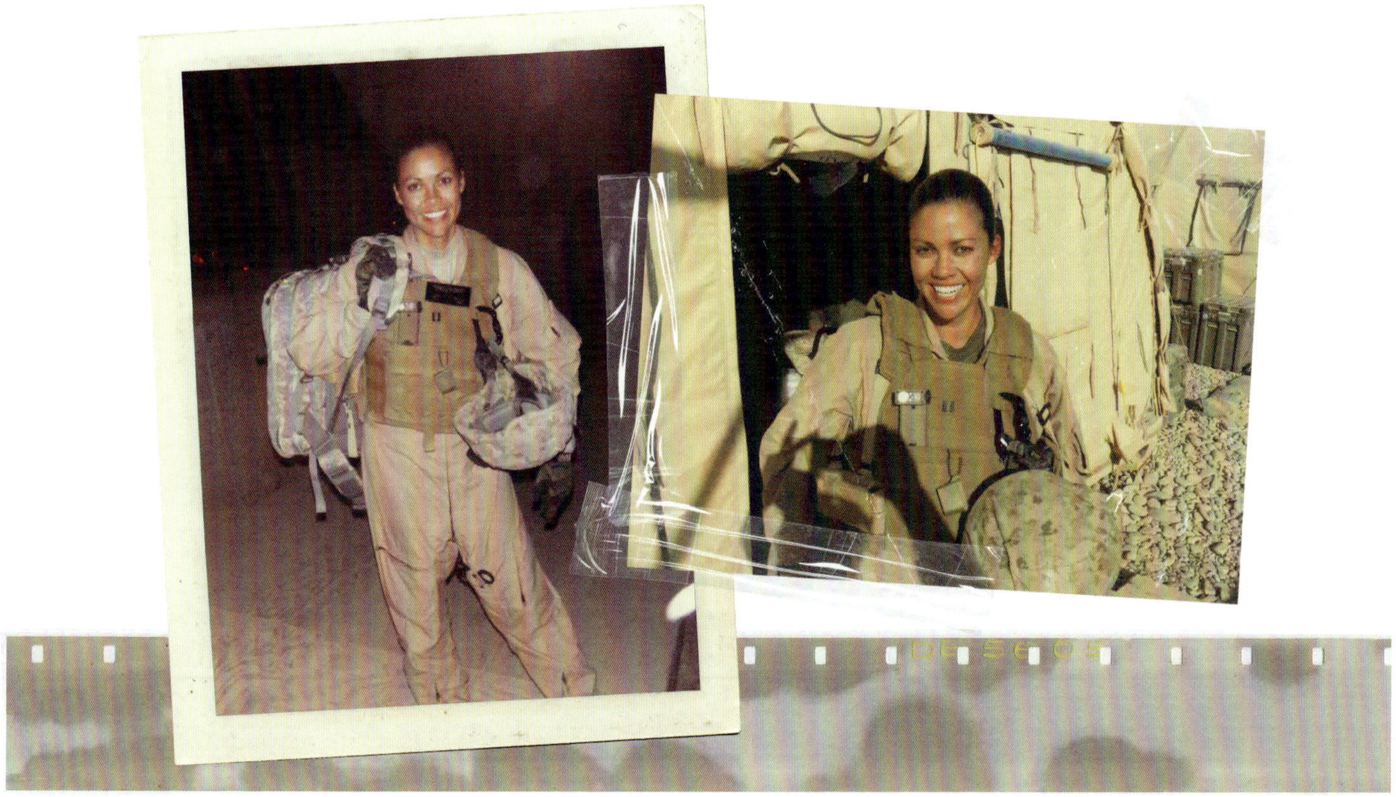

My name is Nikki Selby. I enlisted in the United States Navy as a Search and Rescue Corpsman because I screwed up in high school. As I progressed in my career, I got my degree and transitioned to the officer side, becoming a critical care ER trauma nurse. I retired as a Lieutenant Commander (O-4). I deployed in 2010 to help with Haitian earthquake relief before going to Afghanistan in support of Operation Enduring Freedom at the end of 2010. I also deployed in support of Operation Iraqi Freedom in 2012.

After I retired, I started working for the Joint Trauma Services, which is the military's central hub for trauma records and training. I worked in a job and on a team similar to what I did at the end of my career, which was trying to get the clinical or hospital-based medical component of the Armed Forces more prepared for field environments. It's a big change, a big adjustment, to go from the hospital to the field, especially in a combat environment. I am lucky enough to have been part of the programs that try to help make those changes and help clinicians be better prepared. Now I work for a company called Combat Medical Systems as a Department of Defense accounts manager.

I support HunterSeven and think they're doing amazing work for our veterans. I'm also part of the Global Surgical and Medical Support Group. They are an organization that goes to Kurdistan and teaches the Kurdish Peshmerga Tactical Combat Casualty Care (T-CCC) concepts to improve their readiness. We've saved hundreds of lives on these missions in Kurdistan. It's really rewarding work. I'm a bit of an adrenaline junkie. I skydive, do long range shooting, and I fly helicopters for the San Bernardino air rescue teams as well. I also work out a lot. My mantra in life is: live hard, and enjoy what you're doing!

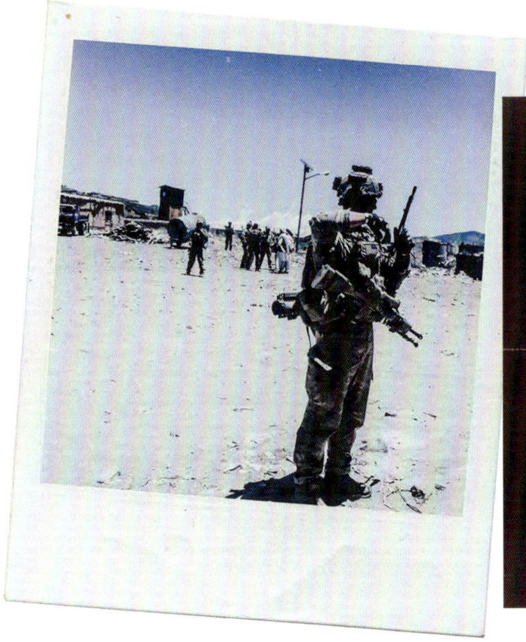
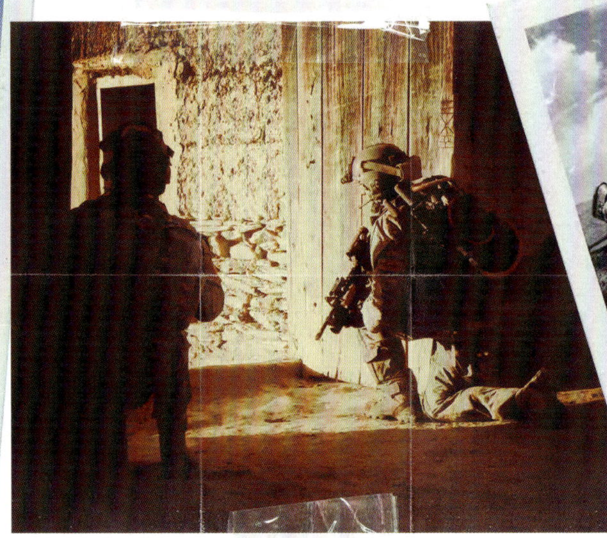
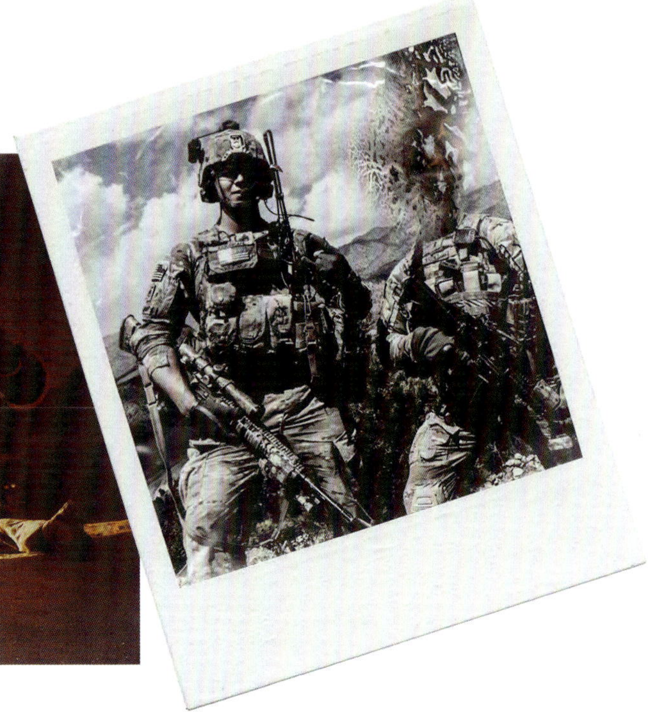

My name is Coby Cochran, and I was an 11B, Infantryman, in the United States Army. I joined because, like many young men, I thought I wanted to go to war and be a hero. I enlisted in 2008, and spent ten years in service to this nation before being medically retired as a Sergeant First Class (E-7). I had multiple deployments, including Operation Enduring Freedom as a member of the 1st Battalion, 75th Ranger Regiment.

Transitioning out of the service was the hardest thing I've ever done. I had no real plans. I was blindsided by my medical retirement. Even though I had also been a successful entrepreneur while on active duty, supporting ten non-profits through those endeavors, I didn't want to reach out for help. I left the service hooked on prescription drugs and in the deepest hole I've ever been in. I had a come-to-Jesus moment, with family and friends intervening. I knew I needed help. The prescriptions just numbed everything. It wasn't just the bad feelings — it was all my feelings. They numbed my entire desire to live. Opioid addiction is a direct contribution to military suicide, especially with the culture of partying the military can foster. The search for something that worked led me to CBD oils and the founding of my company, Uncanna.

We created CBD products that we have found to be effective. We have thousands of reviews and testimonials from First Responders, LEOs [law-enforcement officers], and veterans. It's life-changing testimony to hear that our products are reducing their anxiety, helping them sleep, and reducing their dependency on addictive prescription drugs. The Motocross racing community in Colorado is amazing. The first day I met a dude who was also a med-retired vet. He immediately introduced me to other trail riders, and I was hooked on racing again. The race community here is a family, similar to the military in many ways.

My life mantra is: "Find good people doing good stuff in a good environment." Good people in good environments are saving lives. A solid network and positive environment are key to being able to move forward in life.

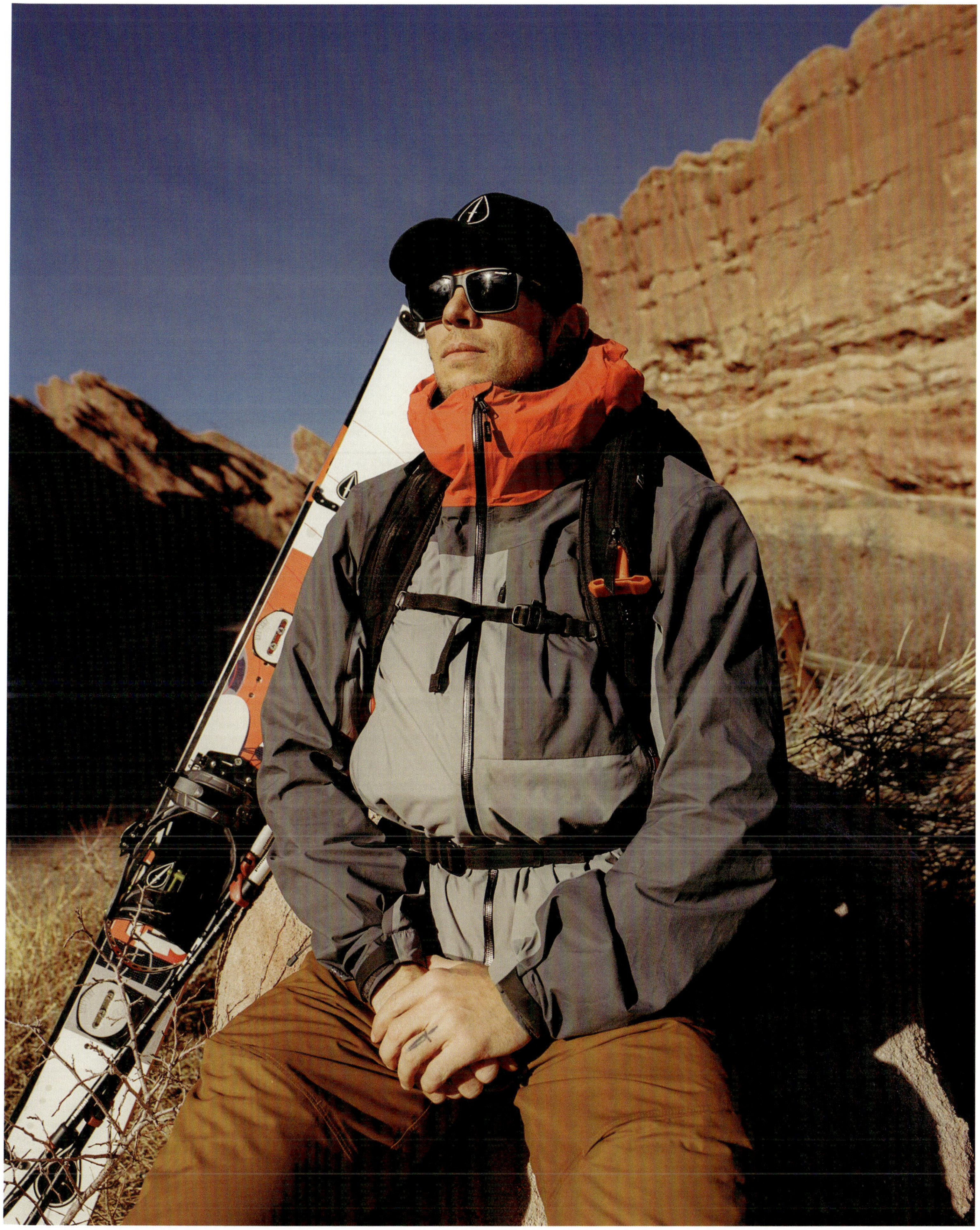

Coby Cochran, Sergeant First Class
Army - 11B / Infantryman
Years of service - 10 years
Deployments - 3 [Operation Iraqi Freedom & Operation Enduring Freedom]

My name is Tim Kolczak and I was a 92G food service specialist in the United States Army. I enlisted in 2005 and picked my job because they had the best bonus options. I had been a baseball player in high school and could've played in college, but baseball scholarships suck. Since I'm from San Antonio, and it's a hugely patriotic city, joining the military to help with college money made sense. Also, my dad had served, and he didn't want me to be a ground pounder. Then I ended up going overseas with the infantry in 2009 in support of Operation Iraqi Freedom. So, I ended up there anyway.

I am now the founder of the Veterans Project. I started it because I had this hippie, Berkeley-type professor for my master's degree program. She had protested Vietnam, the whole nine. She pulled me aside and said, "Tim, you are catching on, and I really like it." She went on to pitch me this idea about using my photography to tell veterans' stories, which is a little ironic. At the Veterans Project, it's not about any one vet; it's about vets around the world. Brits, Aussies and all of our allied veterans. It's about making this the way it should be, about the legacy of service members.

Photography has brought me some of the greatest peace in my life, along with my faith in Jesus Christ. I want to be an active testimony in the world and photography has helped me do that. Through my work I have to focus on the other end of the lens. I have to focus on the people outside of me. My transition was pretty easy, since I was serving in the National Guard and still playing baseball. My dad had also prepared me by instilling in me that no matter what, don't disgrace the two name tapes on my chest. Our legacies are the mission.

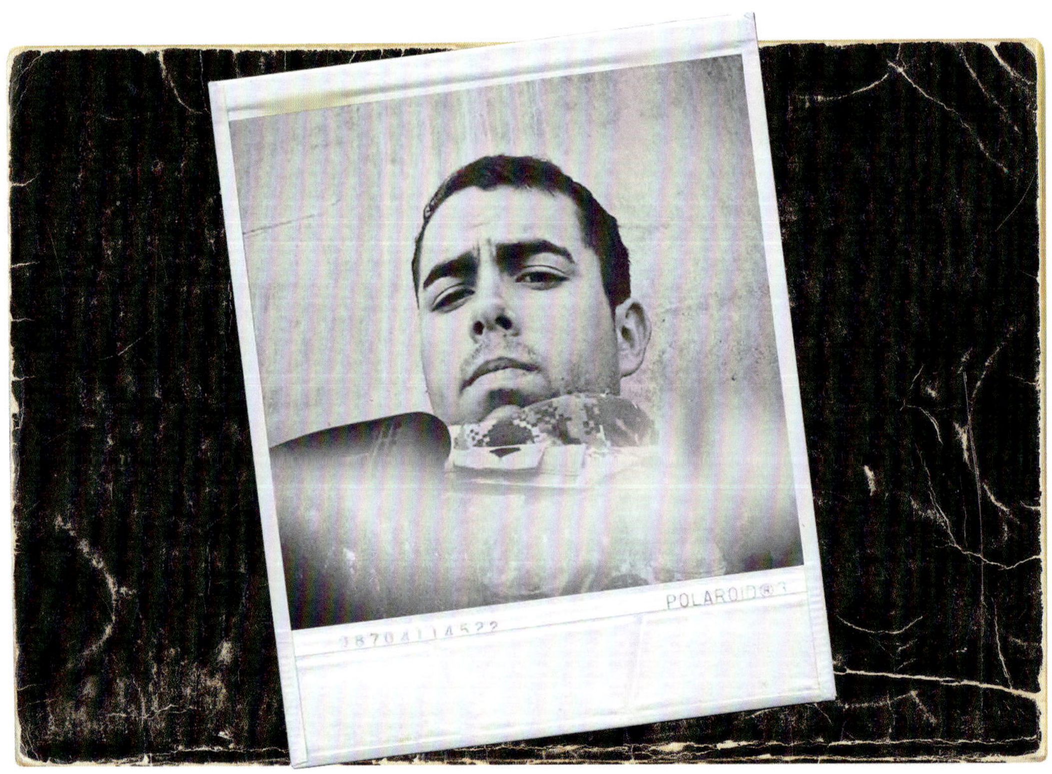

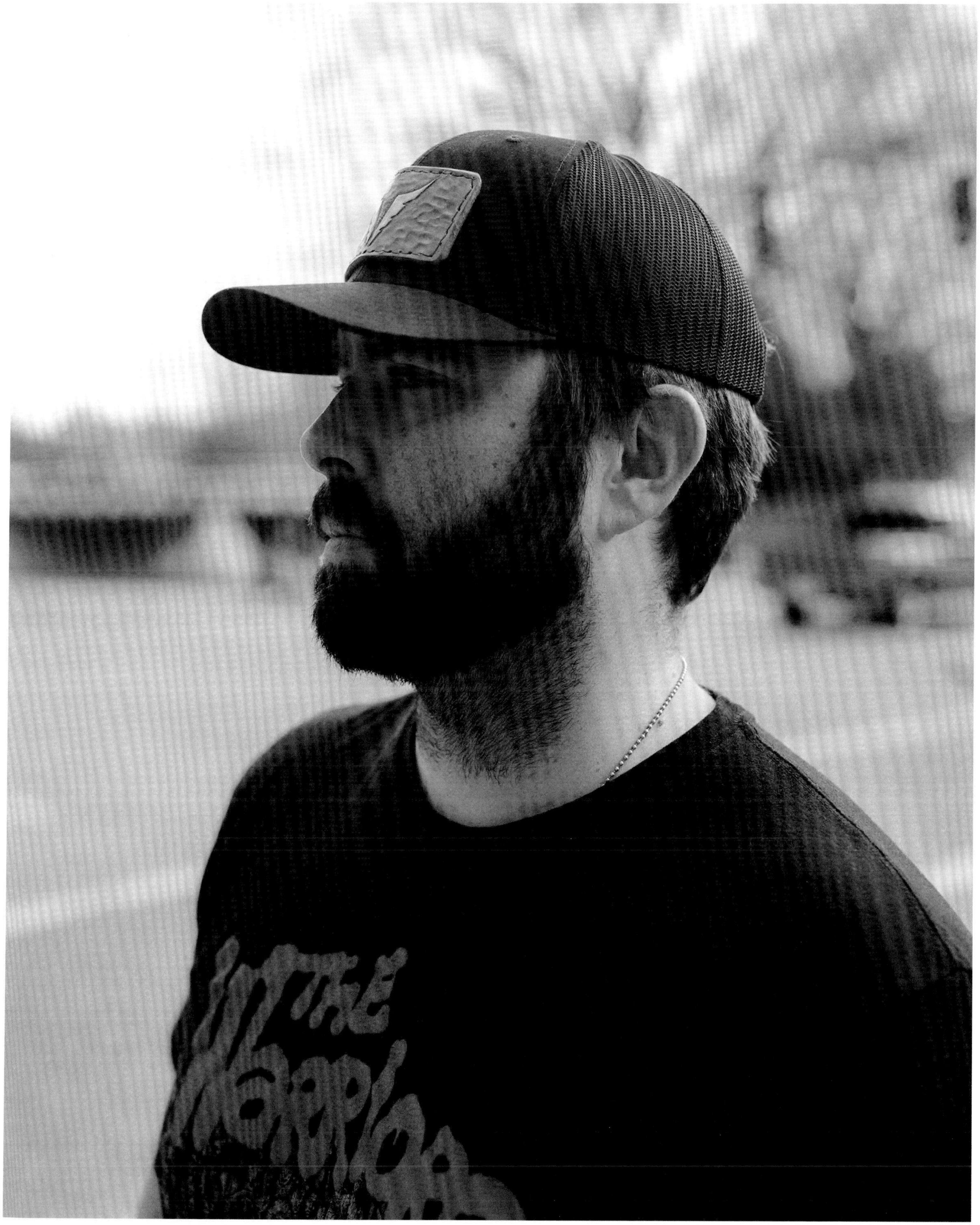

Tim Kolczak, Specialist
Army - 92G / Food Service Specialist
Years of service - 6 years
Deployments - 1 [Operation Iraqi Freedom]

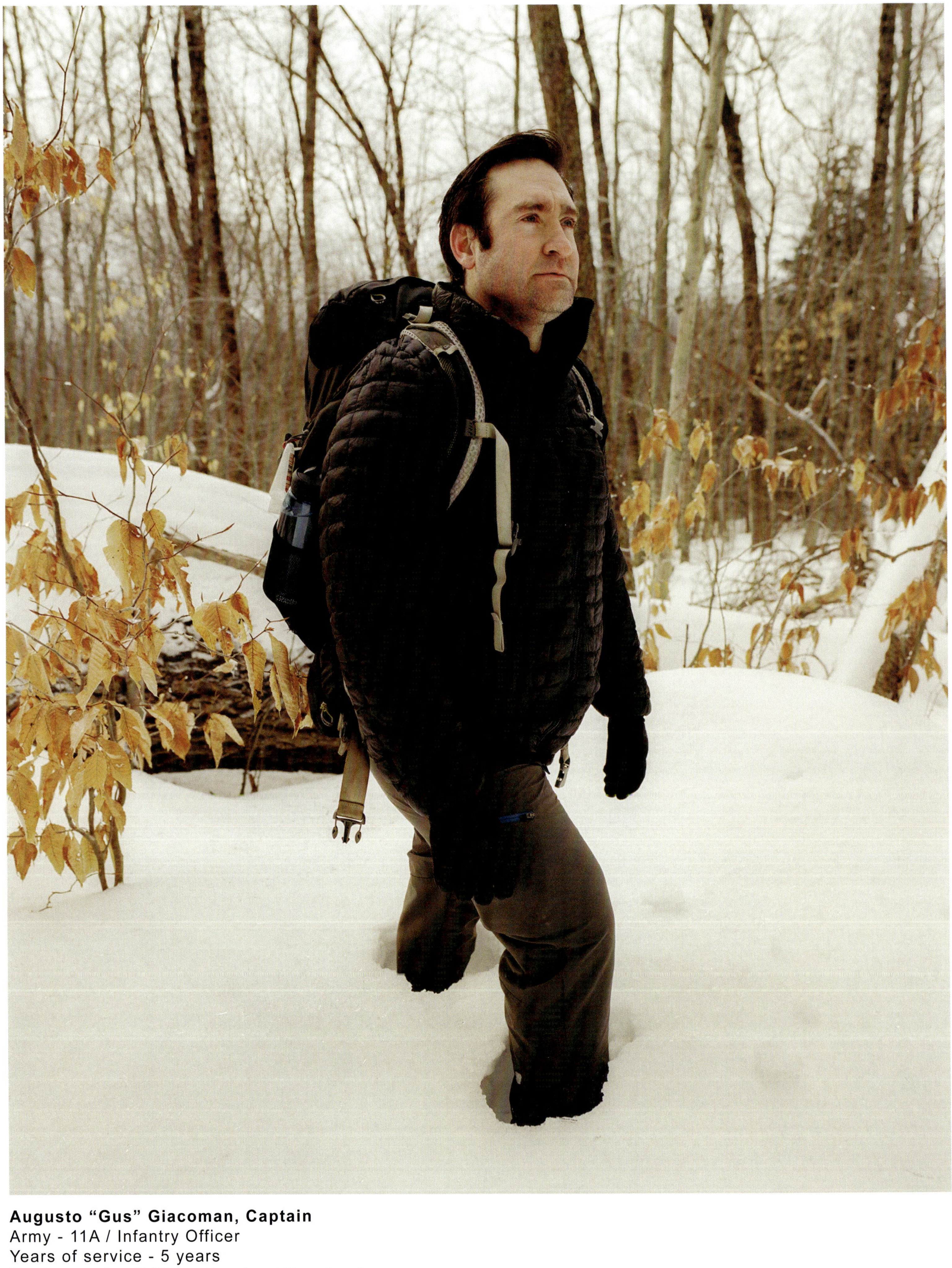

Augusto “Gus” Giacoman, Captain
Army - 11A / Infantry Officer
Years of service - 5 years
Deployments - 2 [Operation Iraqi Freedom]

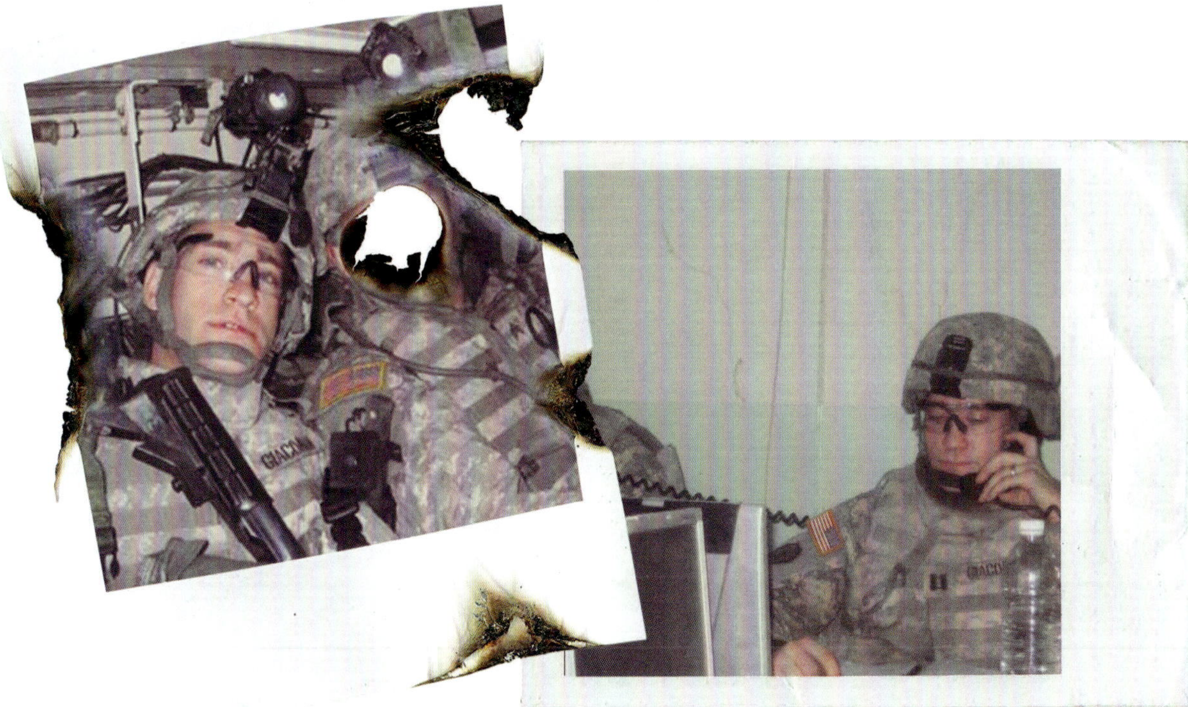

My name is Augusto Giacoman, and I was an 11A infantry officer in the United States Army. My family immigrated to the United States from Nicaragua. The idea of democracy and capitalism, of the American dream, is very important to my family. My mom and dad were in the Army, and my uncles and aunts served, as well. Initially, I wanted to enlist but my dad encouraged me to go to college, and I got appointed to West Point. I separated from the military as a captain, with two deployments to Operation Iraqi Freedom.

On my first tour we were aggressively trying to find Zarqawi. He was really sowing chaos, just blowing up everyone all around Mosul. A lot of IEDs, a lot of snipers. Every sixth patrol or so we'd get an IED or sniper fire. We were in High Philistine and we saw a lady breaking curfew. She said she was going to the hospital and ... boom. I don't remember the explosion. I just remember fire and the instant headache. My ears were ringing, and my squad leader was bleeding all over. The other LT started to bandage him, I collected my wits, and we got everyone back to the hospital. Our Stryker was all messed up, but we got home.

During readjustment I had a ton of help. I had a great resource in the Vet Center. Vet Centers were set up by the VA post-Vietnam. It's outside of the normal VA system, usually run by veterans, and it's a great place to get help safely and confidentially. Since separating, I have also gotten involved in supporting Service to School. We are a 1:1 program to help vets get into the best colleges we can get them into. We've helped vets get into the Ivy League, just top-notch schools. I'm on the board, and very active in fundraising for the organization. I love to read, and I love to write and get my workout in! The advice I have for any fellow veteran is to serve others. Be of service, and everything falls into place.

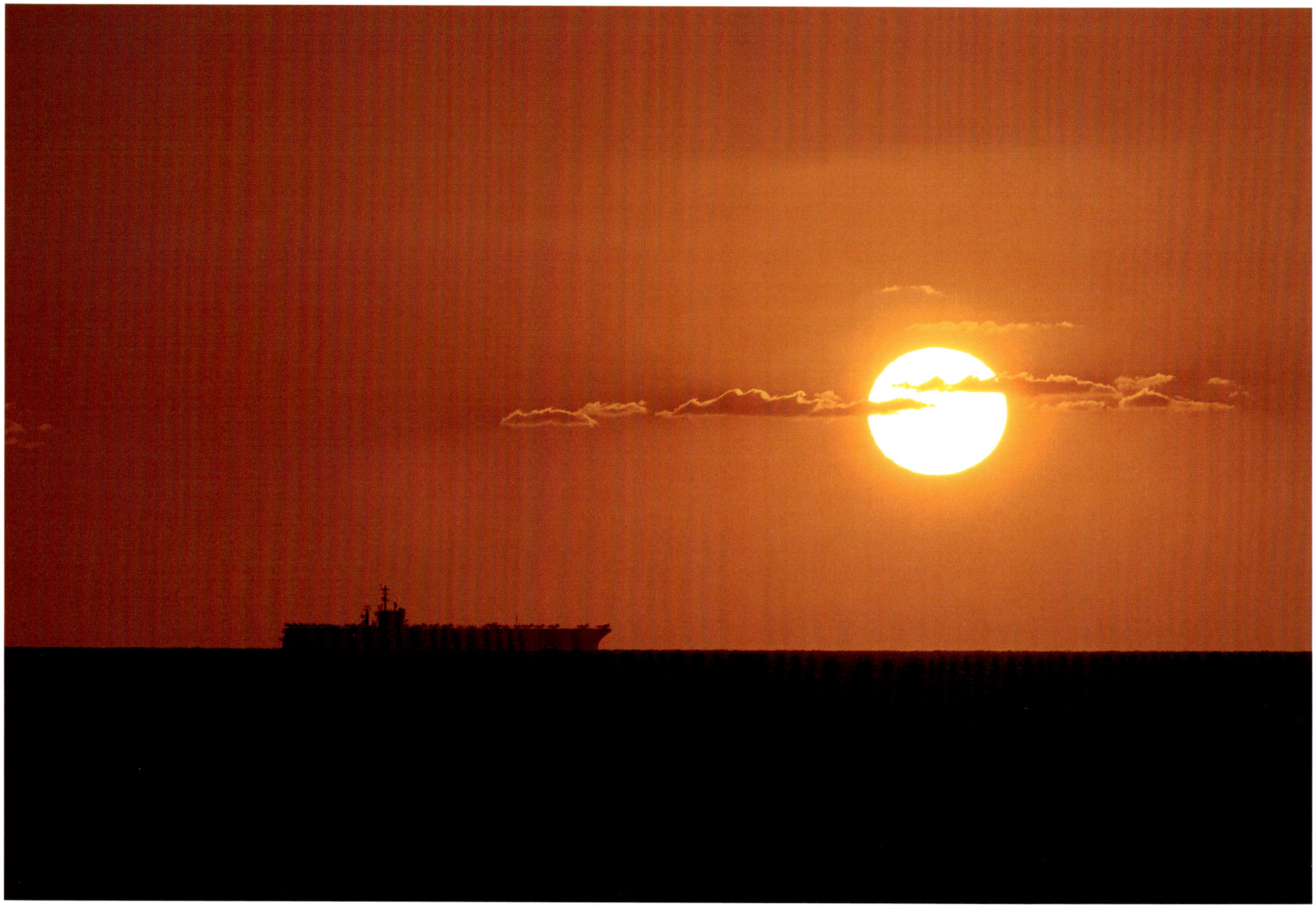

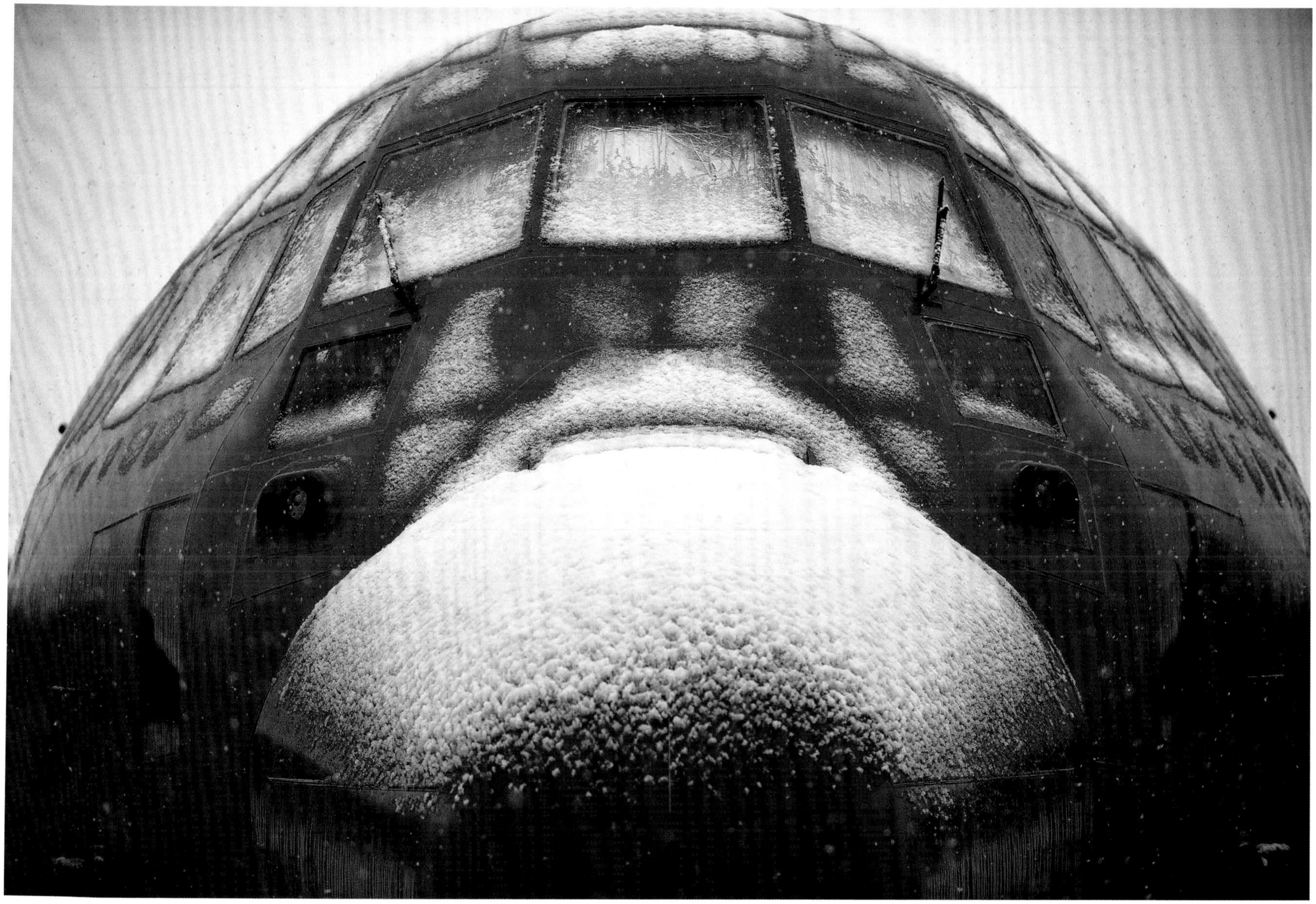

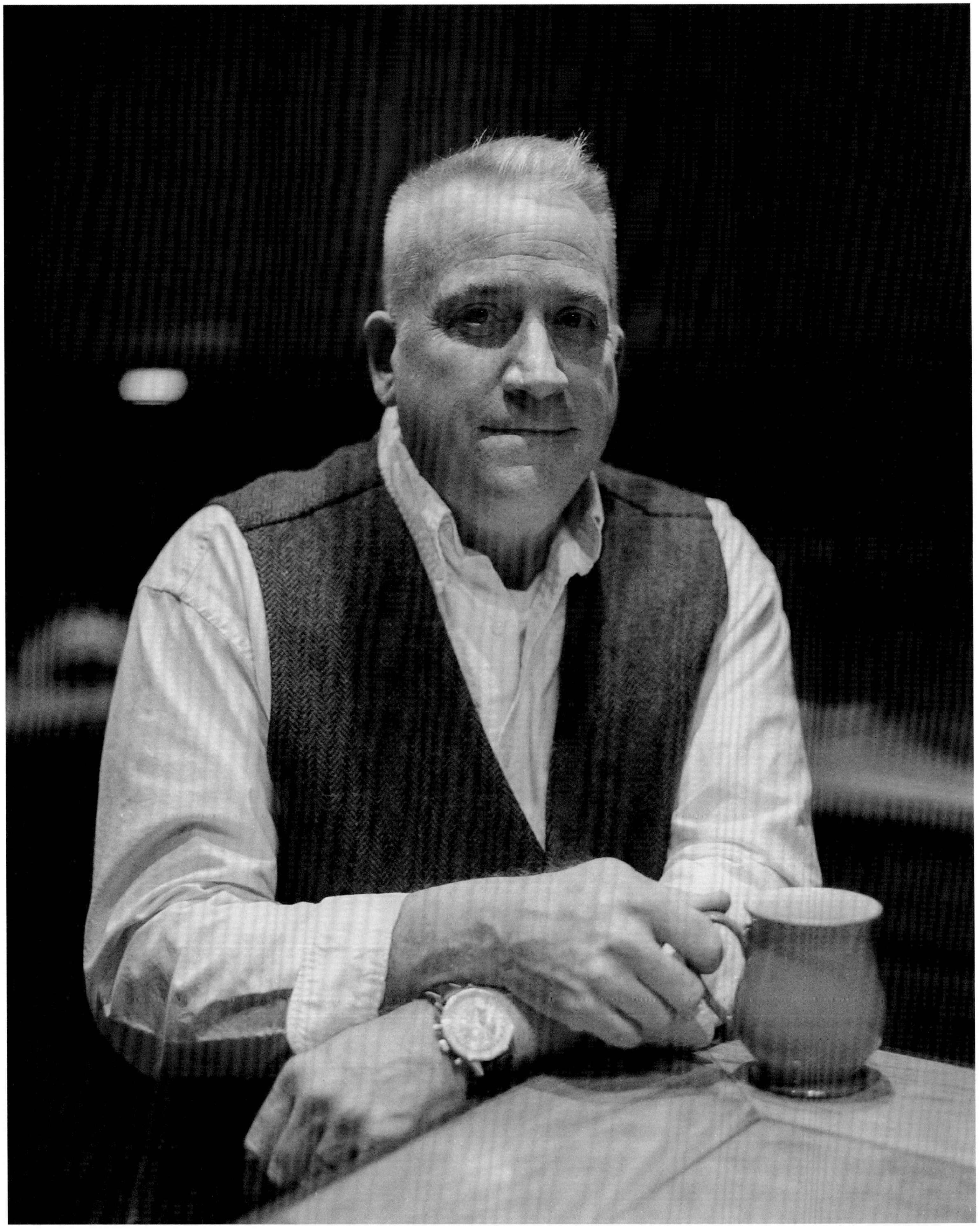

Bradley "BK" Hammer, Colonel
Air Force - MC-130H Talon II / Special Operations Pilot
Years of service - 32 years
Deployments - 3 [Operation Iraqi Freedom & Operation Enduring Freedom]

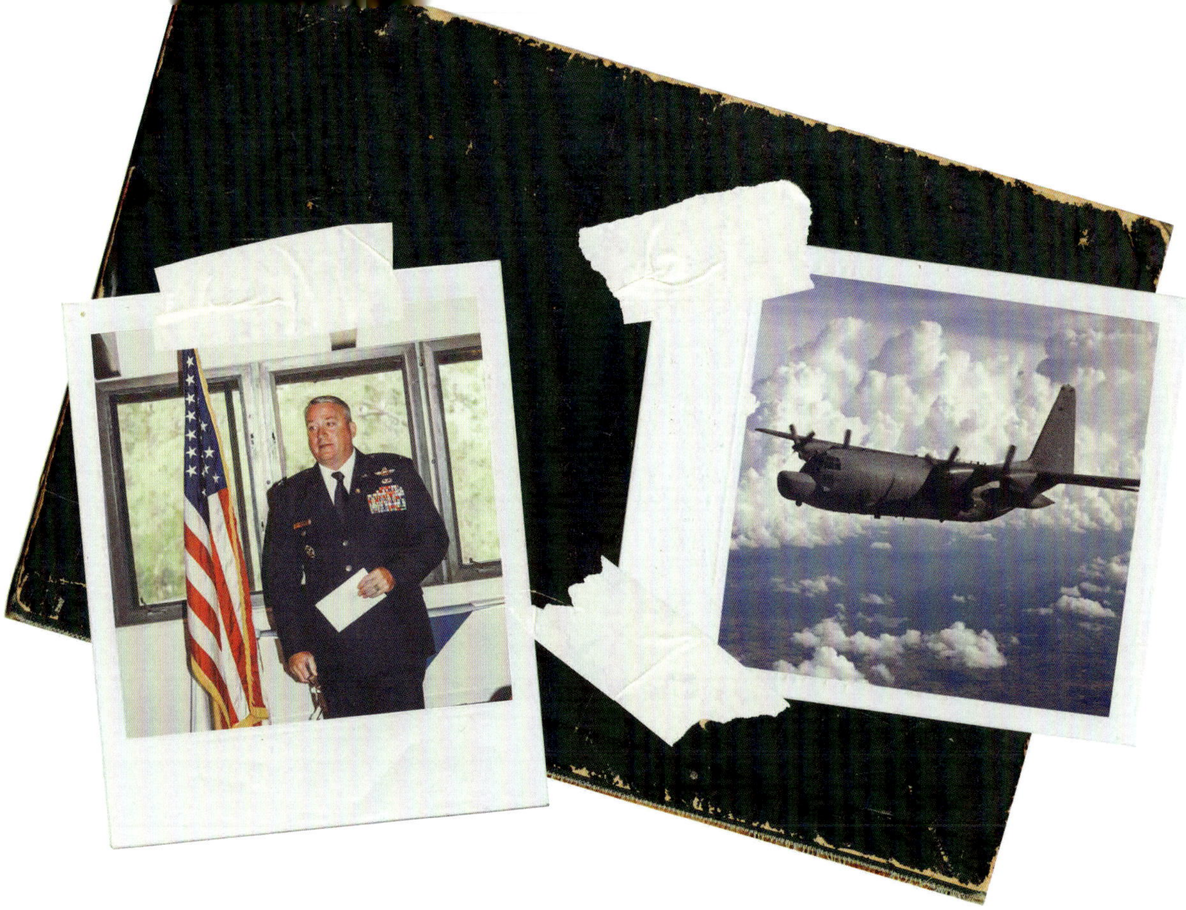

I'm Bradley "BK" Hammer and I was a Special Operations pilot for the MC-130H Talon II and AC-130 Specter Gunship for the United States Air Force. I was in tenth grade, at my grandmother's. Dan Rather was showing the burnt-out MH-53 and C-130s from the failed Iran hostage rescue. It was the type of team I wanted to be part of. I joined the Air Force right out of high school. My dad signed for me, and after thirty-two years between enlisted and as an officer, I retired as a Colonel (O-6).

My first operation was flying Panama in 1989. I had to get a waiver to take my revolver. It was a family affair. My little brother was in Chalk 13 for Third Ranger Battalion, taking Rio Hato airfield. Then it was Haiti, and after that, Desert Storm. Humanitarian missions in the Balkans, then Operation Enduring Freedom in 2002. I also flew in support of Operation Iraqi Freedom and did missions in the Philippines in support of OEF, as well.

Right when I retired, I had a buddy tell me there was a gig in South Korea to help upgrade their C-130's. From there I became a commercial pilot with Frontier Airlines. Readjustment was good for me, because when I went to Korea, even though I didn't initially know the guys I was working with, we all had the same experiences. It was like working with five of your best friends, but with no adult supervision. Also, the DAV (Disabled American Veterans) really helped me with my paper-work and worked me through separation, with educational benefits and such. I love woodworking and hunting, both archery and gun. Every elk season I hunt with my in-laws, who have been doing it on the same ranch in Colorado since the '30s.

I've learned more from my failures than my successes. I've been in crappy places with great people and great places with crappy people. I'd always rather be with great people.

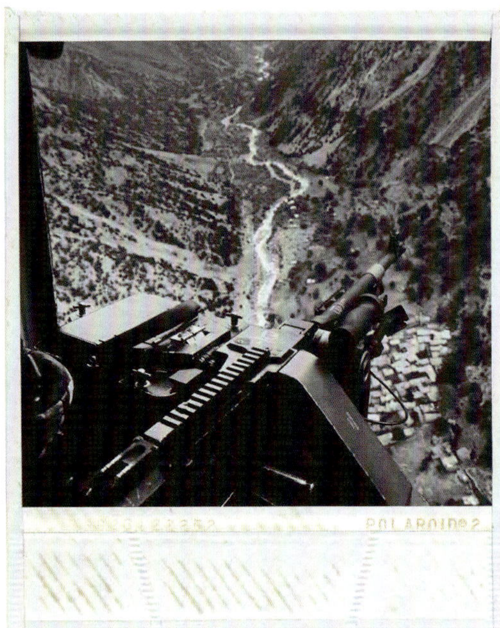

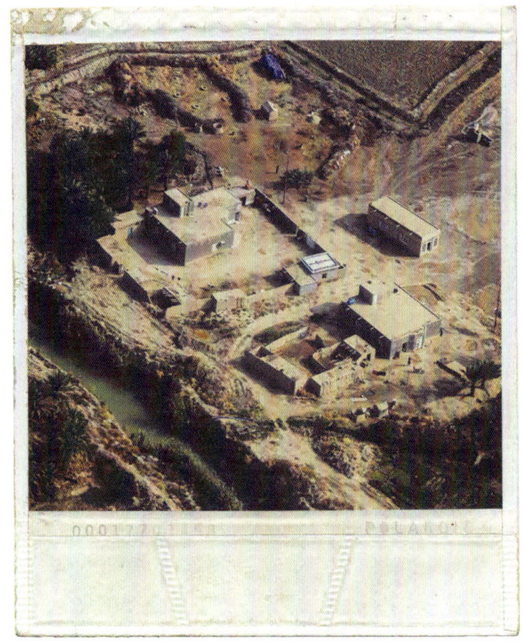

My name is Andrew Thetford, and I was a US Army 11B infantryman. I had this romantic notion that everyone should serve. My dad was a great example, but watching the old World War II movies really instilled that desire to serve. I was medically separated after eight years as a Specialist (E-4) with a deployment in support of Operation Iraqi Freedom and a deployment to Kosovo. I had just gotten back from Kosovo. We were doing a jump at Fort Bragg. I landed, felt some pain and tried to walk it off, and ended up with a broken leg in three places. Rehab and therapy just weren't cutting it. I had wanted to do my twenty and right after that jump I'd been scheduled to go to the Warrior Leadership Course to become an NCO, but it wasn't to be.

I graduated from Fayetteville State University and I am currently working for a defense contractor at Fort Bragg. It's a temp contract, but I'm working to make it permanent. I'm doing business development and operations ... generally whatever they need. Thankfully, I didn't really have a problem separating. I feel like TAPS really helped me, and I had a plan. You are who you are today because of what happened yesterday. And I like who I am today. At times I wonder if I should have pushed harder for an opportunity to keep pace with my peers, but I try not to second-guess it.

This too shall pass. There's been some tough stuff, I won't lie, but I have been consistently realizing it does get better. You can always work to make things better. We don't talk about it enough, but you can have Post Traumatic Growth too, not just the PTSD we always talk about. We're not all broken. In fact, most of us aren't. Most veterans are doing better than their peers.

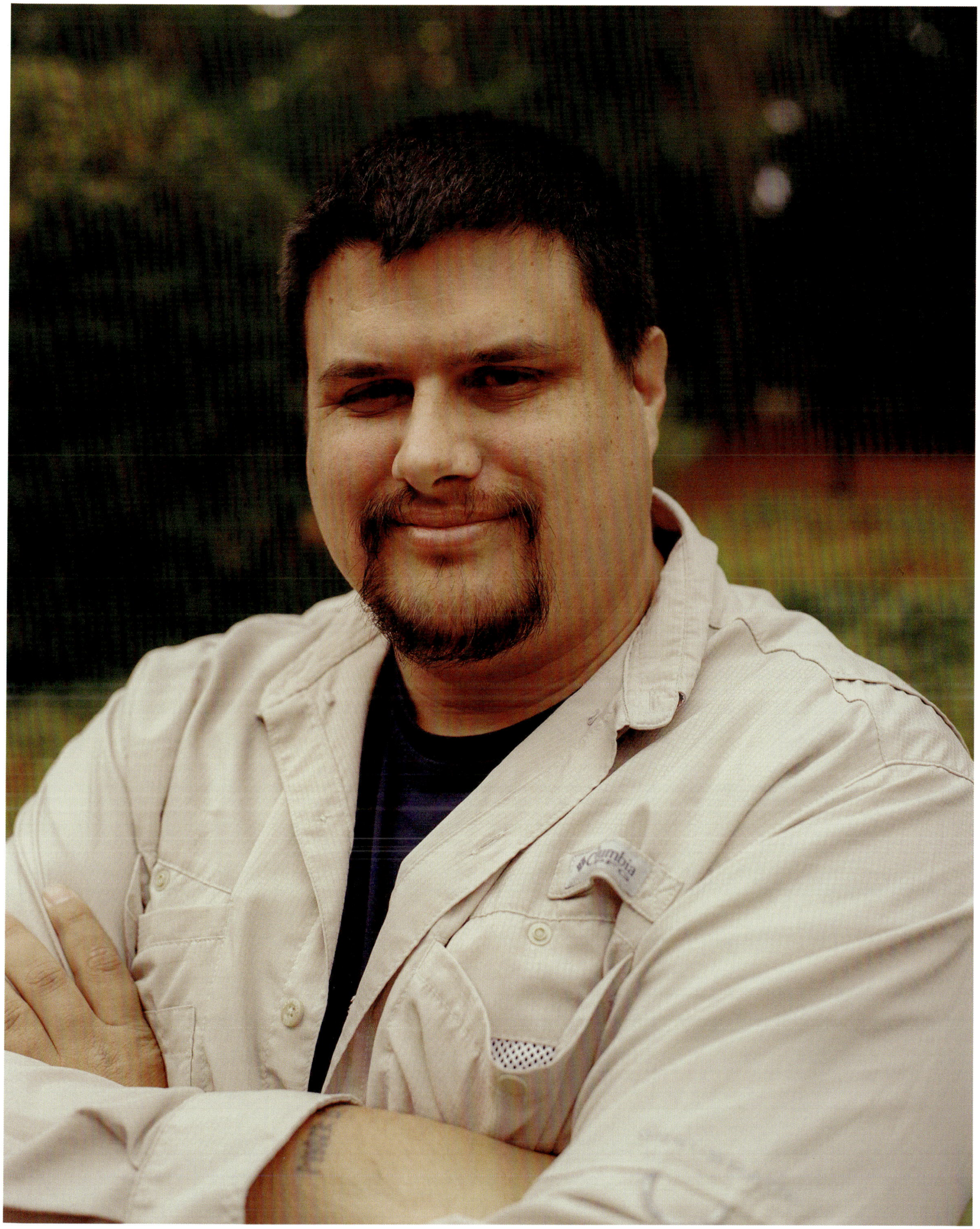

Andrew Thetford, Specialist
Army - 11B / Infantryman
Years of service - 8 years
Deployments - 1 [Operation Iraqi Freedom]

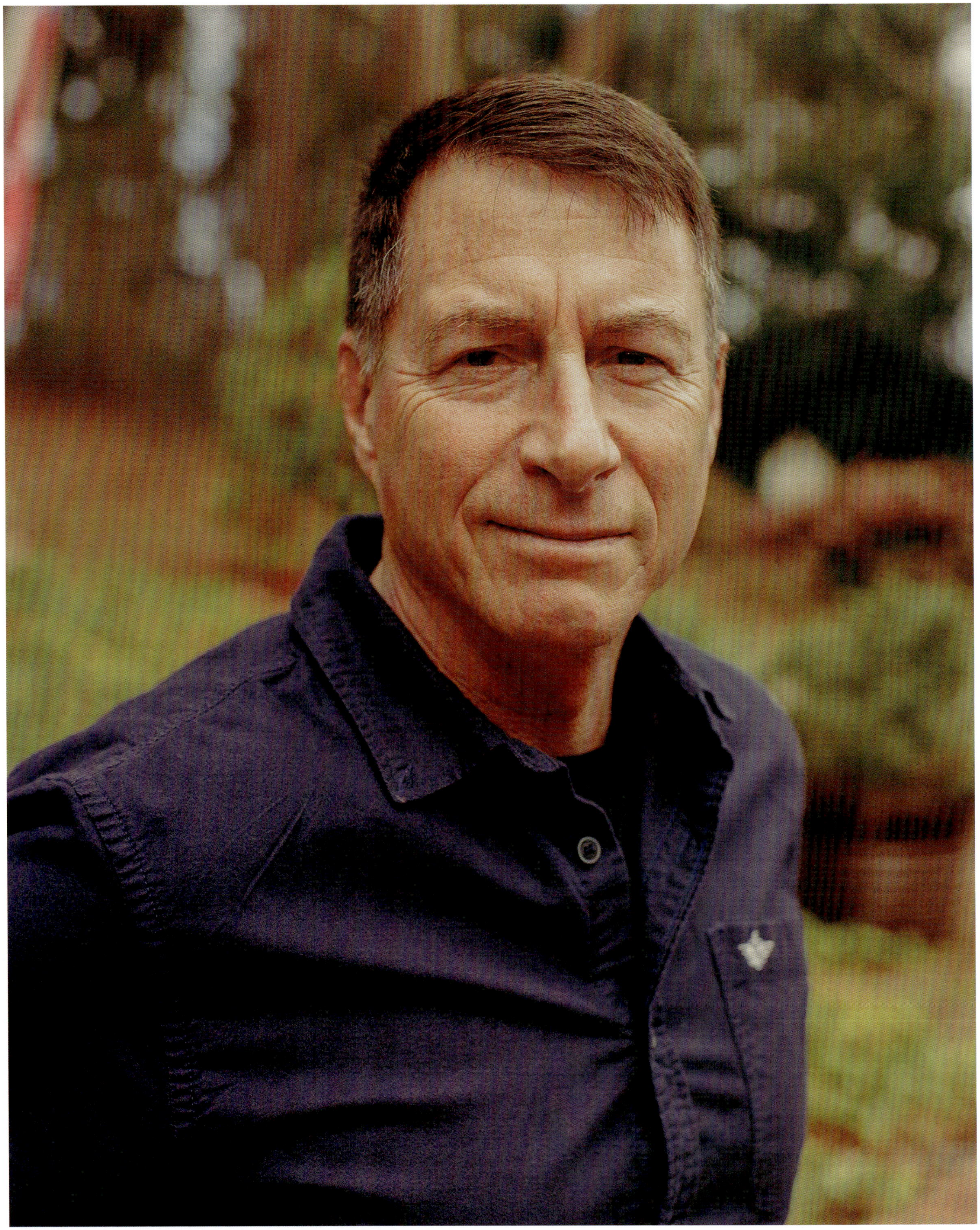

William “Bill” Thetford, Command Sergeant Major
Army - 75th Ranger / Special Forces NCO
Years of service - 38 years
Deployments - Multiple in support of GWOT

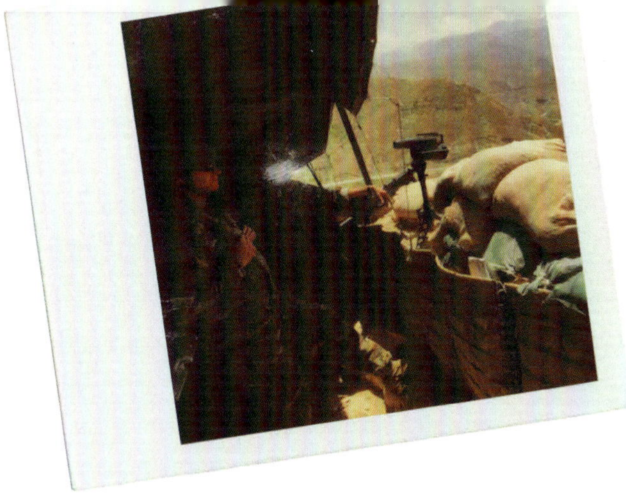

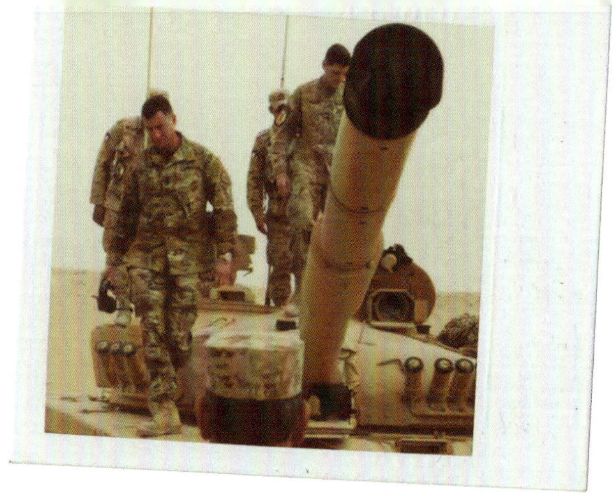

My name is Bill Thetford, and I retired as an 11A infantry leader in the US Army. I started as a Ranger in the 2nd Ranger Battalion, went to The Unit at Bragg and eventually became the CSM of JSOC for General Votel and progressed with him until I was his CSM at SOCOM, and then CENTCOM. I enlisted right after the failed rescue in Iran. I wasn't ready for college, and it wasn't ready for me. My grandmother was the matriarch of the family, and she said being a Ranger would make a man out of me. I deployed first to Grenada, then with TF Ranger in Somalia. I was part of the PIFWC (Personnel Indicted for War Crimes) mission in the Balkans. I deployed for the first time to Afghanistan in 2002 when Bagram was just a tent city. I was also part of the initial invasion force into Iraq. In the end, I did multiple deployments in support of the Global War on Terror. I enjoyed making Battlefield Circulation trips with General Votel to check on our troops and make sure they had the resources they needed and understood General Votel's guidance and campaign plan. We'd go and see the guys and check in on them to see how they were, helping to serve them rather than be directly in the fight. I've been blessed from the time my grandmother said the Ranger thing. Then to end up in The Unit, and become The Unit CSM (I was there almost twenty years) ... and then to have the good fortune to join General Votel and spend eight years with him, well, I wouldn't change anything.

Alison, my wife, took a break from being a teacher when we went to Tampa to serve at SOCOM. She had been doing some FRG with Michelle Votel and we decided we were going to get back to NC so she could continue to teach. I was able to go to the Wexford Group; they snapped me up. Now I'm a consultant for USASOC to help with a contract proposal. I'm really excited about helping them get ready for the next thing. I'm surrounded by SOF veterans and other like-minded people. It's exciting. Project Transition USA is a great organization that I highly recommend for transitioning service members. I had seen something about a one-day workshop on LinkedIn during my TAPS process and thought, "That could be handy." I took what I learned there about networking and about the elevator pitch and how to use LinkedIn for Project Transition USA ... it is just amazing.

Work hard and good things will come. Hard work pays off.

I think TBI and PTSD get overplayed a little bit. There are a lot of veterans who saw a lot of things and a lot of combat and are doing okay. For all of the PTSD, there is also Post-Traumatic Growth. Sometimes you expose people to stress and trauma, but you can use it like the gym, growing your muscles. You see that in SOCOM. You can put people in hard places, but as long as you give them some rest, recuperation, therapy, and space, they come back even stronger. Some will struggle, of course, but I do think some people over-leverage the broken vet when really, most of us aren't that at all.

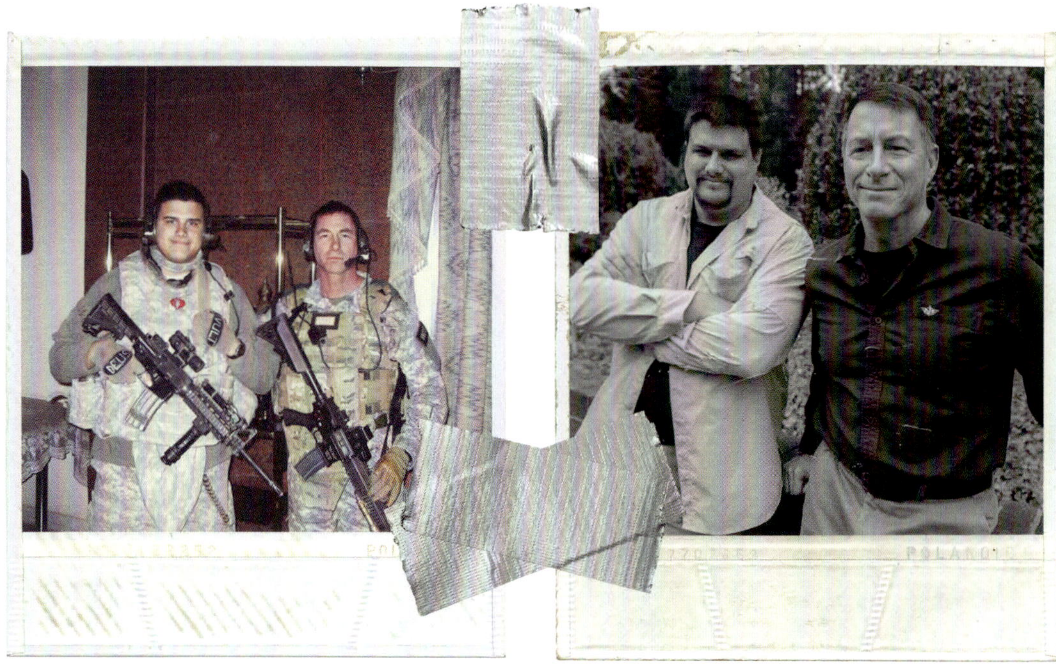

Andrew: I didn't join because of my dad, but it was a truly once-in-a lifetime experience to serve with him overseas. I really enjoyed getting to do that.

Bill: Andrew was in Iraq and it coincided with me as The Unit CSM. In that role I had a lot of flexibility. My commander was really excited that my son was deployed, so I'd go see Andrew for a day or two. I'm sure he had different feelings ... (*Andrew chuckles*) But I always coordinated with his 1SG before I came out to see him. We pulled guard duty and ate MREs together and sometimes he'd get to come with me.

Andrew: Serving made my connection with my dad stronger. He was a CSM at the time, super responsible, and I'm an E-4 with two years in. We're still kicking the same door down; we're equals. We could both have to shoot a guy or catch a bullet; it was just the mission.

Bill: One time, Andrew came to Baghdad with me, and we were hanging out with SEAL Team 2. The team guys thought that a father and son deployed together was cool, so they offered to let us go out on target with them. I initially told them, "We'll follow behind your command, make sure we stay out of your way." Well, they were kicking doors, we were following behind, pulling security together. The SEALS got on the radio and said, "We're getting spread thin, can you guys help us clear this room?" And we did. Target was secured, we found the guy we were looking for, and the SEALs took a picture of us after a successful night. It was a great memory for Andrew and me. We've also jumped together here at Bragg. One of them was in December when we jumped with the Brazilians for a set of foreign jump wings. Great memories.

Andrew: Look, my dad's done far more than I have with his career. I'm both disappointed and extremely grateful I haven't had to do some of the things he has, but at the same time there's all the training that we did. I spent eight years proudly serving my country. Did I really get used properly? I wonder sometimes. But in the end, our service together helped make a stronger connection. My dad and I have a stronger bond, and I'm really grateful for that.

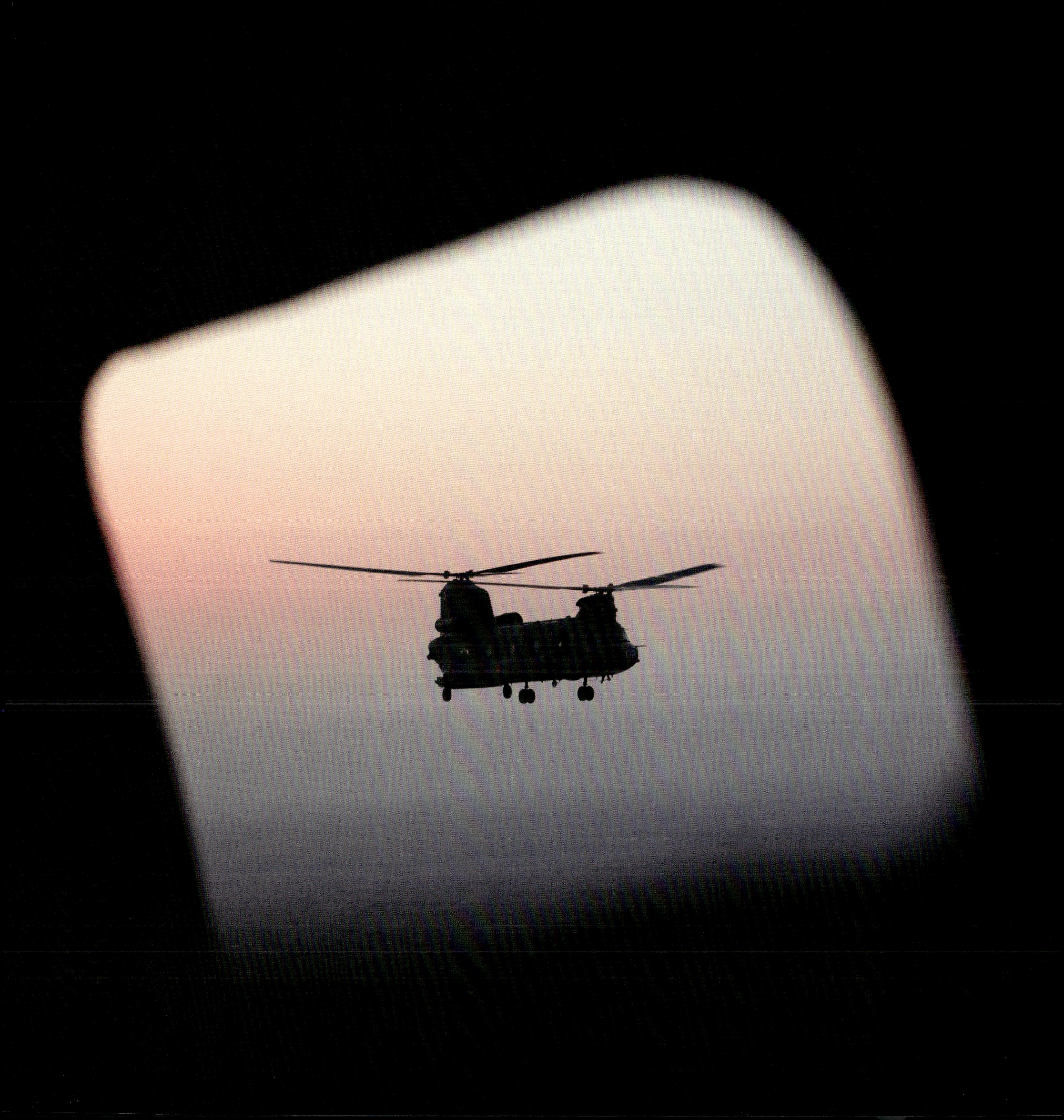

Airborne Angel Cadets of TX

Kandahar, Afghanistan
10 OCT 2012

Dear Angel Cadets,

Thank you very much for your supporting efforts in sending items over here to us in Afghanistan. I think organizations like yours don't get thanked enough. When I saw all these girl scout cookies, it was really impressive! We here at the Headquarte[illegible] Regional Command-South enjoyed them a[illegible]t a treat!

Life is not [illegible] folks like yourselves se[illegible] really good for morale [illegible]d or support or disappro[illegible]hing is clear - that you suppo[illegible]member - and for that we are very thankful! It is comforting to know that after all these years, we still have groups that still think of those serving overseas - in far places.

A big Texas thank-you to the Girl Scouts of the Metroplex from all of us here in Kandahar! Your support is touching and will be remembered for a long time to come!

[illegible]
MSG, USA

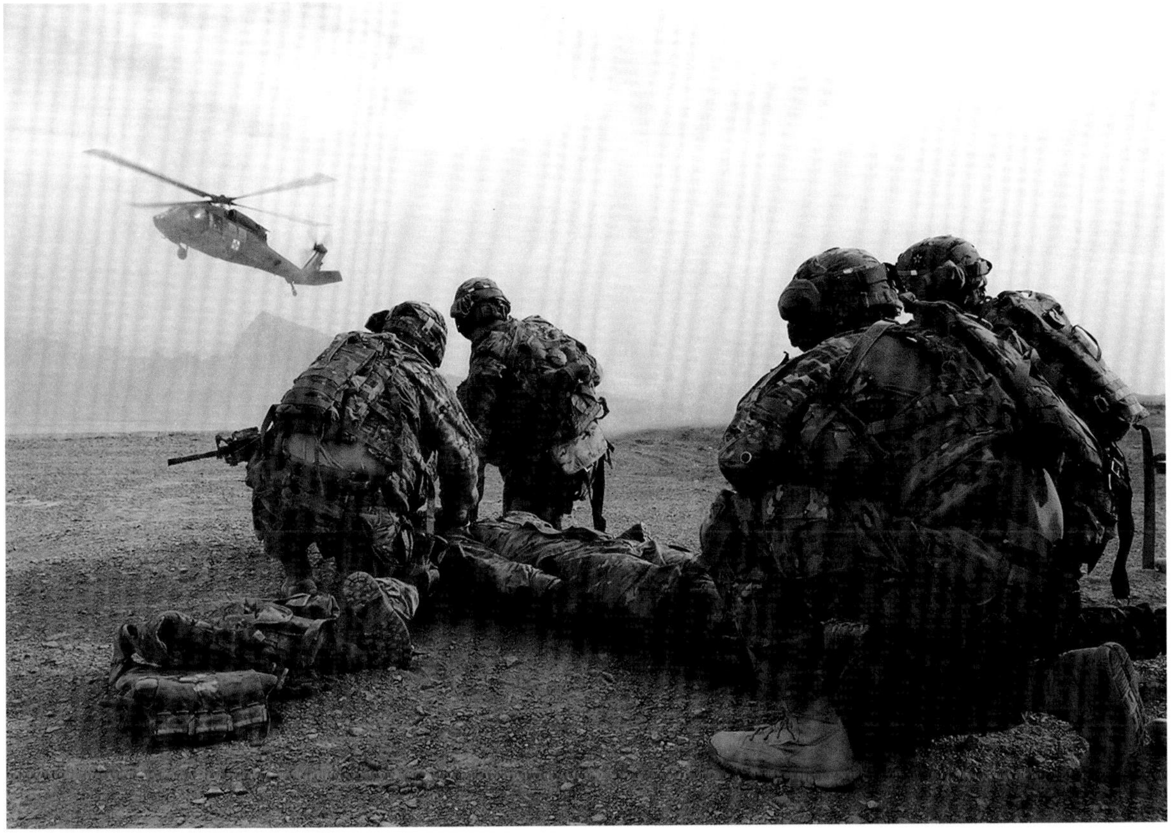

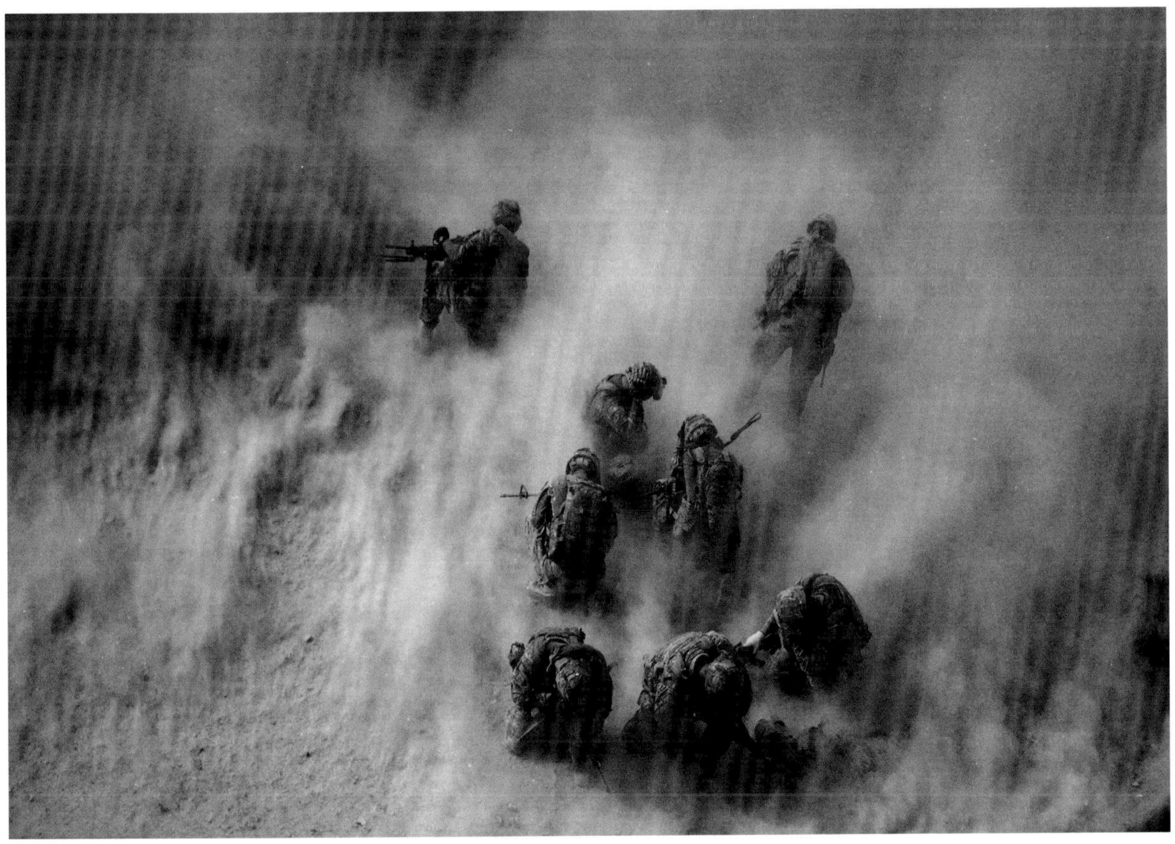

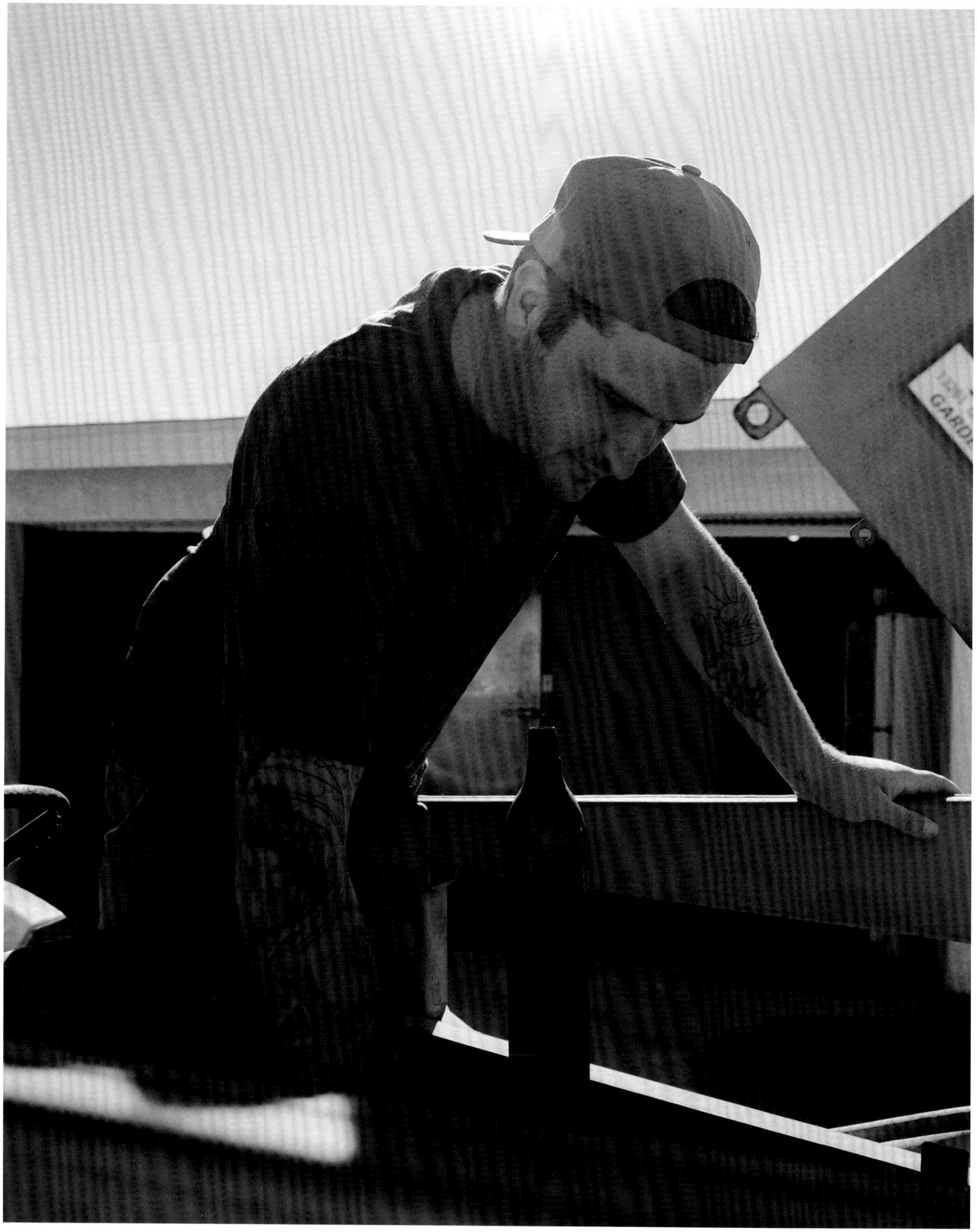

Eric Wilson, Specialist
Army - 68W / Combat Medic
Years of service - 6 years
Deployments - 1 [Operation Enduring Freedom]

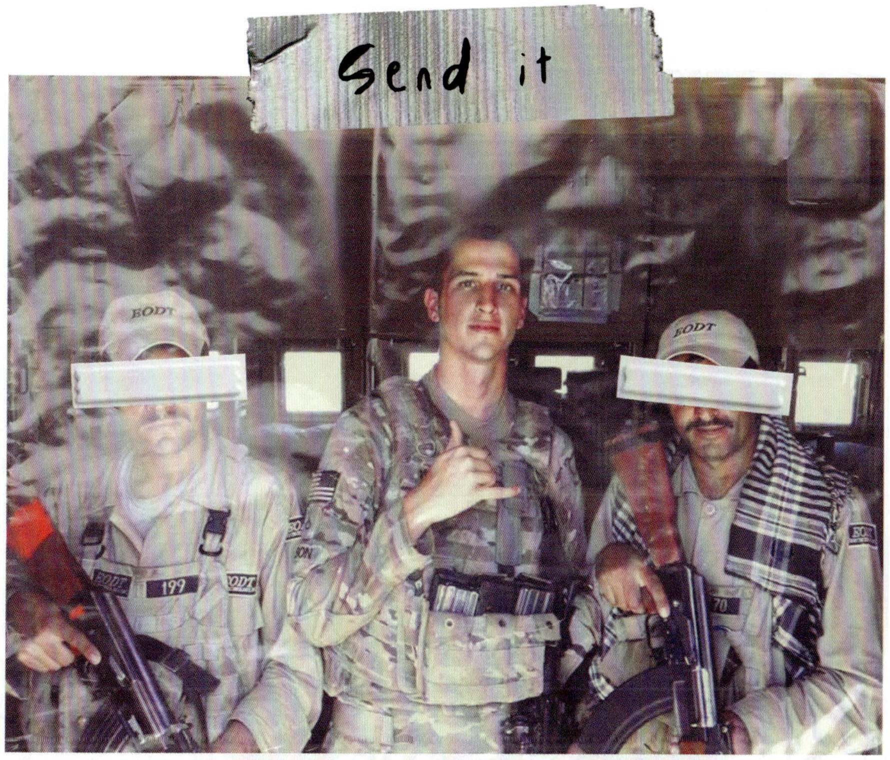

I enlisted in the Army as a 68W combat medic, as a way to get out of the desert in the Palm Springs area. I'd been an EMT before I enlisted, and wanted to be more adventurous than that. I left the Army as a Specialist (E-4) and I deployed in 2011 to the Kunar province of Afghanistan in support of Operation Enduring Freedom. I hurt my back and had spinal surgery a year after I left active duty, due to an injury I suffered while deployed on mission. I flopped around for a while after I left the Army. First, I went to school for photography but was like 'I don't need a degree for this,' and left the program. Then got linked up with the Farmer Veteran Coalition and have been there ever since.

The purpose of the Farmer Veteran Coalition is to help veterans transition into agriculture. We give them the tools, from start to finish, to learn how to farm. Gardening was really therapeutic for me and has helped my transition. Readjustment was a trip ... it was really difficult. When I was leaving Afghanistan, an old crusty Command Sergeant Major said, "Drink this up, 'cuz you'll never be this badass again." He was kinda right. In a war zone, as a medic, everyone relies on you. Everyone needs you and you have a deep sense of purpose to support and care for your guys. Coming out, people don't need you in the same ways. It's a weird transition. If I could change anything about my life ... man, I don't know. Win the lottery? Nah, I got it pretty good. I'd just say: in life there are certain parts you simply have to face no matter what, so send it.

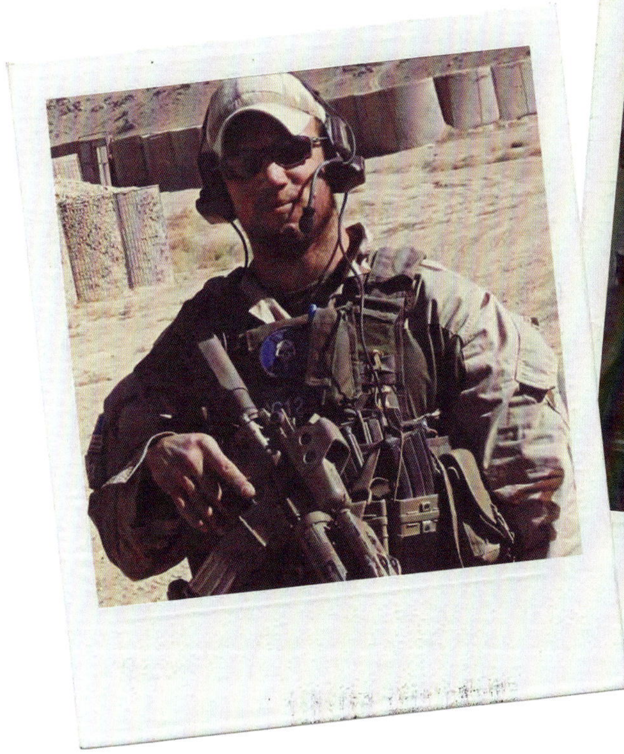 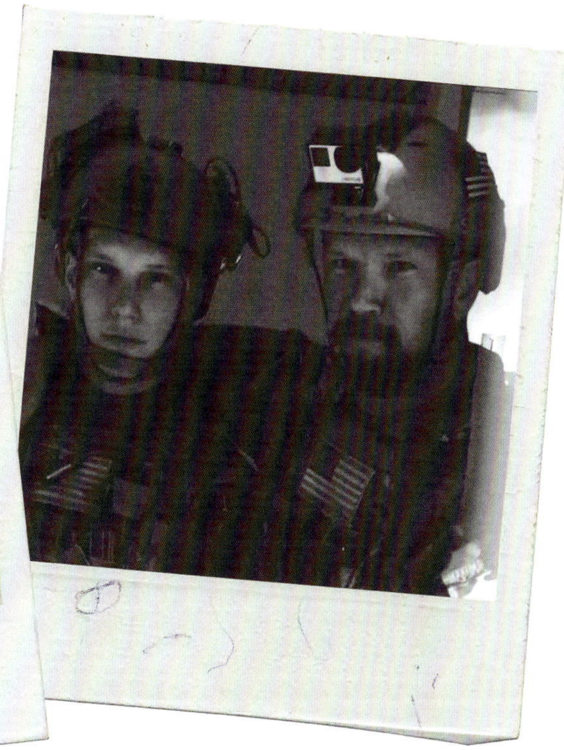

"Be nicE"

My name is Jeff Adams, and I am a retired 18Z, formerly 18D, Special Forces Team Sergeant. I enlisted initially to get a free college education and see the world. I retired as a Master Sergeant (E-8). My first deployment overseas was in 1996 to the Balkans. I also deployed in support of Operation Enduring Freedom, Operation Iraqi Freedom, and in support of missions in Northern Africa. After separating from the service in 2012, I became a defense contractor but got tired of taking the devil's money. My last contracting deployment was in 2015. My transition out of the service was terrible. Combat was like a black tar heroin addiction I was fighting. I had to wean myself off it. I couldn't watch war movies or read war novels; they were like another hit off a crack pipe. I was in a really dark place, post-military. PTSD, suicidal thoughts and the like. Worse, I saw a lot of guys were killing themselves and it hurt me. These men were someone's father, son, grandson. No matter what they did overseas ... a cook or an infantry guy or a Green Beret ... they mattered.

I started to work with various non-profits. One is Mission Six Zero, which does leadership and team building. For veterans who are hurting, there's the PTSD Foundation of America. I also support the Green Beret Foundation and the Special Operations Warrior Foundation. I'm a firearms instructor now in Colorado, teaching everyone from civilian beginners through high-level competitive shooters. It's really important for me to make sure that people who don't fit my genre of lifestyle still can understand what I do, and they can use firearms safely. I see myself as a firearms ambassador and go the extra mile to make sure everyone who shoots with me has a positive experience. My life mantra is: Be nice. It's always better to love people than hate on them.

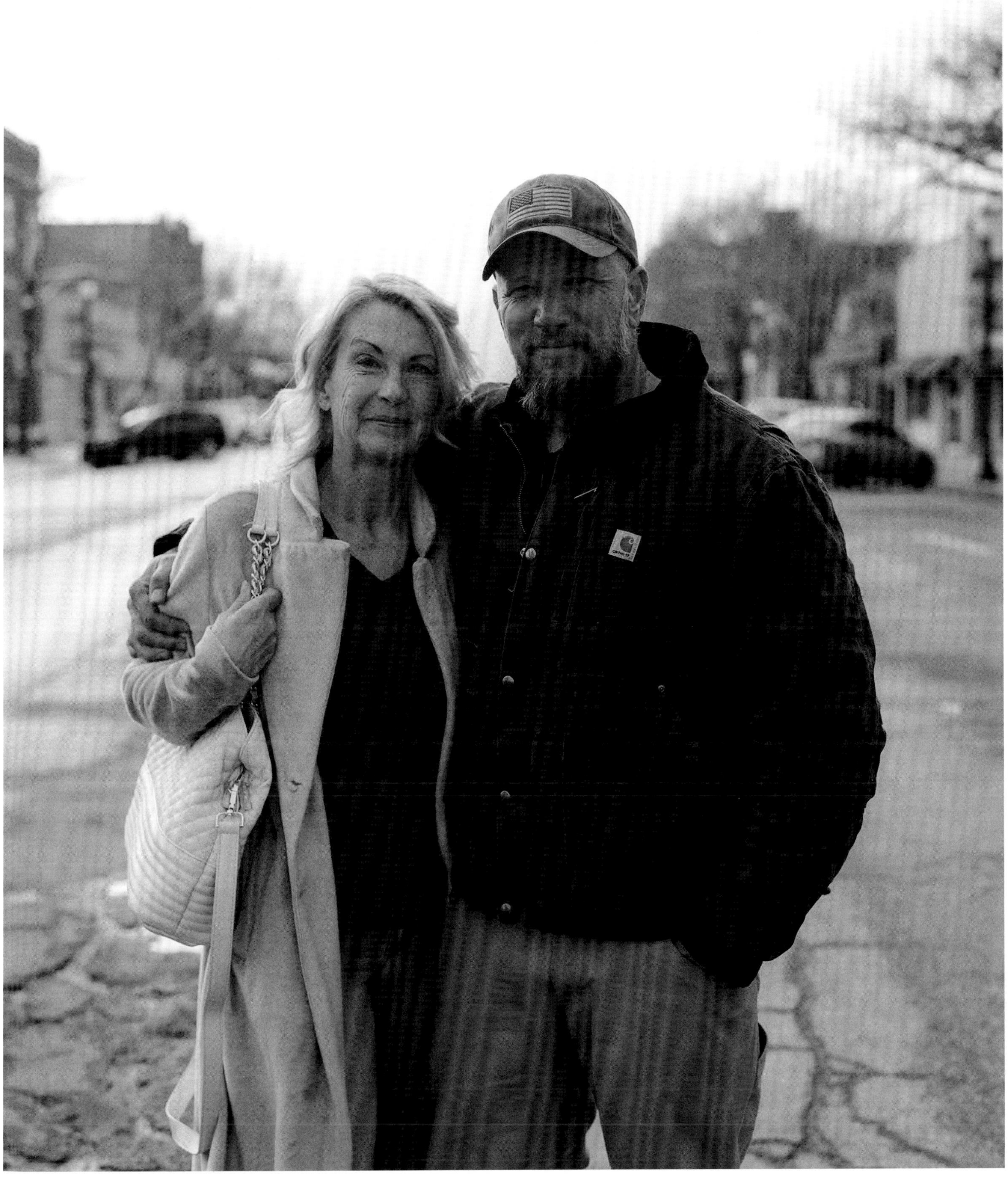

Jeffrey Adams, Master Sergeant
Army - 18Z / Special Forces / Delta
Years of service - 21 years
Deployments - 12 [Operation Iraqi Freedom & Operation Enduring Freedom]

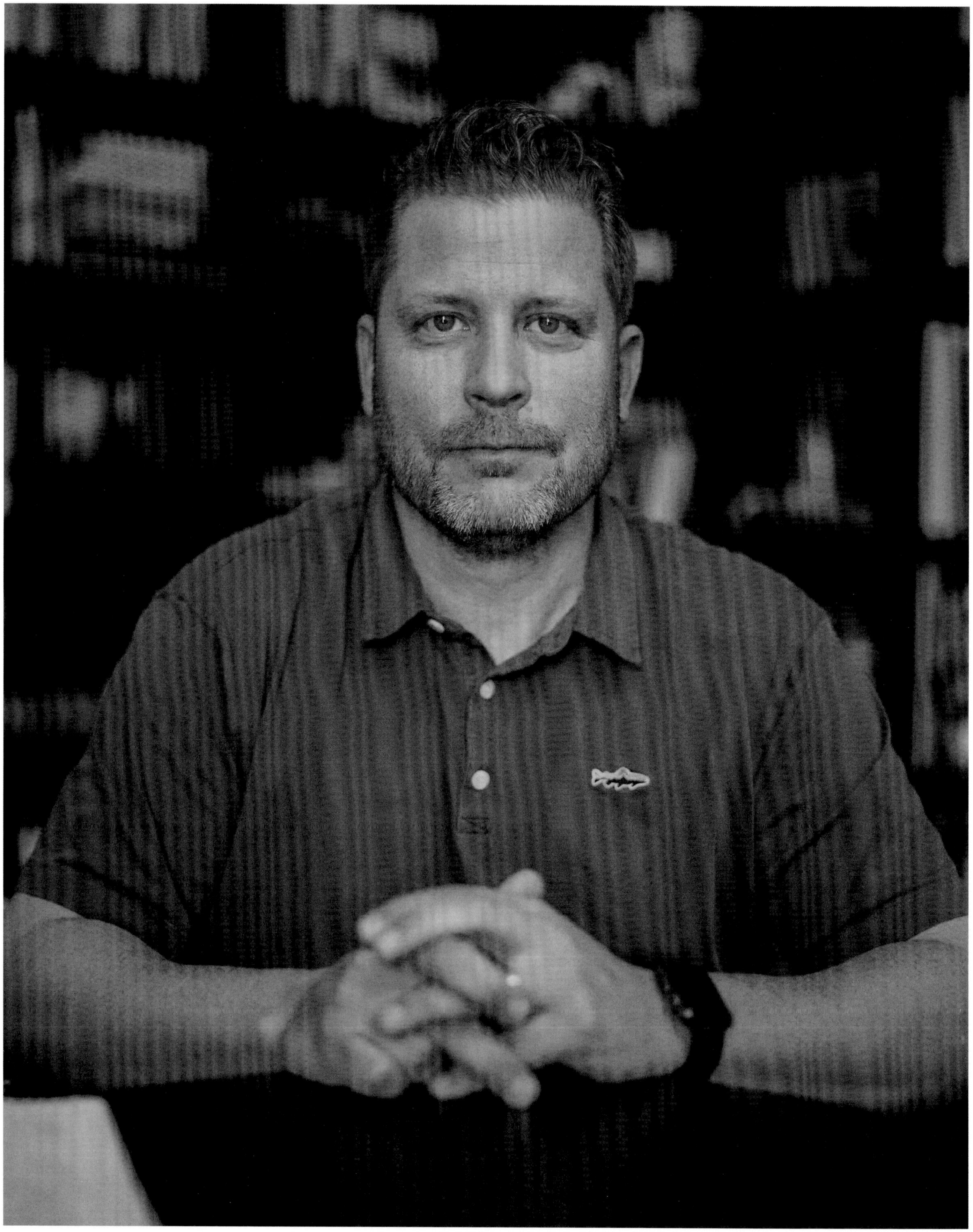

Nate Self, Captain
Army - 11A / Infantry Officer / 1/75 Ranger
Years of service - 6 years
Deployments - 2 [Operation Iraqi Freedom & Operation Enduring Freedom]

My name is Nate Self, and I was an infantry officer in the United States Army. I got a postcard in the mail from West Point when I was in high school. There wasn't anyone close to me that was in the military, but it looked like a unique challenge, so I sent it back and got the catalog. I talked to my parents, and the more I got into it, the more I felt like it was something I needed to do.

I was initially stationed in Germany and was one of the first groups into Kosovo in 1999. I spent New Year's Eve 1999 in Kosovo. New Year's Eve from 2001 to 2002, I was flying to Afghanistan in support of Operation Enduring Freedom as a platoon leader in the First Ranger Battalion, 75th Ranger Regiment. I had already pinned my Captain (O-3) rank before we deployed. Usually, I would've been moved to a staff job, but I was kept in place as a platoon leader since they knew we were going overseas. I had been with my guys for well over a year at this point, and we were a great team. The Rangers were a part of Task Force 11, which was basically all of the Special Operations units looking to capture or kill bin Laden in the very early days of the war. Intelligence was starting to notice the build-up of Taliban and Al-Qaeda forces in a specific area, and the mission became Operation Anaconda. The conventional forces went into the region heavy while we waited to see what shook out. It got hot really fast, and conventional forces started wearing down.

As the battle developed, they decided to insert a SEAL team onto a mountain called Takur Ghar, literally the "Tall Mountain." The bird took heavy fire and as it was pulling off the mountain, Neil Roberts fell off the back of the helicopter. He fought alone for as long as he could and eventually was killed. As TF 11 spun up a Quick Reaction Force (QRF), they also were heavily engaged and took serious casualties. They realized they couldn't hold on and retreated off the mountain ... except for their Air Force Combat Controller, John Chapman. Neil Roberts was already dead. John Chapman was now up there fighting alone as well. I was tasked with leading two helicopters, with ten Rangers apiece, to try to save John Chapman and recover Neil Roberts's body.

We came in at daybreak and took intense fire. We basically got shot down, but it was only a hard landing, because we were already so close to the ground when the RPG hit my bird. The other peeled off and dropped their Rangers down at the bottom of the mountain and they climbed up to us. We lost five men from the helicopter to gunfire as we were running off the back. We secured a perimeter and fought all day through knee-deep snow at high altitudes. A lot of bombs dropped as we tried to secure our position. We suffered more wounded throughout the day, but we continued to fight on. [Editor's note: Captain Self himself was one of those wounded. He was awarded a Purple Heart.]

In the face of all of this, we retrieved Neil Roberts's and John Chapman's remains.

Continued on next page

After my tour to Afghanistan, I went to Fort Campbell, Kentucky, to be a member of the 101st Airborne Division, where I also deployed to Iraq in 2003. After separating, I wasn't really connected with people from my time in the military. When I left, maybe I was running away a bit, and I started to struggle. I felt like I had run away, that I had quit. I separated from service just three years after it all started. I knew what ended up happening was going to happen: the guys were going to just keep deploying. I felt this extreme guilt. I wanted to hurt myself. I've worked through it, and acknowledge how ridiculous that is, but that was a very real thing for me.

I was contacted by a retired general, one of my classmates' dad at West Point. He said to me, "I'd like you to come to tell your story." I didn't want to, but it was a part of a Campus Crusade program focused on veterans and their local communities. I started talking at churches, telling my story to help transitioning veterans, and that really got me through. Writing my book, Two Wars: One Hero's Fight on Two Fronts--Abroad and Within, also helped. I really didn't want to do it at first. But once I started, I found a lot of catharsis writing it. I wanted to be able to honor the guys who served in the battle. No one who served there had written anything about it. Some journalist pieces, sure, but no one who was there. The last third of the book is about my transition. A lot of people felt it was really important to hear and read. Looking back on my transition, I would say to any fellow veteran: don't be a loner. You need to find guys and gals who have already done it and have them support you. Find connections, find camaraderie. Even if you want to get away from some of the military cultural aspects, have someone help you find the path ahead. Again, keep from isolating.

I started my business and after some fits and starts, I started working with the Army, and with cadets at West Point. We started bringing stories from the frontlines, primarily from young officers. My company, Praevius, Latin for "lead the way," is a consulting company with primarily government contracts. We also do some intel fusion work and fielding of equipment to deploy to the frontlines. Resilience training has also been a core piece of our business model. About five years ago, the medical community approached us and said they were having problems with doctors burning out. We're working with them in similar ways as we do with the military to promote resilience and self-care.

Everyone needs help. You don't get anywhere on your own, ever. Also, you need to stay connected. Not just professionally but personally. Don't ever get isolated. Our mission at Takur Ghar went from finding bad guys to going to find good guys. We had a SEAL and combat controller alone on a mountain. We needed to go get them. After the military, I felt like I was up on the mountain, and eventually, people did come to help me too.

Constantly look to learn and grow. Growth is a lifelong thing and never stops. Be good, be generous, be grateful.

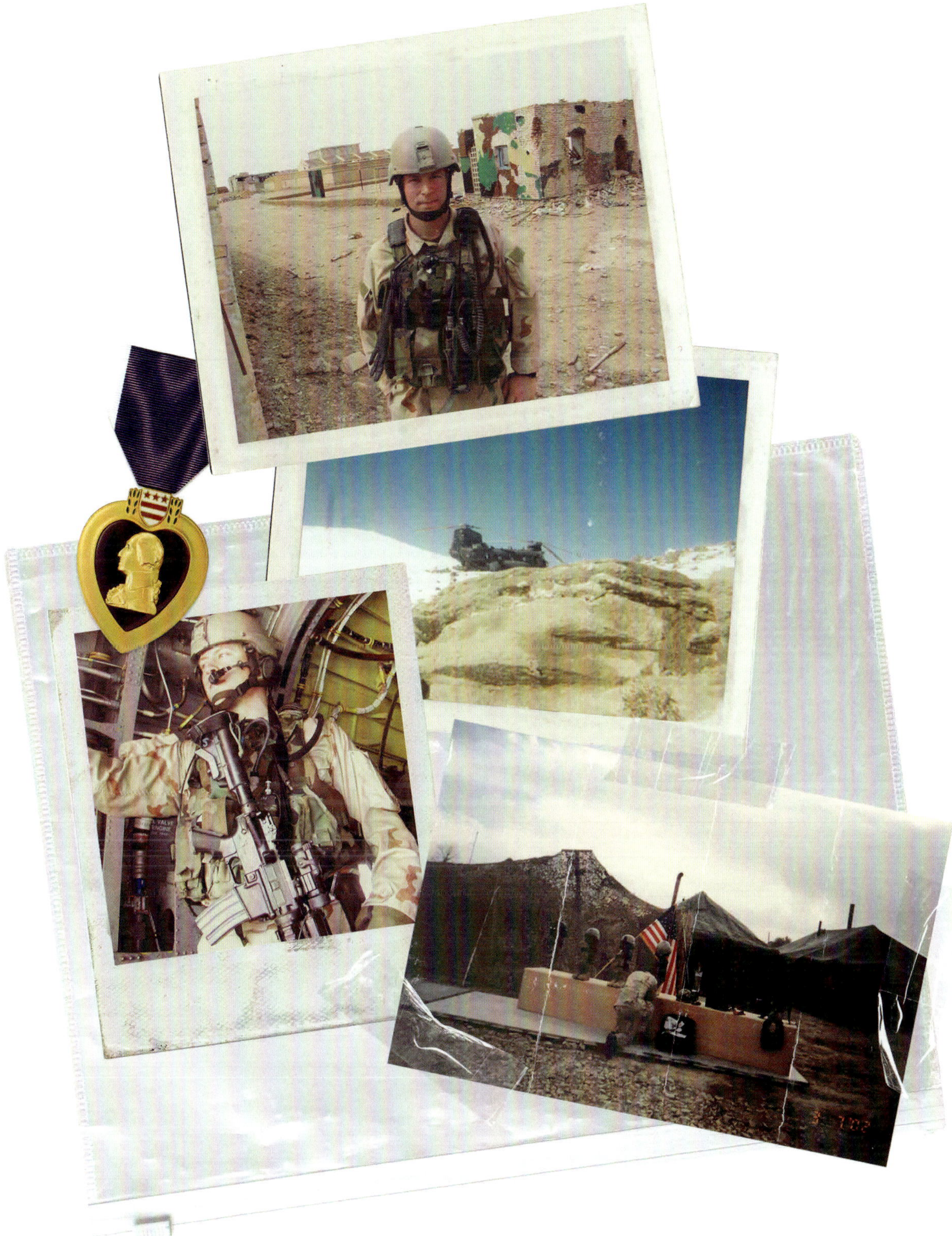

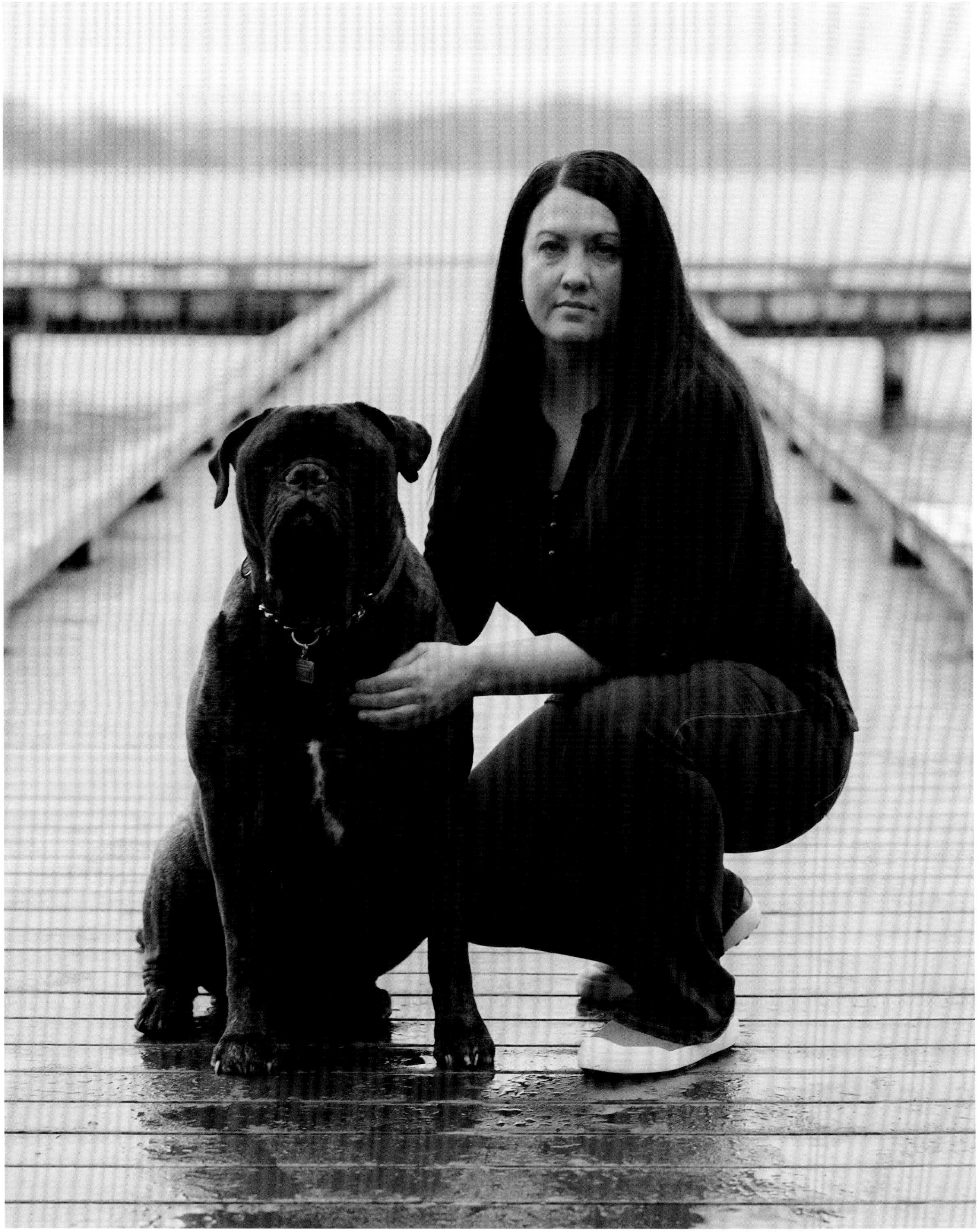

Jessica Harris, Sergeant First Class
Army National Guard - 68W / Combat Medic
Years of service - 20 years
Deployments - 1 [Operation Iraqi Freedom]

My name is Jessica Harris, and I was a 68W Combat Medic in the Washington Army National Guard from 1995 to 2015. I joined the military my senior year of high school and initially wanted to go on to medical school and become a surgeon. I never planned on making the military a career, but I ended up serving for twenty years in the National Guard, with the last nine on active duty status, which was spent mostly with the Counterdrug Task Force. I deployed in 2008 in support of Operation Iraqi Freedom, where I served as the senior medic for the 1/303D Cavalry Regiment. I retired as a Sergeant First Class (E-7) in 2015.

My initial adjustment post-service overall was somewhat easy and exciting. I was ready to leave the military and I had secured a great job that took me from Washington state to North Carolina to work at the University of North Carolina as the Administrative Director of Cardiac and Vascular Surgery. I started this new job just two weeks after coming off of Counterdrug operations, so I didn't really have a chance to process all of the changes at the beginning, as well as getting settled in a new state. After about six months however, it hit me this isn't really what I wanted to do, but I was going to stick it out and give it a fair shot. As luck would have it, my position was eliminated and I was laid off ten months after moving to North Carolina, and just six weeks after I purchased a house.

It was during this time that I struggled to find employment and started dealing with the issues of regret after leaving the military and wondering if I made the right decision. I didn't have a network or a support system in North Carolina the way I did in Washington, but knew I had to find one, as I was determined to stay and make things work. I ended up discovering Bunker Labs, an organization that helps veterans and their families start and grow businesses. This was my turning point in finding the right people I needed to be around, and ultimately led me to starting K9 Salute in 2016, a natural dog treat company that honors K9 heroes and gives back to working dogs. That same year, I won a local business contest, which allowed me to get funding for my business and access to local mentors. In 2018, K9 Salute was named an American SmallBusiness Champion by SCORE and Sam's Club, and I won the Bob Evans Farms Heroes to CEOs contest, securing a grant and mentoring from Daymond John of Shark Tank.

As challenging as those initial years were after retiring from the military, I don't look back at them with regret. I would have regretted not going for it. If I hadn't taken the opportunity, K9 Salute might not exist. It can be difficult to let go of our identity in the military, but my hope is that veterans know there is so much more outside of military service that we're capable of. Take those chances and find new and meaningful ways to give back and serve.

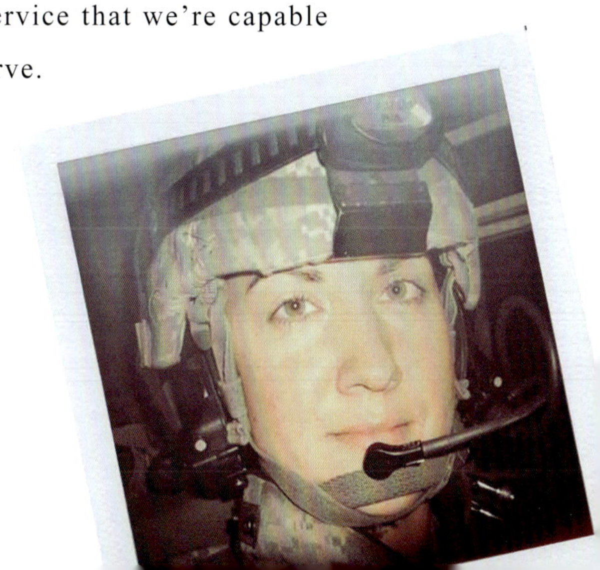

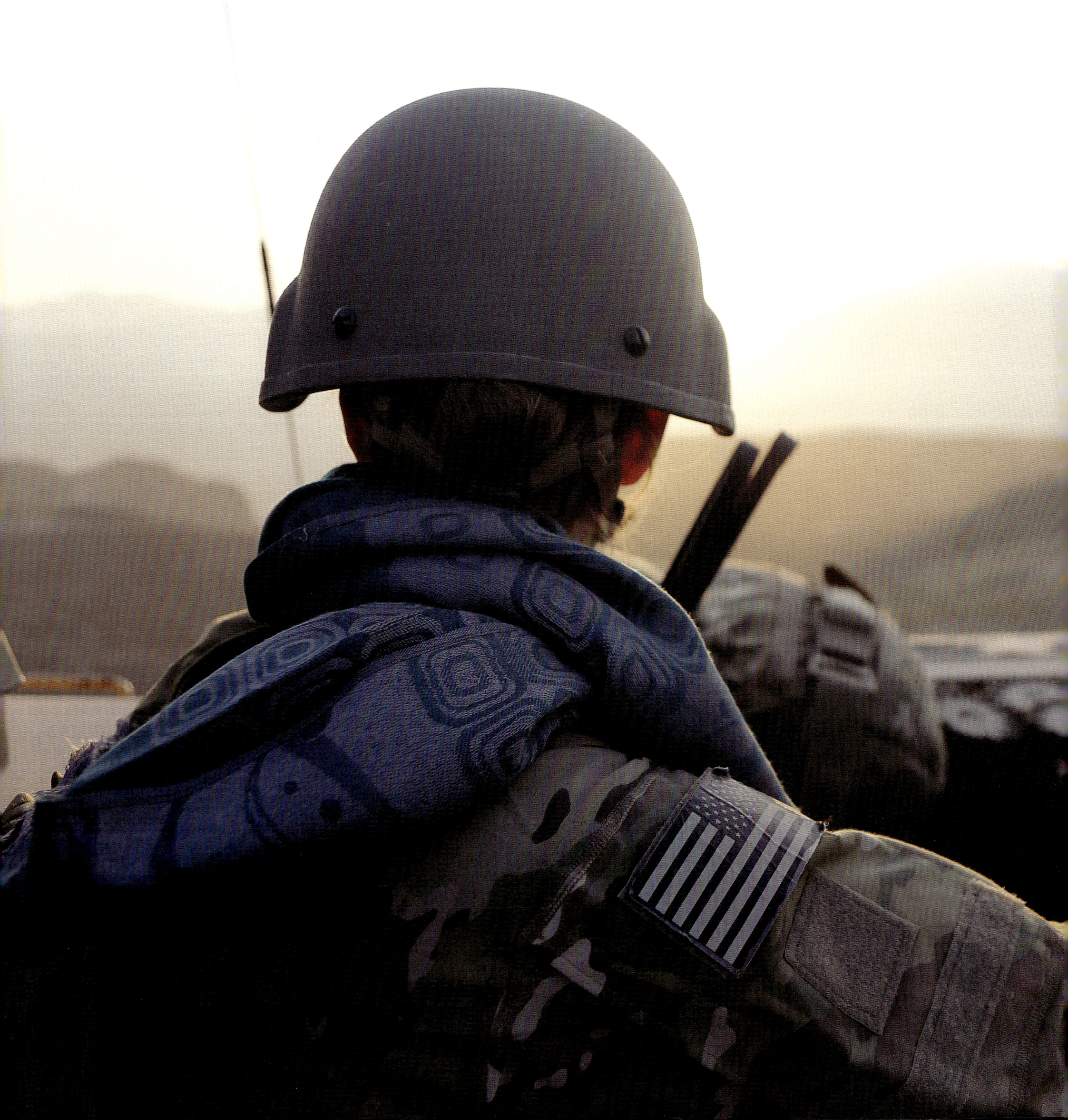

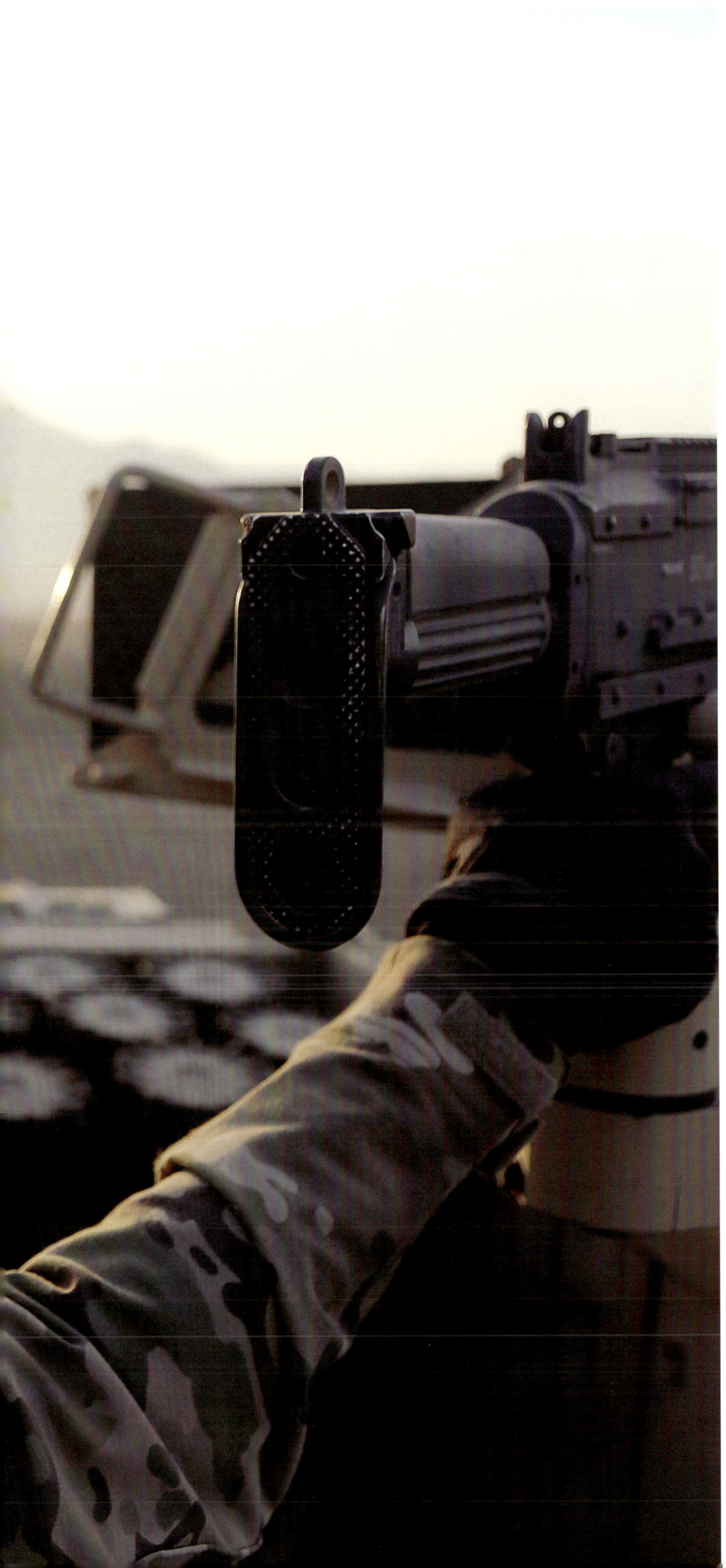

My name is Robert Bentley. I started off as an 11B infantryman and jumped into Panama in 1989. From there I became an MOS 29M tactical satellite repairman and got picked up by the 10th Special Forces group as an attachment. I went with them to Desert Storm. I had a seven-year break in service, but reenlisted in the National Guard as an 11B to deploy to Afghanistan in support of Operation Enduring Freedom.

Since I left, I've been working as a farmer and with the Herd 2 Human program. Horses don't lie. You have to build a bond and they help you understand what you're going through. Civilian versus military trauma isn't handled the same way. EMS and firefighters see some stuff, but they go home to their loved ones and can find ways to relax. The military goes back to the FOB. You spend a whole year in the military with your adrenaline going. You're never off. It isn't until you come home that you can turn it off, but by then you're so conditioned to be on. There's been so much change in how we handle soldiers coming back. When we got back from Panama and the first Desert Storm, you didn't talk about it, you just sucked it up. The VA has gotten better from those days, but obviously they still have a lot of room for improvement.

In the civilian world, it feels like there isn't basic accountability. When I was in, I always wanted to move up. But out here, people seem to deflect and not hold themselves accountable. Part of the reason I became a farmer is that I can relate to at least 80 percent of farmers. I do custom hay baling and other farm-related jobs. I was a cowboy growing up, and when I got out, I had to find something to do that was away from the general public, and away from that lack of accountability.

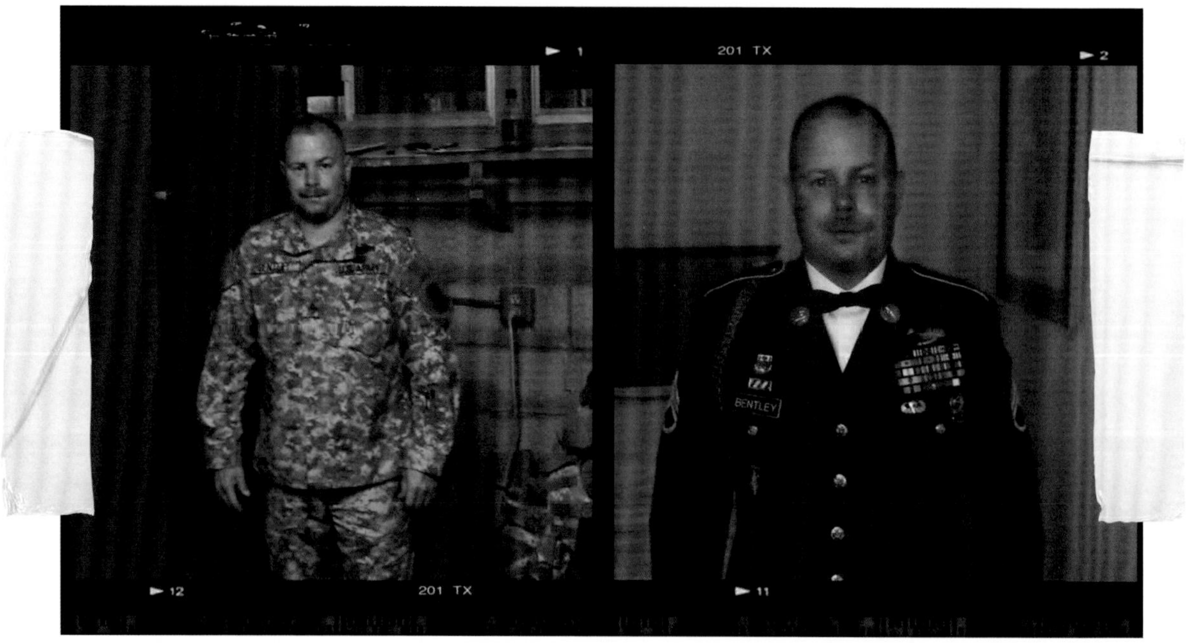

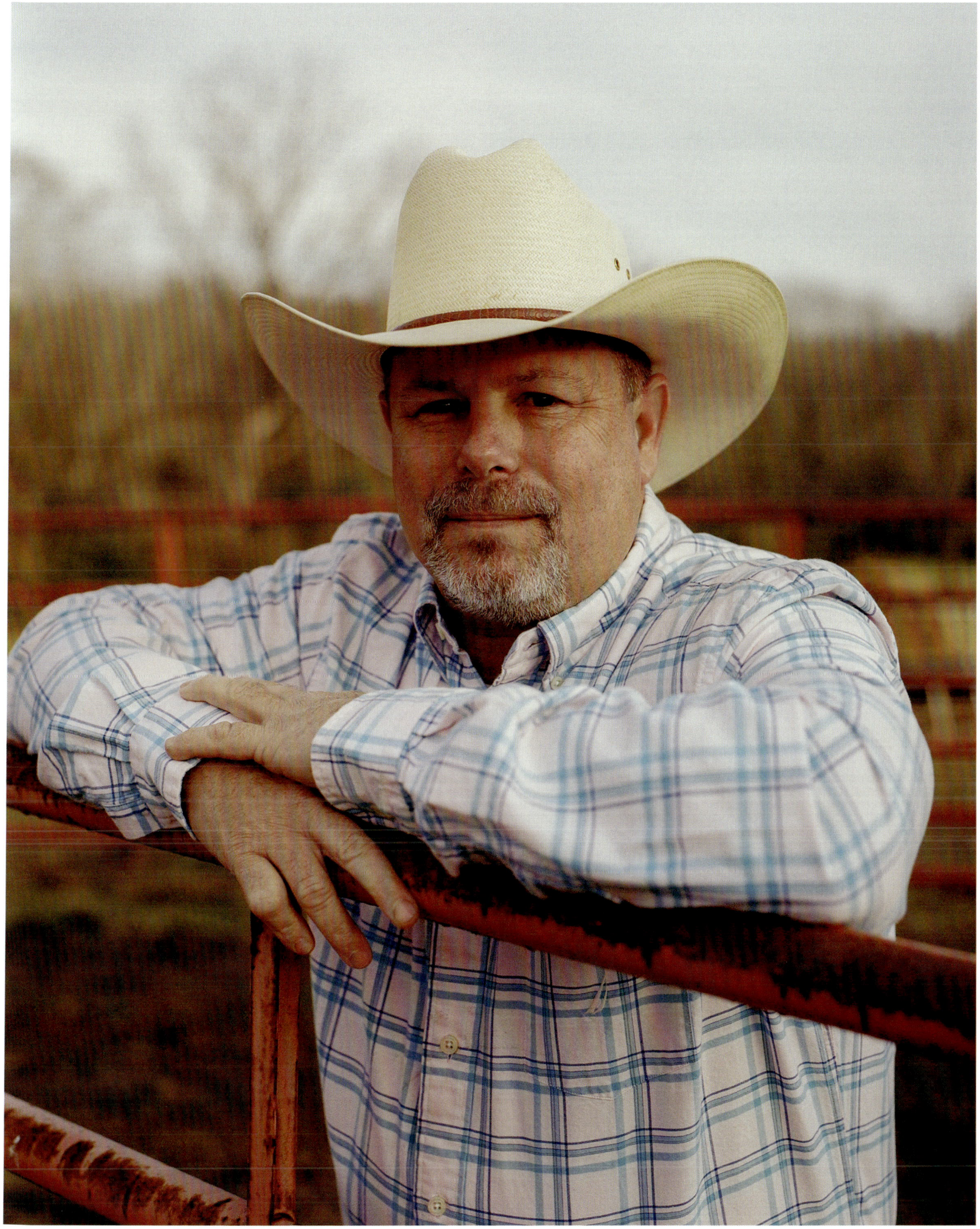

Robert Bentley, Staff Sergeant
Army - 11B / Infantryman
Years of service - 23 years
Deployments - 1 [Operation Enduring Freedom]

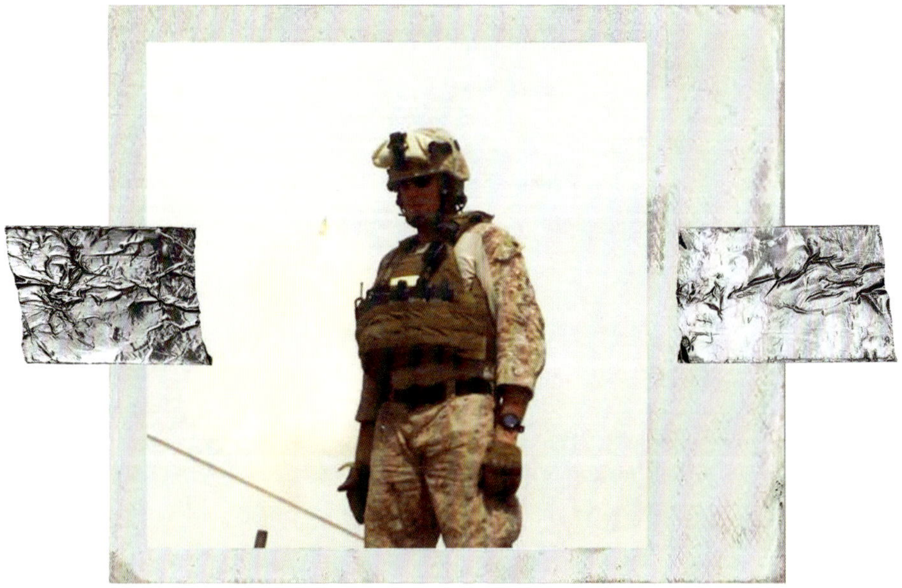

I joined because of my dad. He was a Marine in Vietnam and that meant a lot to me growing up. I deployed regularly, both during Operation Iraqi Freedom and Operation Enduring Freedom. There were some pretty dark times and close calls which took brothers and sisters from us too early. I also took some hard hits which led to multiple surgeries and a road to recovery that has proven to be a great challenge. It took about two years just to get back to zero. Everything. Eat, sleep, walk, enunciate words correctly, even to learn how to trust and embrace love again. Eventually, I took a breath and stepped forward outside the Marine Corps. I started a handyman company while putting myself through college. I felt like the odd man out, not knowing what to say as a twenty-eight-year-old in a college class full of eighteen-year-olds. Fortunately, during my time in school, I had some remarkable professors that just got it and understood my position and really took the time to ensure I was successful. To them, I will always be grateful for such kindness. Once I graduated, I stepped into the corporate world and enjoyed the commercial landscape industry. Made my way up to the VP level for a landscape firm, but separating from the military was the hardest thing I'll ever do.

For me, family is everything. They got me through and are still what gets me through. When the nightmares come, my wife is there with me. When I have a bad day, my kids always save the day. Despite the ups and downs, I wouldn't change a thing. I carry the memories of lost brothers and sisters lost forever, and I will always honor the promises made over beers prior to deployments. So if I may leave you with this simple quote, it may just make all the difference: "If you can risk it all on a game of pitch and toss, and lose, but stand again, that's when you'll know you're a man."

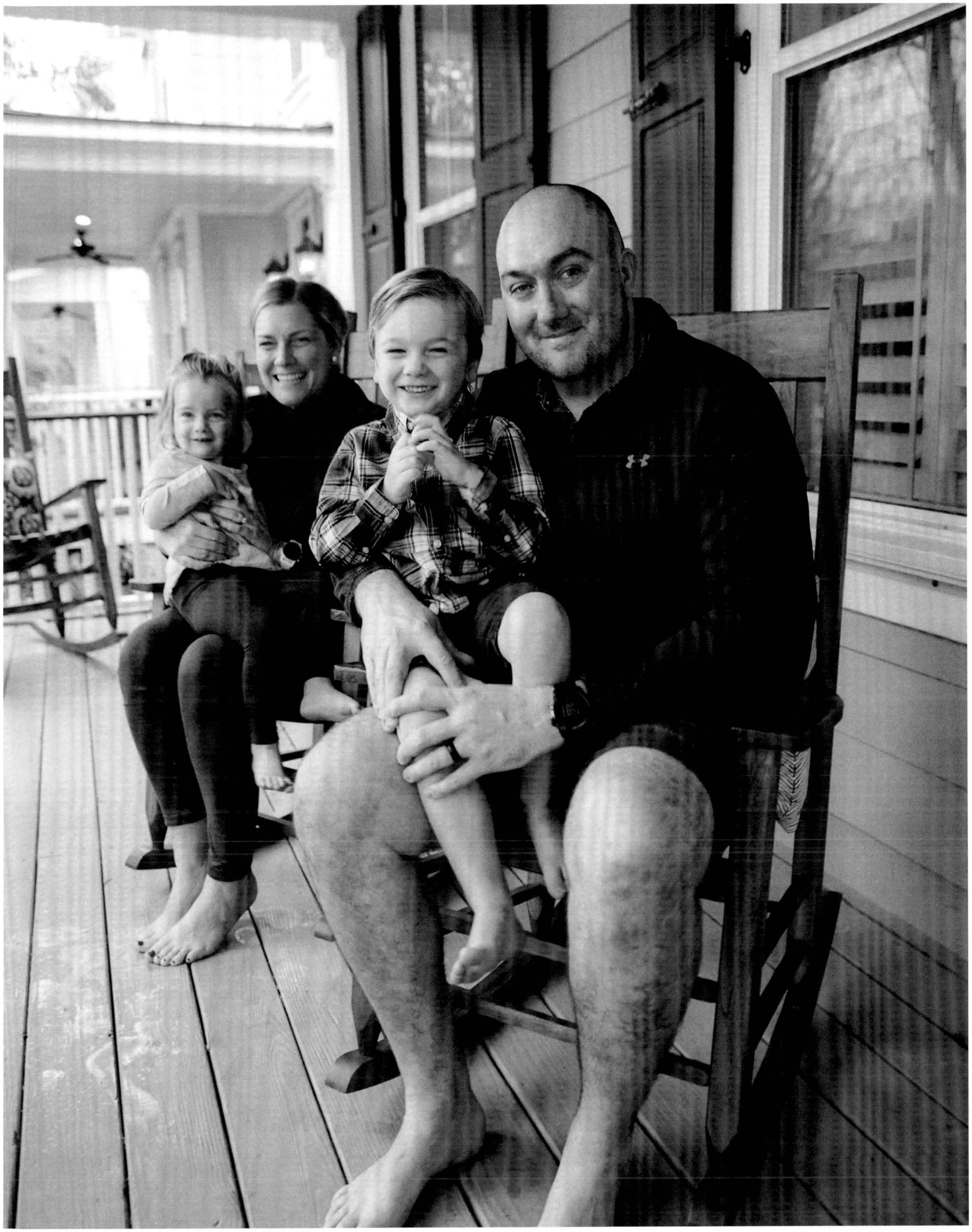

Ryan Siler, Staff Sergeant
Marine Corps - 8152 / EFSS CHIEF
Years of service - 9 years
Deployments - 8 [Operation Iraqi Freedom & Operation Enduring Freedom]

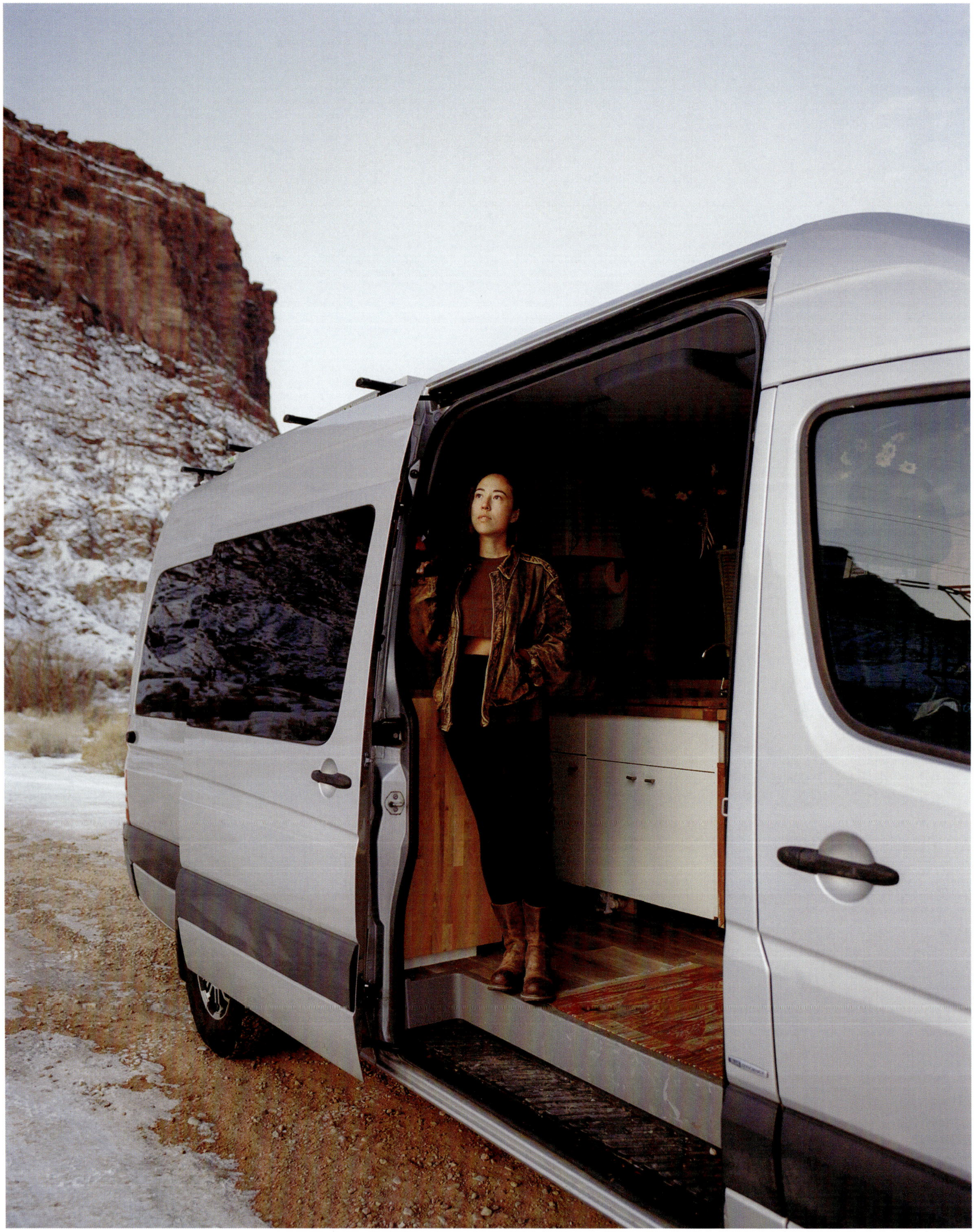

Sophie Hilaire, Captain
Army - 90A / Ordnance Officer
Years of service - 5 years
Deployments - 1 [Operation Enduring Freedom]

My name is Sophie Hilaire, and I was a 90A Ordnance Officer in the United States Army. I deployed in support of Operation Enduring Freedom from 2012-13. I left the military as a Captain (O-3). My father introduced my siblings and me to the idea of the Army — which was peculiar, since we weren't a military family. I had never considered the Army, but my dad took us on a visit to West Point, and the rest was history.

I had a busy first summer after separating from the Army; I took a solo trip across Mongolia, volunteered at a Syrian refugee camp, and became a yoga instructor in Costa Rica. I also crossed off my fiftieth state by climbing Denali in Alaska and started to get into mountaineering. Professionally, my next stop was getting a Wharton MBA, which led to my first civilian job, at McKinsey & Company, where I've been a management consultant for four years. My readjustment post-Army was tough. I'd just left an organization that felt like a family; some of my soldiers wept during my farewell, which almost made me question leaving. It felt strange trading that environment for the parties at business school, where I initially struggled to connect with my peers in ways beyond alcohol.

While at McKinsey, I pulled the trigger on a longtime pipe dream and climbed Mt. Everest. A lot of people tried to talk me out of it; it cost $70,000 that I didn't have, would take me out of work for two months, and required intense training. But I had an employer that encouraged sabbaticals, and I'd climbed the prerequisite mountains over the past few years. Summiting Everest was a major turning point in my life. I noticed how much better I'd felt and slept while living out in nature, since my body produced so much natural melatonin. I decided I wouldn't renew my lease in NYC in the fall of 2019. When the pandemic hit, I still wasn't sure where I wanted to live. Since I couldn't make up my mind, I went straight for something easy that I assumed would be short and temporary: a Sprinter van. I named her Sage. It's been almost a year with Sage, and I have no plans of moving out.

Being separated from society has been priceless. I've finally had the time and lack of distractions to work on myself. To process the places I've been, the things I've seen, and the experiences I've had. To understand myself and why I'm wired the way I am. My entire outlook has changed. To some, van life appears to be rock bottom. The truth is, I'm flying higher than Everest's summit. I've learned that time in nature, minimalism, and becoming my own best friend makes me enduringly happy. And I'm not waiting until retirement to cash in on all of my "investments for the future." I'm giving myself space and permission to enjoy life every day.

Continued on next page

I've also learned that my compulsion for the extreme isn't always rooted in something healthy. It's a trait I've recognized in many fellow service members. I'm working to untangle my own negative patterns of behavior so I can enjoy my "second lifetime" with ease and happiness. Through all this, I'm honoring the younger version of myself, who gave ten years to our nation and the Army and became self-reliant and resourceful enough to someday climb into this van to do the most important work of all: inner work.

In society and in the military, we're programmed to think and react in certain ways. Some of it has been helpful to me, but some has prevented me from actualizing my true self. Once I realized the armor I'd been wearing throughout my life was just a coping mechanism, I knew it was time to take it off, and finally find levity. That's the mission I'm on in my second lifetime: doing the inner work.

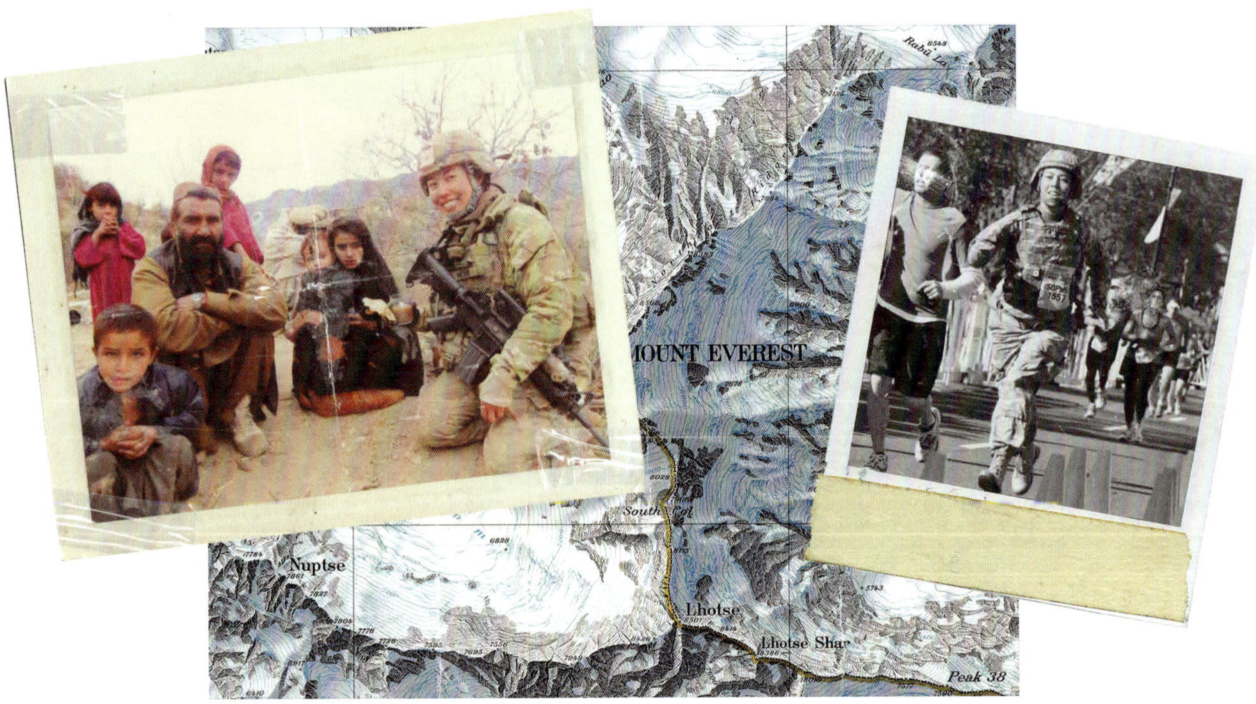

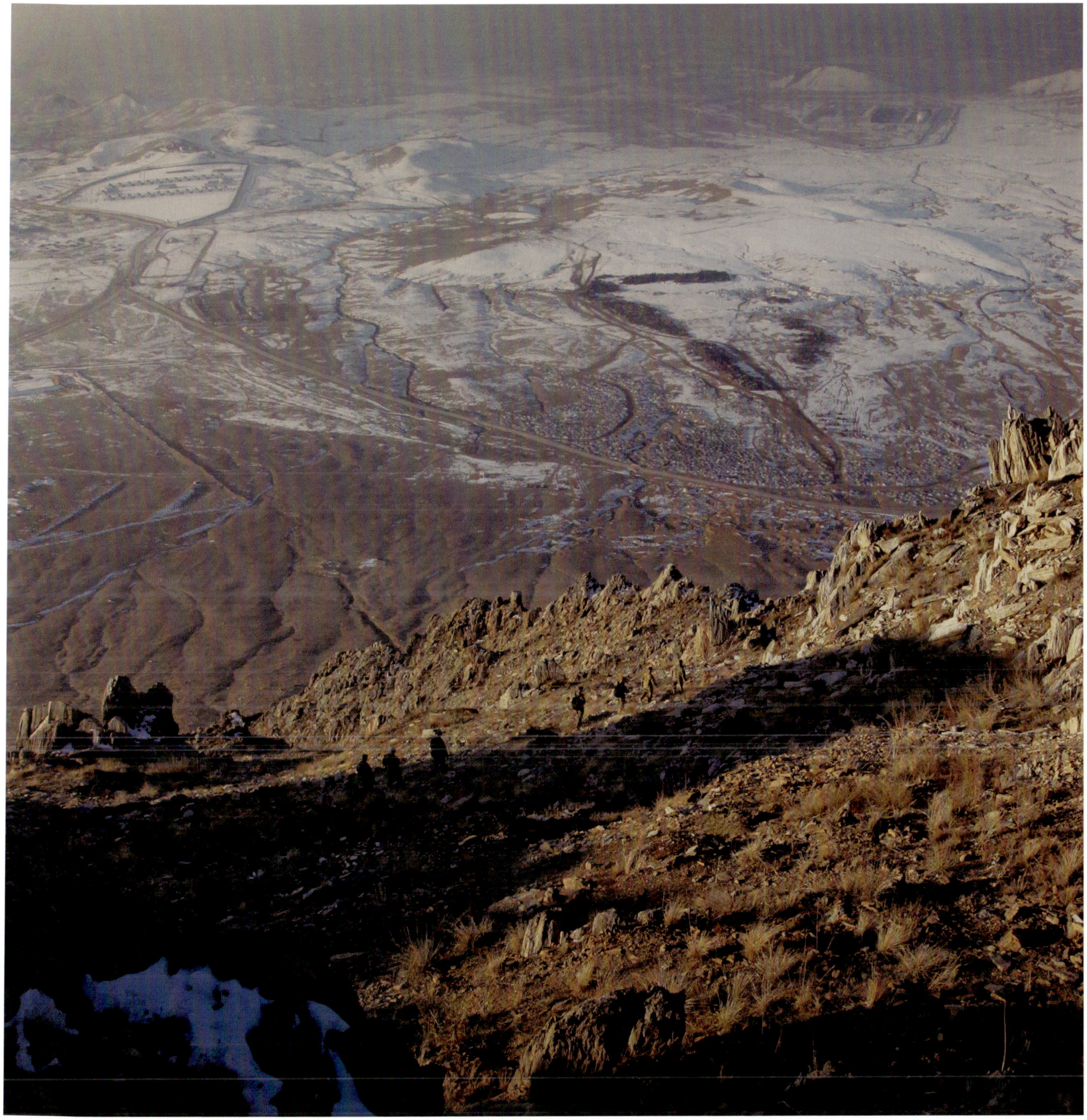

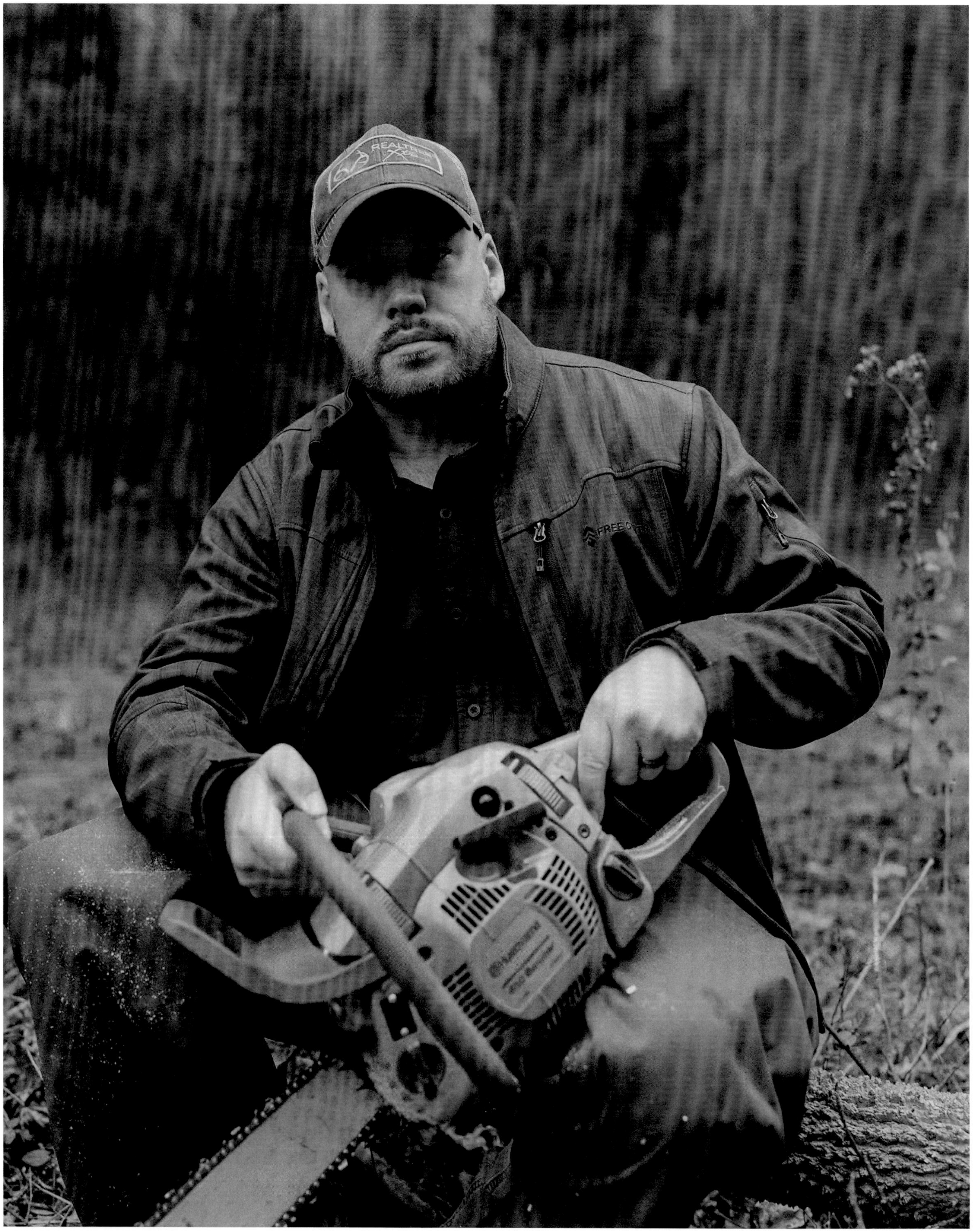

Joshua Grim, Specialist
Army - 11B / Infantryman
Years of service - 3 years
Deployments - 1 [Operation Iraqi Freedom]

My name is Joshua Grim, and I was in the US Army as an 11B infantryman. My younger brother had joined the infantry about a year before me. We'd always grown up playing Army and always had a passion for it. Then, 9/11 happened my freshman year of college. That planted a seed. I dropped out and joined up. I was a Specialist when I got out after my three years of service and one deployment to Iraq. I deployed immediately after completing basic training. We went right overseas to support Operation Iraqi Freedom in 2005. We were supposed to be there, as a unit, for a year. Then we got extended and the tour was ultimately sixteen months, of which I was in-country for thirteen.

My readjustment was really difficult. I ended up getting out with an honorable discharge and I really should've gotten a med discharge. The VA wasn't much help when I separated in 2007, but eventually I was able to appeal and get the rating with the VA. I struggled with the usual stuff. I didn't know how to deal with or understand PTSD. I drank a lot and fought a lot. I was really not myself and for the first five years, I was really struggling. Since getting help and readjusting, I found the love of my life and remarried. I also moved to Virginia. Life is amazing now. We have four beautiful kids and we're getting ready to buy some land and build a home. In addition to playing golf and working with Fairways for Warriors I also enjoy scuba diving and camping.

Find a purpose. I know that sounds clichéd, but it's so true. If you don't have something to put your attention towards, life becomes really rough. You need to find the things you have passion for.

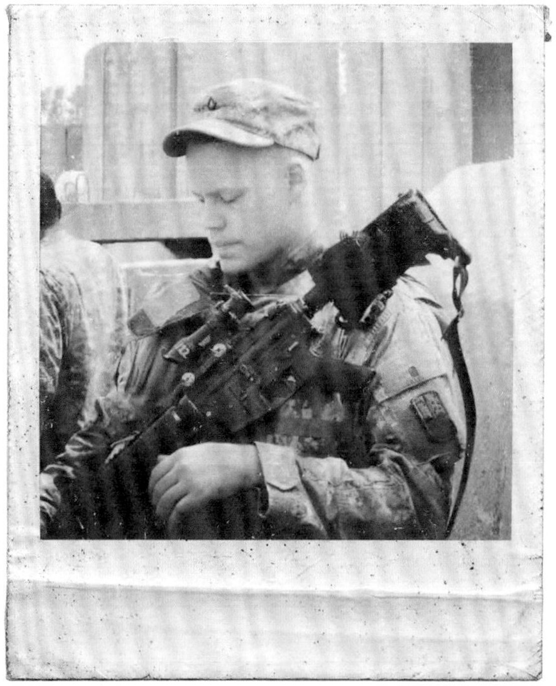

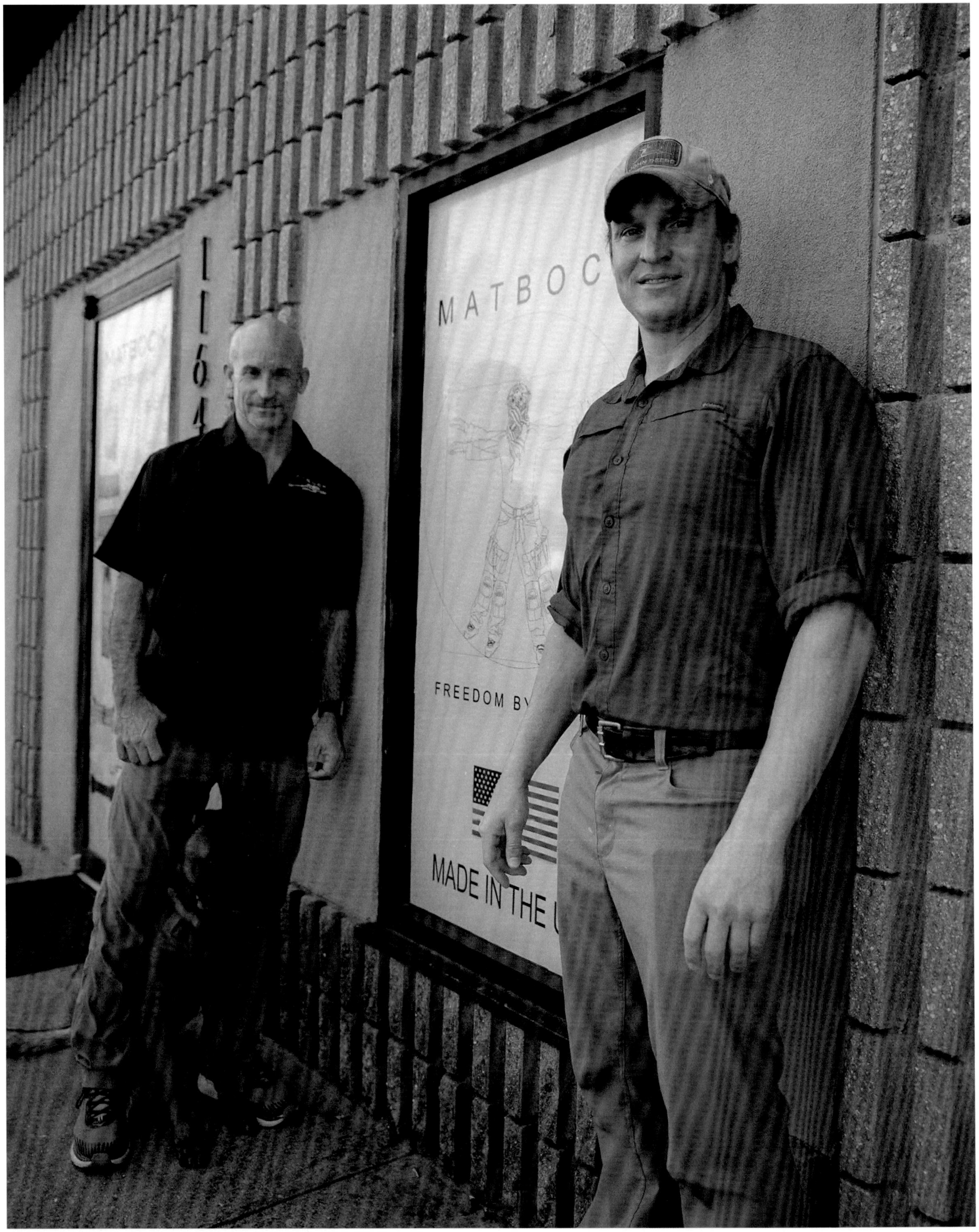

Jim Foreman, Command Master Chief
Navy - Seal Team 1 / DEVGRU
Years of service - 27 years
Deployments - 18 [OIF, OEF, Bosnia]

Zach Steinbock, Lieutenant Commander
Navy - Seal Team 4 / SBT-20 / Seal Team 18
Years of service - 7 years
Deployments - 4 [OIF & OEF]

My name is Zach Steinbock, and I was an officer in Naval Special Operations. Growing up, I wanted to be a fighter pilot, ever since I saw Top Gun. That sparked my curiosity about the Navy. In high school, I learned about the SEALs from a family friend, and that was it. After graduating from the Naval Academy, I was AOIC for SEAL Team Four. I then moved over to Special Boat Team 20 as a training officer. After separating from active duty, I went to medical school and was in the SEAL reserves on Team 18 as a troop commander. I left the service as a lieutenant commander after deploying three times to Iraq in support of Operation Iraqi Freedom and once to Afghanistan in support of Operation Enduring Freedom. My transition was difficult for the first few years, but I feel like I had it easier than a lot of other guys. Anyone coming back with PTSD or other problems, we need to make sure they understand they are not broken. There are people out there willing to help, as long as you are willing to put in the work. Mental Health Therapy has been a big part of my life, ever since I recognized there was a need to change. As I was separating, Disabled American Vets were great helping me get my disability packet done. American Corporate Partners was an initial key support program as well, with their mentorship program. I have an engineering degree but didn't know much about business, so ACP was really helpful. Now, I'm the president of both MATBOCK, a tactical equipment company where we sell primarily to the Department of Defense, and Aceso Plasma. At Aceso Plasma, our core product is an innovative medical device that creates high concentration, low temperature oxygen plasma for disinfection of wounds, rapid hemostasis and accelerated wound healing. We're trying to get things to where if you have a traumatic amputation, you can put a tourniquet on it to stop the bleeding, use our device, and then take the tourniquet off in a few minutes because the arterial bleeding has clotted so quickly. Go to work, that's always my mantra. It's time to put in the work and do something.

My name is Jim Foreman, and I served twenty-seven years as a Navy SEAL. I left the service as a Command Master Chief (E9) with eighteen deployments. The most difficult part of my transition from the military was finding a new career that was fulfilling and enabled me to stay somewhat connected to the Teams. Part of the unique aspect of being in a SOF unit is that you generally stay with the same group of people your whole career, forming strong bonds. Leaving the military also means leaving your brothers. Another struggle is going from having a high level of purpose to not having a mission. Working as the sales team lead at MATBOCK has allowed me to stay connected with military units across the country, while assisting in their missions by providing some of the lightest waterproof tactical gear on the market. MATBOCK has also allowed me to design and create new gear based on specific requirements for units, which has been fulfilling to me.

ZACH STEINBOCK

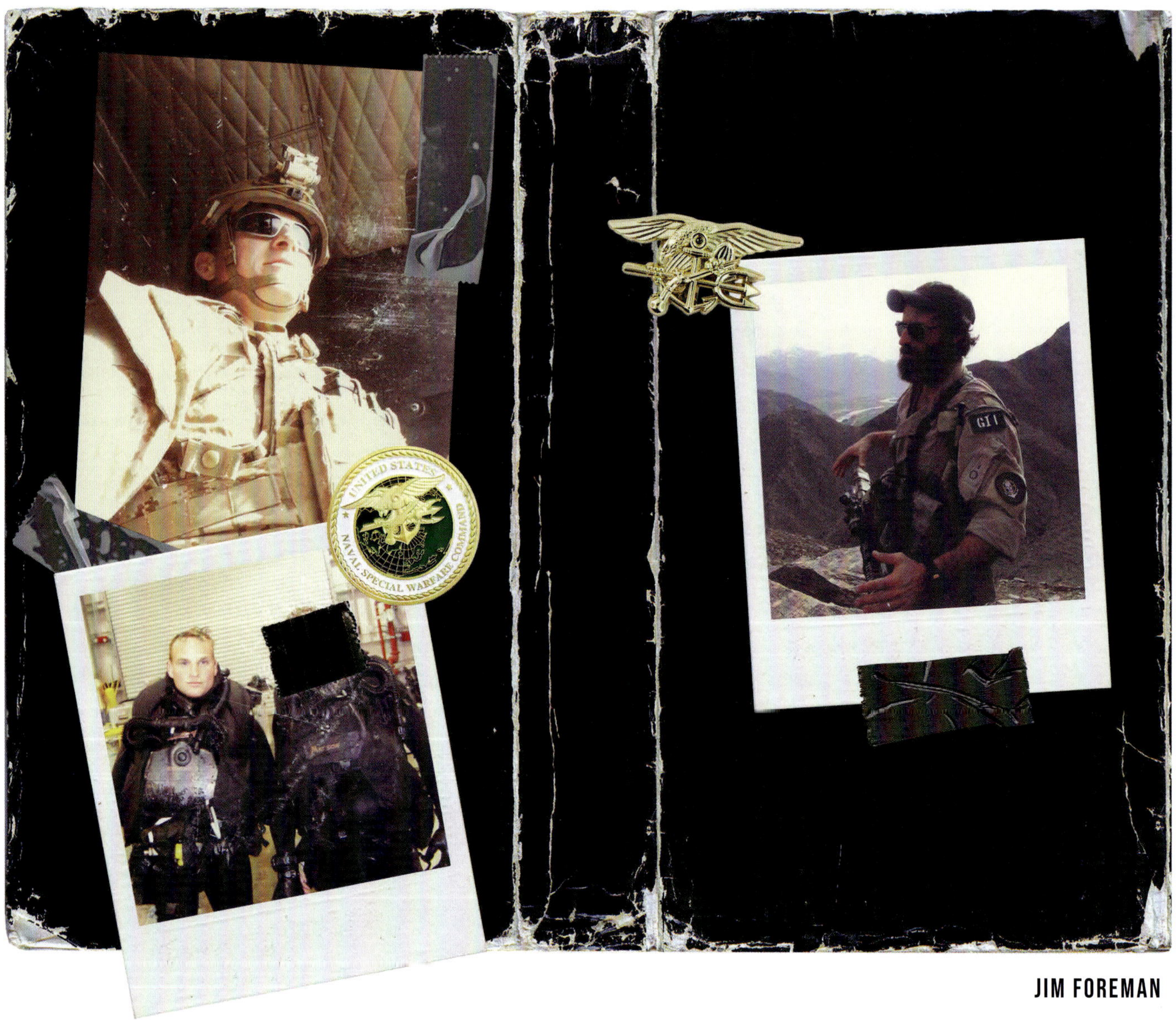

JIM FOREMAN

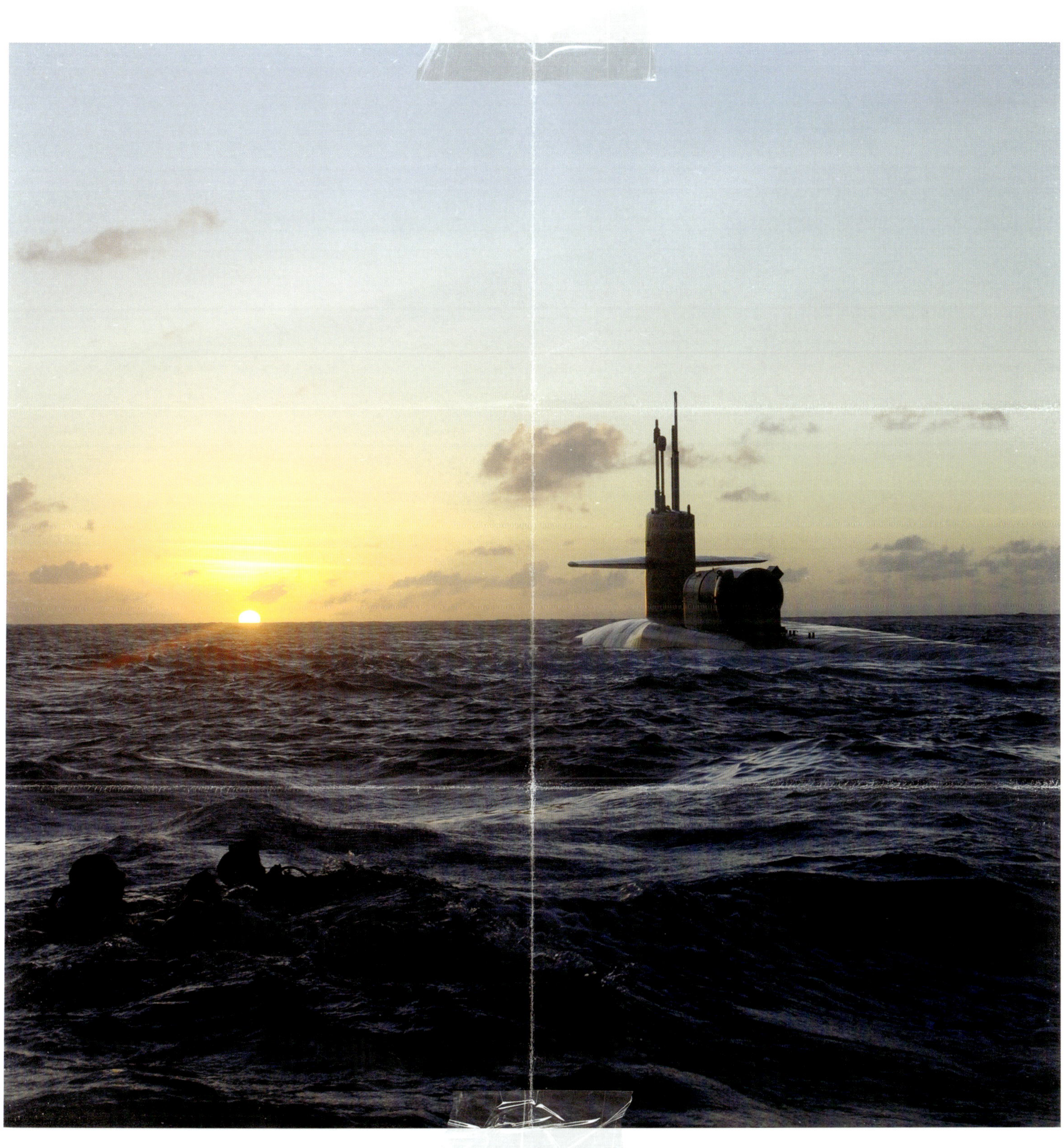

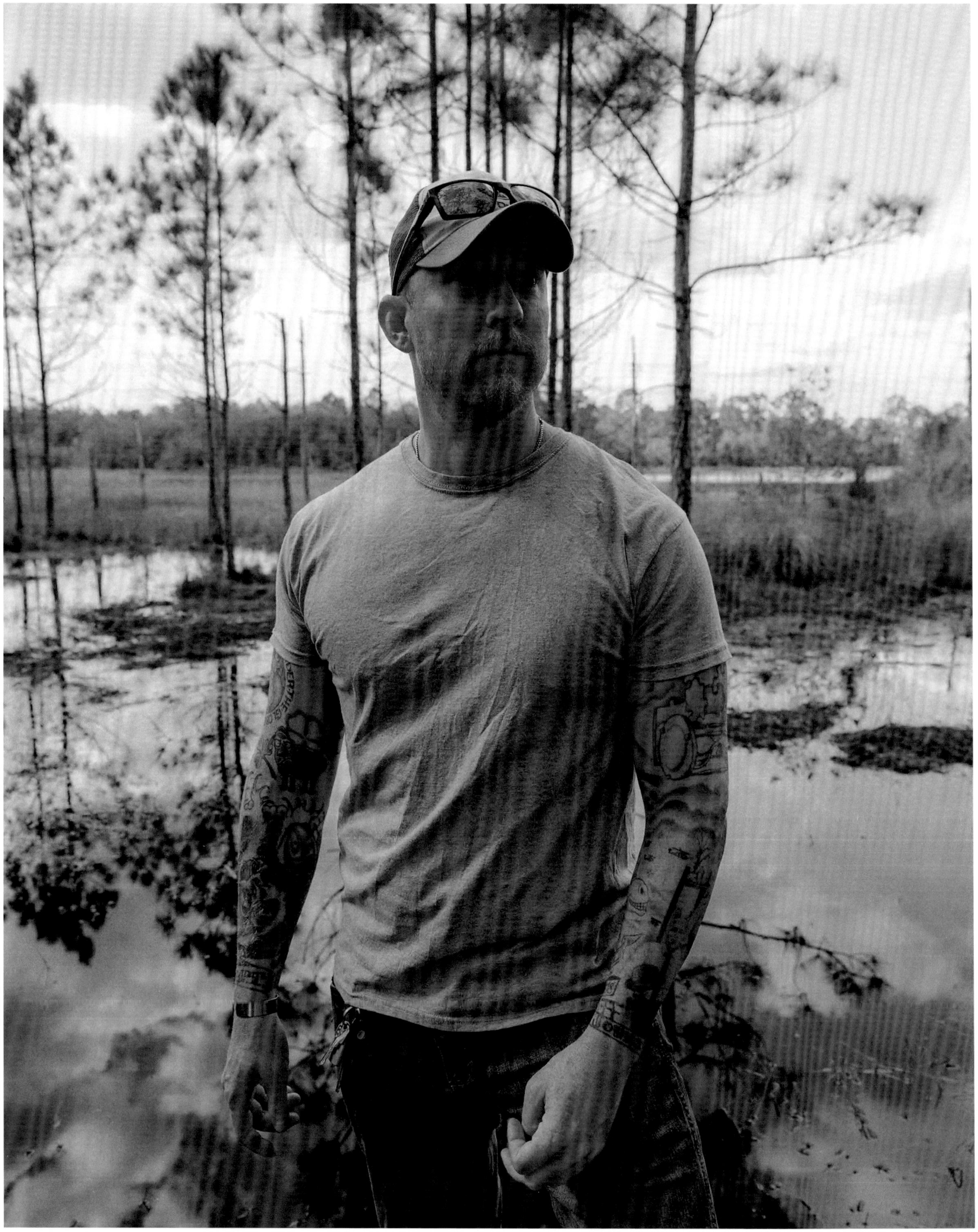

John Edwards, Staff Sergeant
Army - 11C / Mortarman
Years of service - 6 years
Deployments - 3 [Operation Iraqi Freedom]

My name is John Edwards, and I was an 11C infantry mortarman in the United States Army. I joined because it was something I wanted to do as a kid. Growing up near an army post, it was always in me. Initially, I had promised my mom I would go to college, and had started in September of 2001. I dropped out to enlist in January of 2002 and deployed to Operation Iraqi Freedom in 2003, 2005, and 2007. I got burnt out on the big army mentality and all the deployments. Looking back, I wish I had tried to go to a special ops unit or try to be a pilot. Right before I left, I had come up on Drill Sergeant orders, so I could have done that and used the time to reevaluate and maybe have tried to be something else, other than be a grunt. Instead, I separated from the Army as a Staff Sergeant (E-6).

After I separated, I took a year to travel all over the United States. I realized America isn't so big. I credit the road trip with making my readjustment to civilian life much easier. I was a professional vagabond, and that intermediate step really helped. If leaving the military was "hot" and civilian life is my now "cold" situation, the trip was my lukewarm period. After my trip, I got a degree at Appalachian State and have been working in law enforcement since 2013 in Orlando, Florida. I love just getting out into nature: biking, hiking, kayaking, and some hunting as well. My mantra is: Have a no-easy-days mentality. I never want to be known as the person who quits. I just keep pushing and trying to make myself better.

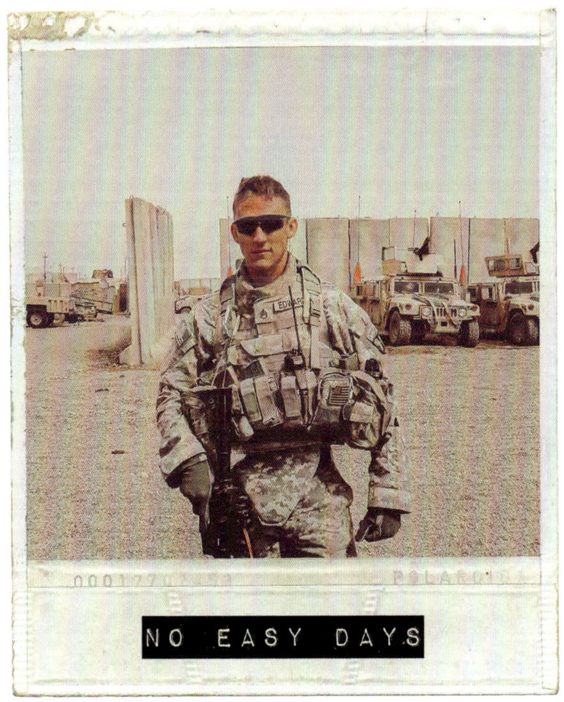

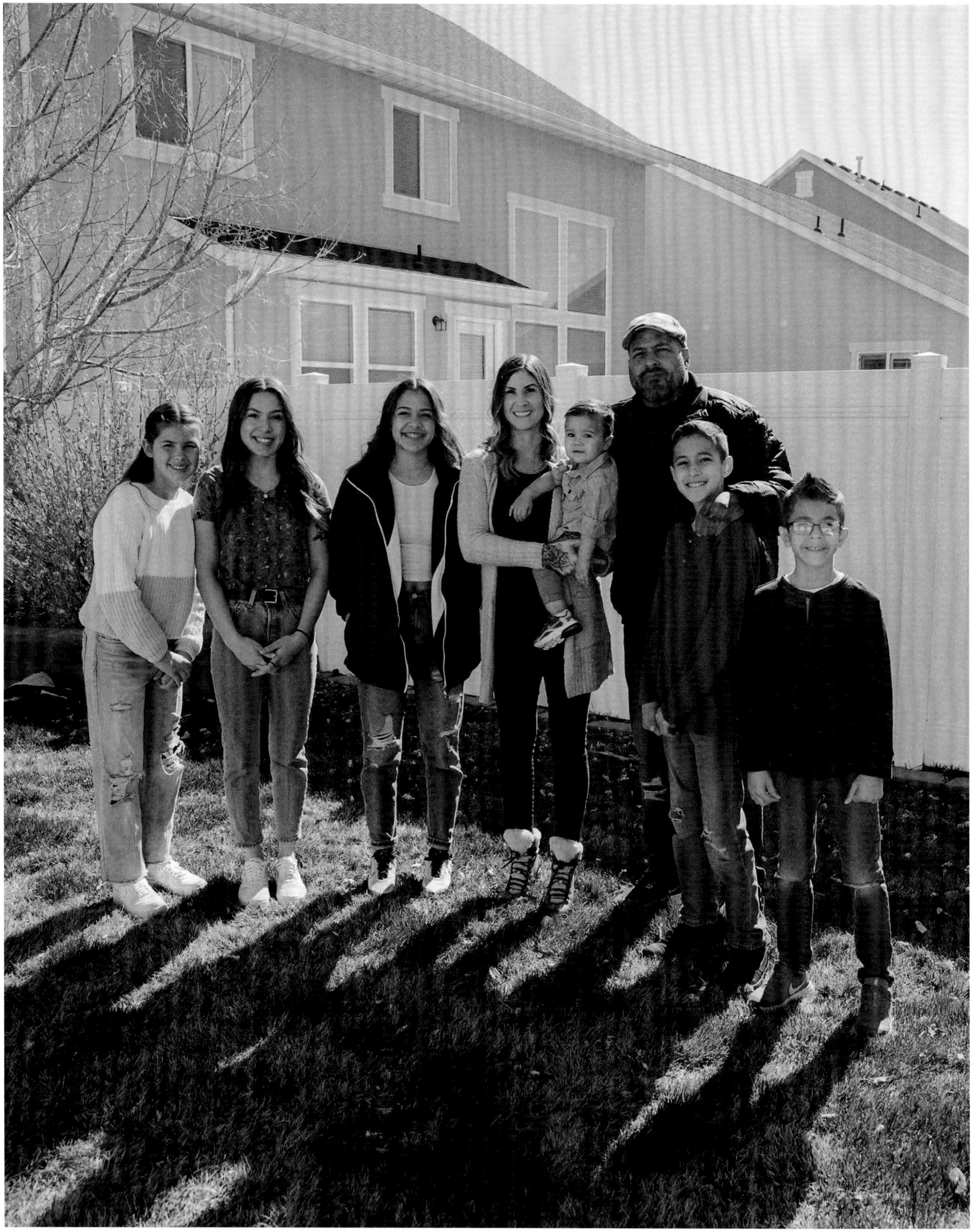

Vincent “Rocco” Vargas, Sergeant First Class
Army - 11B / Infantryman / 2/75 Ranger
Years of service - 18 years
Deployments - 3 [Operation Iraqi Freedom & Operation Enduring Freedom]

My name is Vince Vargas, but most people call me "Rocco." I am currently in the Army Reserves as a drill sergeant and was an 11B in the 2nd Ranger Battalion before that. I'm about to medically retire after eighteen years of service. I deployed in support of Operation Enduring Freedom twice and Operation Iraqi Freedom once. Growing up, I hated the military. My dad was a Marine and I thought he was a dick. Very stern type of guy. As I grew up, I started to respect and understand where it came from. He was scared I would get into gangs like he and my brothers did. Instead, he got me into baseball. For so long I had no life outside of it, and eventually went to junior college to try to get drafted. I got into a fight with a coach at my first junior college. I had gotten into another fight off campus and the coach made me run laps. He was talking to me disrespectfully as I ran, and eventually I'd had enough. Immediately, I drove to the next juco at Glendale. There I got conned into cashing a bad check. It was an accident but now I was toxic to the MLB because I was such a high risk.

I'm also dyslexic and had avoided education because I was embarrassed. At this point, life was hitting way too hard. I got a girl pregnant, and I knew I needed to get away from LA. I was watching the war on TV in a bar next to some Navy guy. They'd just put the flag over Saddam's head. CNN was interviewing the family of that Marine and they were so proud. I couldn't remember the last time my family was proud of me. Baseball was done and I had a daughter to provide for, so I enlisted. I maxed out two credit cards overseas, drinking. I was thinking I was gonna die. I was young and angry at the world and like, "Fuck it, if we're gonna do this then let's fucking do it." So, I decided to do Rangers.

I didn't have any veteran organizations or anything after active duty until I linked up with Mat Best and the Article 15 crew. It was really cathartic for me. It was healing to be around guys who understood me. Mat and I had served together in the Rangers. It was healing with humor; through that I realized I needed counselling. I always, from day one, wanted to do more than just YouTube skit comedy. I think twenty years ahead in life. I wanted to make movies, to really act. Acting was always the thing growing up in LA that's always around you. I also did two years of theater in college to try to keep my grades up. Eventually we leveraged Article 15's success to partner with another company to make the movie Range 15. We hit number one on Amazon for three weeks in a row. I was like: I can do this! I also won an award at the GI Film Festival for a project called The Long Way Back. Jamie Keller was the host and a Navy vet. After the festival, we did some skits together and it was his most-viewed video ever. He really encouraged me. So, I walked away from Article 15. I met my wife and I said, "I really want to pursue acting." She was really encouraging, and I started to gain a little steam. Through Jamie I met Steve Howey, who is on Shameless. He mentioned Kurt Sutter's new project, Mayans, was looking for someone. I sent in my stuff. A couple hours later, I got an audition. I recorded the script to audio and I was listening to it all night. Steve's first piece of advice was, "Don't use the script, be off paper. If you lose it ... whatever, stay in character."

Continued on next page

It seemed like the audition went well; the casting director even hugged me. I flew back to Texas and the same day I landed, I got the email I needed to go back to audition again. This time with show creator Kurt Sutter himself. My dad, who's a biker, drove me to the audition, all hyped. Steve's final piece of advice, and I still use it, was, "You've been to war, you've done harder shit, you don't need to stress." I kinda blacked out in the audition. I walked out to my dad, who I have a great relationship with now, and we were high-fiving and he was jumping on my back, just so pumped up. Kurt was watching from the window and said, "That's true, that's honest." I got the call the next day. Initially I was gonna die in the first season. I had a single line in the pilot, but I was still pumped: this was the first step to what I'm going to do. I had my first gig. We filmed the pilot, I went home and learned a few weeks later they threw that pilot out. They were going to reshoot, but I might be out. They called and said they were sending the new pilot script. I got the script and I suddenly had seven lines! I was one of the main guys! It went from me being a guy who was going to die, to now I had the biggest role in a TV show of any veteran in Hollywood. It was because of honesty and authenticity. I'm a creative. I just pour myself into my acting and my art and my music and my desire to empower other vets and other fathers.

I don't regret what we did with Article 15. We were growing as businessmen, just trying to be good at marketing. What most people didn't see was the "Zero-to-sixty, nothing-to-legit-business-studs" process we went through. They only saw face value — and the face value was just drinking and shooting and slapstick goofy shit. When the laughter's done, what's next? And there was nothing there. It wasn't fulfilling and it wasn't what I wanted to do. So, in addition to my acting career, I created Beteran. I make my money being me. As an actor and in my small businesses I will always be me, but I made Beteran with the goal of making more inroads for the community. Bettering veterans, Beteran, is a change for the positive. To be pumped for the good in what we did and what we do. Veterans are not broken. Nobody gets out of college thinking, "I'm going to commit suicide," so why are vets getting out thinking that?

Grow. I have grown past the friends I thought were my closest from my youth on down, and I'm going to keep putting people in my life that support me growing and going where I want to go. Fuck all these walls people put up. Fuck the doubts and insecurities and impostor syndrome. It's not what you did, it's what you're doing now. The question is: did you grow?

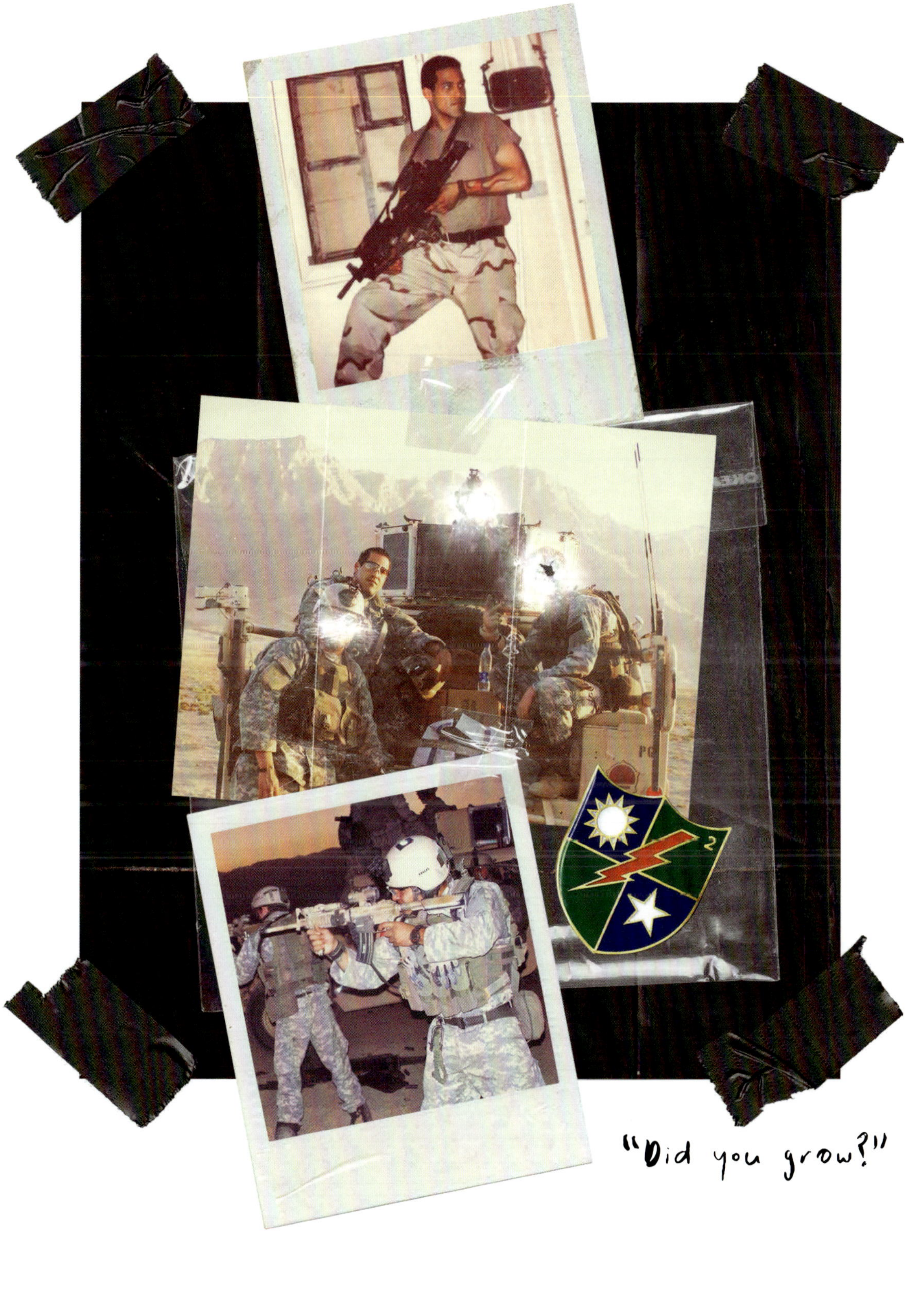

"Did you grow?"

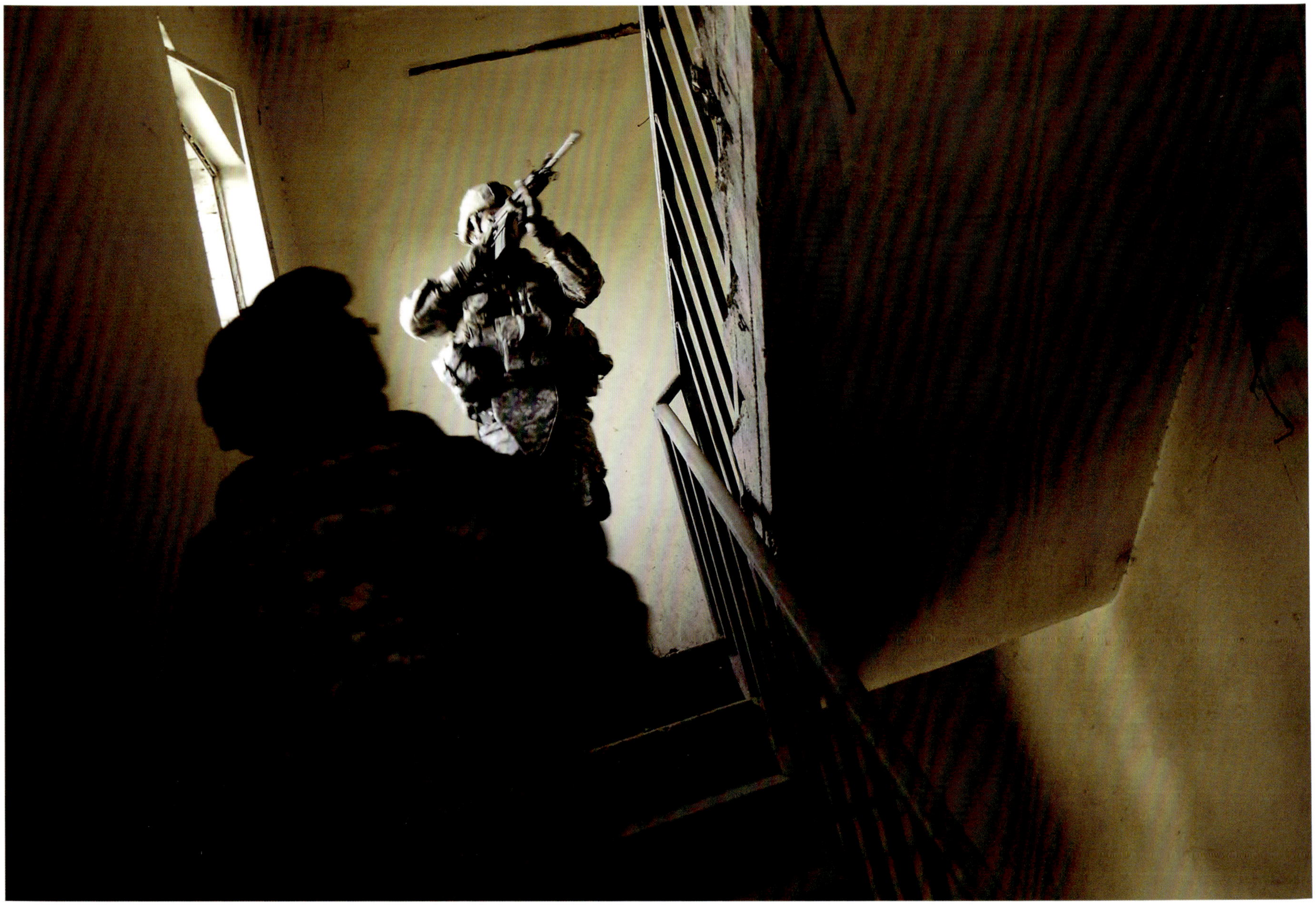

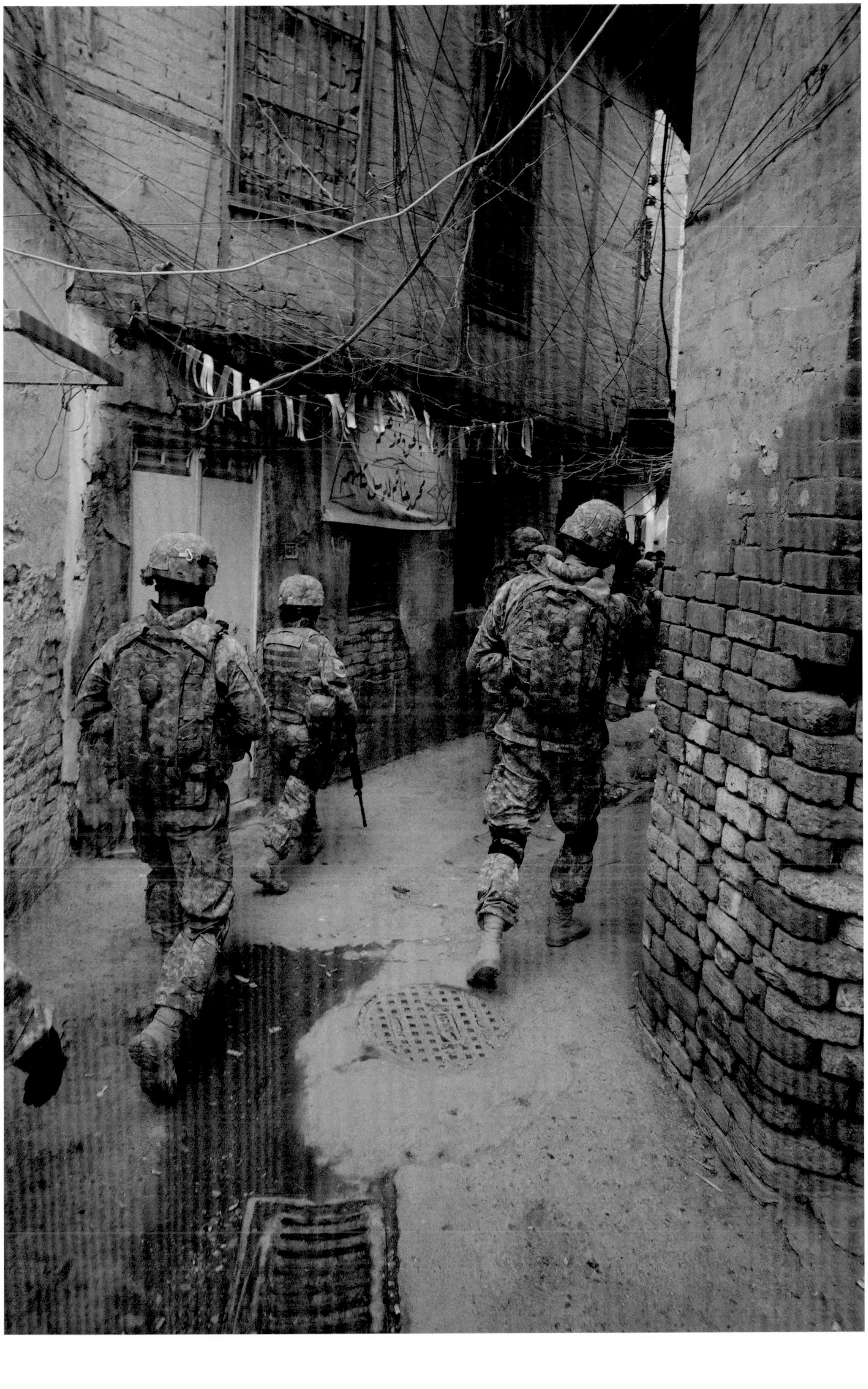

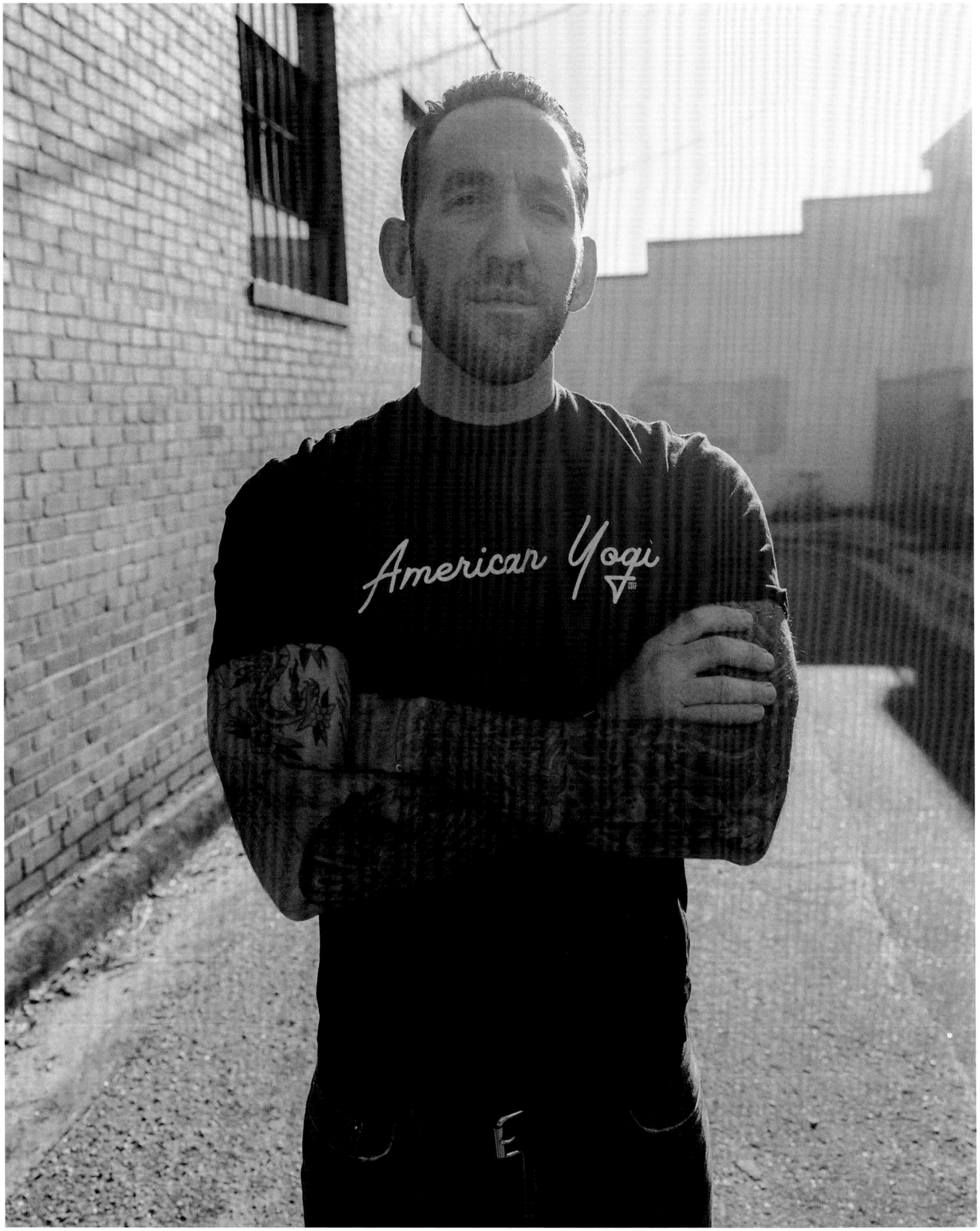

Phil Sussman, Captain
Army - 19C / 38A
Years in service - Active
Deployments - 3 [Operation Inherent Resolve & Operation Spartan Shield]

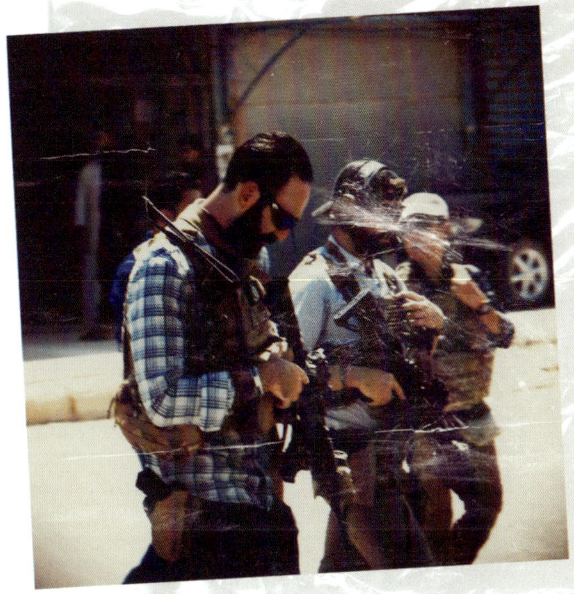

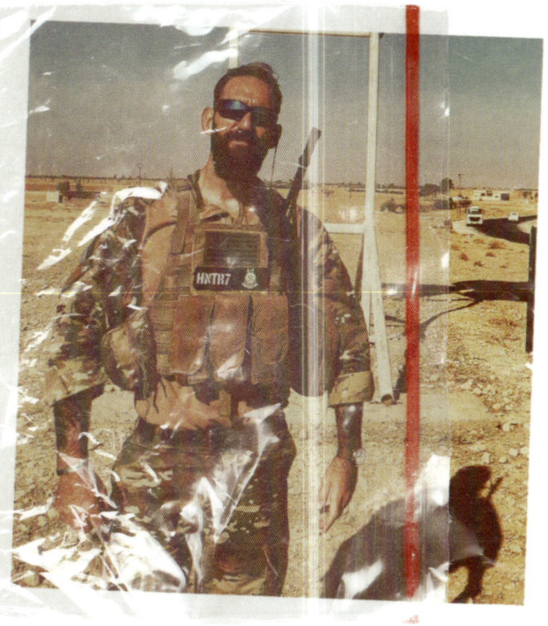

My name is Phil Sussman, and I am a Civil Affairs Officer in the United States Army. On 9/11 my life changed. Suddenly, there was a larger world than the one I'd known. I had always wanted to serve, especially in the Middle East, and like many others in uniform, I wanted to be a part of something bigger than myself. I deployed three times in support of the Global War on Terror. I returned from my last tour in December 2020.

The job of Civil Affairs is truly interesting. We are tasked with engaging the civilian population at the lowest levels in order to be the eyes and ears of our military in any given country. The opportunities afforded to me as a Civil Affairs Officer in Jordan and Syria allowed me to work with refugees, internally displaced personnel, and local leaders affected by poor economic conditions and the conflict in Syria. I prided myself in creating and maintaining deep connections with our local partners and venturing to areas where Americans were scarce. I'm currently medically separating from the Army, and it's a truly difficult process, as I'm finally starting to deal with a decade of mental and physical trauma from a life of service. While transitioning, however, I've been able to work on and find comfort in American Yogi, an apparel company I started with my brother when we saw that there was a population who could benefit from yoga as much as it has helped me. Our goal is to spread a message of healing as widely as possible, especially in the veteran community. We created a fully veteran-run yoga festival here in Southern Pines, North Carolina and use it to break the stigma of yoga in the military community and raise funds to send at least one teacher through teacher training.

My journey in the Army never felt easy. I joined later in life, lost friends and fellow soldiers, broke my back in a training accident, and finally ended my career nearly a decade later in Special Operations. Without all of my hardship, without this tough journey, however, I wouldn't be nearly as ready to help others heal for the rest of my life.

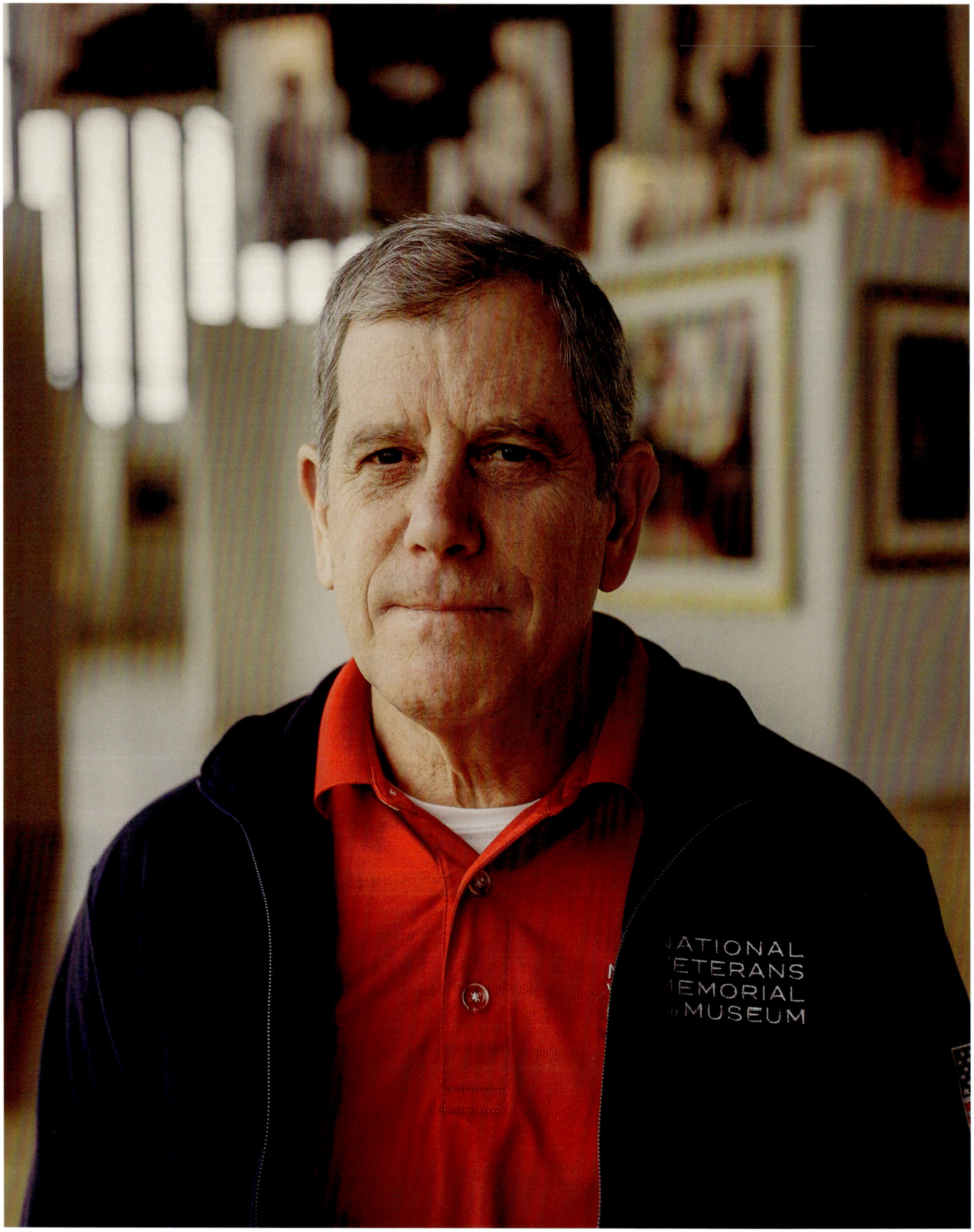

Mike Ferriter, Lieutenant General
Army - 11A / Infantry Officer
Years of service - 40 years
Deployments - 3 [Operation Iraqi Freedom]

My name is Michael Ferriter, and I retired as a Lieutenant General (three stars) in the United States Army. I grew up in an Army family and moved eighteen times as a child. Service is a calling in my family, as three of my four children have served as well. Two of my sons have also served in the Ranger regiment. As a young officer, I deployed to Somalia with the Third Ranger Battalion and I came forward with A Company after the Battle of the Black Sea to support Bravo Company. I also have three deployments in support of Operation Iraqi Freedom. My final combat tour was as commander of the NATO training mission in Iraq. On that final deployment, I was also the deputy commander for the Multinational Force-Iraq. Additionally, I deployed as brigadier general, deputy commander of the 82nd Airborne Division. I didn't go to Ranger School until I was a captain, and it's rare to be able to serve in the Ranger regiment or to command a Ranger battalion getting such a late start. I was incredibly fortunate to serve there.

In addition to my combat deployments, I had the privilege of commanding Fort Benning, Georgia and helped create the Maneuver Center of Excellence. Through this training center, Fort Benning is now home to infantry, armor, and cavalry one-station unit training (OSUT) and the professional development courses of young officers in those career fields. My final posting was as the commander of US Army Installation Command, where I helped save three billion dollars over two years against a twelve-billion- dollar budget. Finally, I had the privilege of helping reestablish the Army Combatives program, which is something I am very proud of.

As any good leader will tell you, you get where you get on the shoulders of the young. My troops carried me. I was nothing without them, and I'm extremely grateful to them for their service. As a military or civilian leader, taking care of your people is going to help with everything. You go into a leadership role with your faith and your friends and your integrity. You hope that this will create a family, and you need to lead that way. You're going to have people challenge that. Keeping ego in check is critical. Losing control of your ego is the fastest way to get off track from your faith and friends and to risk your integrity. When your time as a leader is done, all you want to have are fond memories.

As I progressed in my career to being a senior leader, I had a policy. I would never keep any of the gifts I was given. It's not that I wasn't grateful. I was. But I knew that if I took the gift, I'd like it. If I liked it, I'd want it, and if I wanted it, I'd start to covet. So, I always had a deal with my sergeant major to give them away to worthy soldiers and staff.

Continued on next page →

Transitioning from the military is like jumping out of an airplane without a parachute. In my eyes, the first thing a separating service member should do is to remember how good they are; that it's absolutely critical to continue to believe in themselves. By ability, and through their experiences in service, they are already in the "fast group." By drive and creativity and experience, they are already ahead of the game, but now they are doing something new. When you're new at something, it means you have a lot to learn, just like when you were a private. Do what's right. Improve daily. Start the process to learn the skills to continue to be successful. No one wants to be a private in life forever.

When I got out, one of the first things I did was start Ferriter Group, a for-profit consultancy. We do speaking and leadership coaching engagements, and we also do change management and business improvement work. A way we separate ourselves is with our Hands-On Inspired Leadership program. I'm a Brazilian Jiu Jitsu black belt and there is a lot you can learn about leadership on the mats. If you're a veteran and you want to start a business, you should pursue it! Your big idea is a big idea. Go out and conquer that next objective. Start on purpose, and make sure that you celebrate successes along the way. One of the lessons I had to learn was: if you run a business, you're in sales, period. If you're in sales, you're going to get rejected. Reject that rejection! Don't let it dissuade you.

In addition to consulting, my full-time job is as the CEO of the National Veterans Memorial and Museum in Columbus, Ohio. Our mission is to honor those who served and to preserve their memories and experiences for future generations. I'm also active in the non-profit space. I'm on the board of the Veteran Golfers Association and Wounded Warrior Golf. I work with Down the Stretch Ranch, a place where retired racehorses are paired with veterans to help them learn horsemanship and animal husbandry. Finally, I'm active with the Patriot Foundation.

I always remember a saying by Frances Hesselbein: "Leadership is who you are, management is what you do." Life is about who you are. As veterans, we are leaders. We answered the nation's call to serve. As you transition into being a veteran, you get a chance to make a difference in your community and the lives of those around you. And anyone can be a leader. Human connection is the magic sauce that makes everything happen. You should be asking: how can I be the best leader I can be? Who needs my help today?

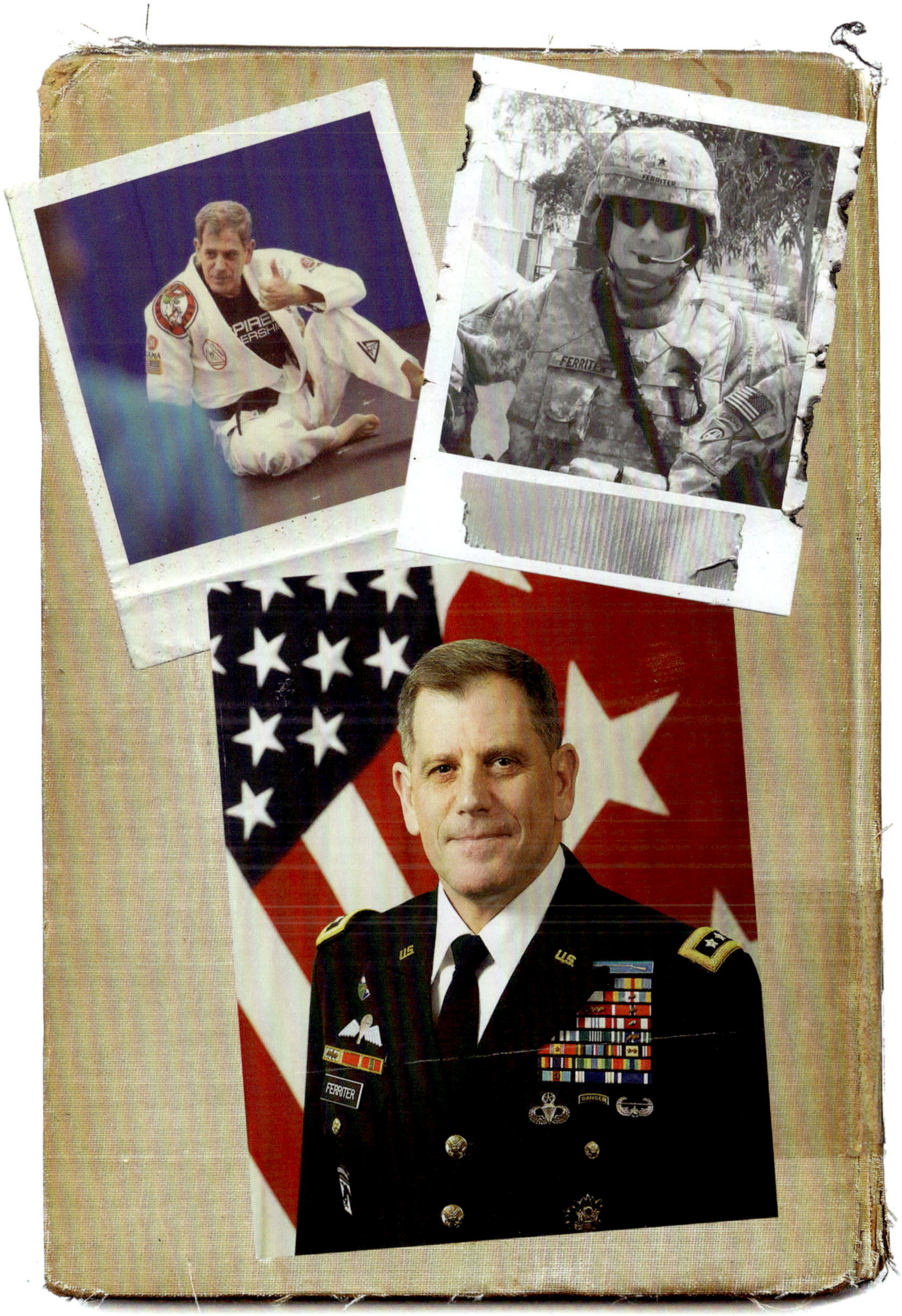
FERRITER

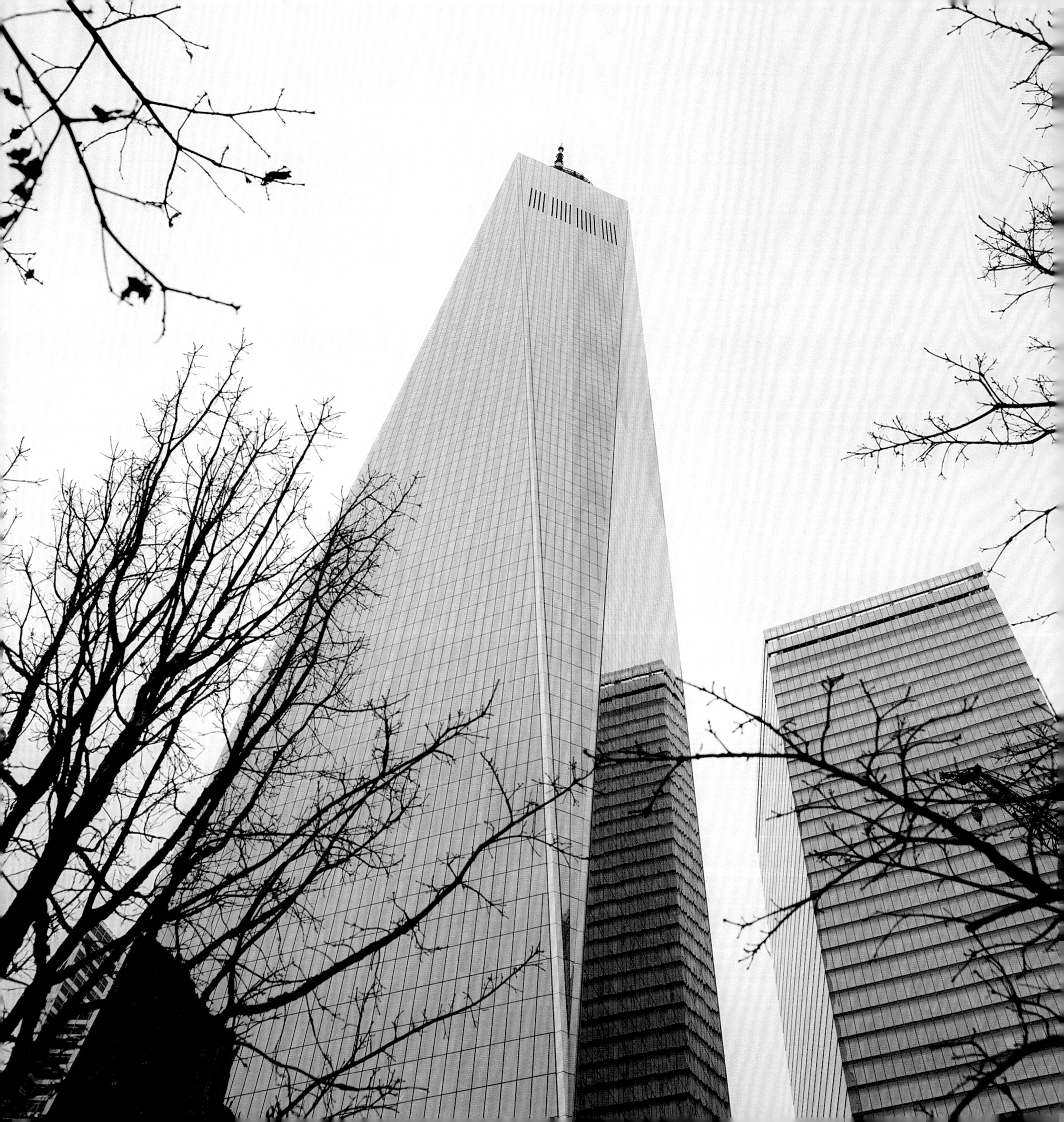

AFTERWORD

In the military, after every mission in training or combat, you conduct what is called an After Action Review (AAR). From the military's TC-25-20 training manual: "An AAR is not a critique. No one, regardless of rank, position, or strength of personality, has all the information or answers. After-action reviews maximize training benefits by allowing soldiers, regardless of rank, to learn from each other. An AAR does not grade success or failure. There are always weaknesses to improve and strengths to sustain."

An AAR asks the question, in the words of Vincent Vargas, "Did you grow?"

Our goal for this book was to show you the American veteran. Through Beau's photographs and our veterans' stories, we hope you are starting to see just how frustratingly normal we are. We're not broken. Most veterans, as you've seen, are thriving. No doubt, many of us have seen ghastly things. Yes, they have left scars. They have also forced a level of growth that was otherwise not possible.

Growth, as Nate Self put it, is a lifelong endeavor. It never stops. Just as the Freedom Tower rose to replace the World Trade Center, so too can trauma fuel post-traumatic growth. Sadly, more people by the day are becoming too comfortable behind a screen and less comfortable looking into another human's eyes. We are starting to forget the wisdom of Bill Butler, who spoke about how much relationships matter. Sometimes, it's hard to talk in person. It can be uncomfortable. Jon Jackson created Comfort Farms to teach people that you only grow in discomfort. A message Jason Van Camp teaches through Mission Six Zero as well. Growth breeds confidence. It helps us gain the courage to make those ripples that turn into the waves Cody Alford talked about. To look honestly at ourselves.

We all have some sort of internal dialogue that says, "I need to be in a place that allows me to change some of the negative patterns in my life," the way Sophie Hilarie has done. It tells us to realize that the waters of the ocean are always different, and one mistake isn't the only thing your life is judged on, like Aaron Laurich did.

Growth requires the humility and honesty to say, "I was a real piece of shit," like Tim Kennedy did, and then to invest in personal growth. To grow from a follower into a leader, as he has with the founding of his school.

Continued on next page

When you sign up to serve, you're required to accept the mantle of leadership — first, among your peers, as in the basic training environments Eric Flesch is responsible for. As your career progresses, be it enlisted or officer, your career continues to be measured by your ability to lead.

Above all, a veteran is trained to be a leader.

To lead, JC Glick believes, requires action. In combat, you learn the only bad decision is not to make one at all. If you want to be a better, more effective leader, you need to keep pushing, keep learning, keep developing. You need to ... keep growing.

As both this war and this book draw to a close, what we hope you get from reading these veterans' stories is: We vets are neither as special nor as broken by PTSD as the media makes us out to be. We aren't heroes for our service, for going to work and fulfilling our contract. That's what you do every day of your life. We did a job for a while, and now every newly minted veteran is looking for a new job. A new chance to grow.

Everyone reading this knows someone who has had a horrible experience in life. We all have non-veteran friends who seek therapy and struggle with mental health. Tragically, we all know non-veterans who chose to end their own lives. The sooner we, as a society, stop sensationalizing the veteran experience and start learning from the wars these veterans served in, the sooner we all, regardless of perceived societal rank, position, or strength of personality can start learning from each other. The sooner we can all grow, together.

Only then can we, as a nation, look at the longest conflict in American history and find the weaknesses we need to improve and the strengths we need to sustain.

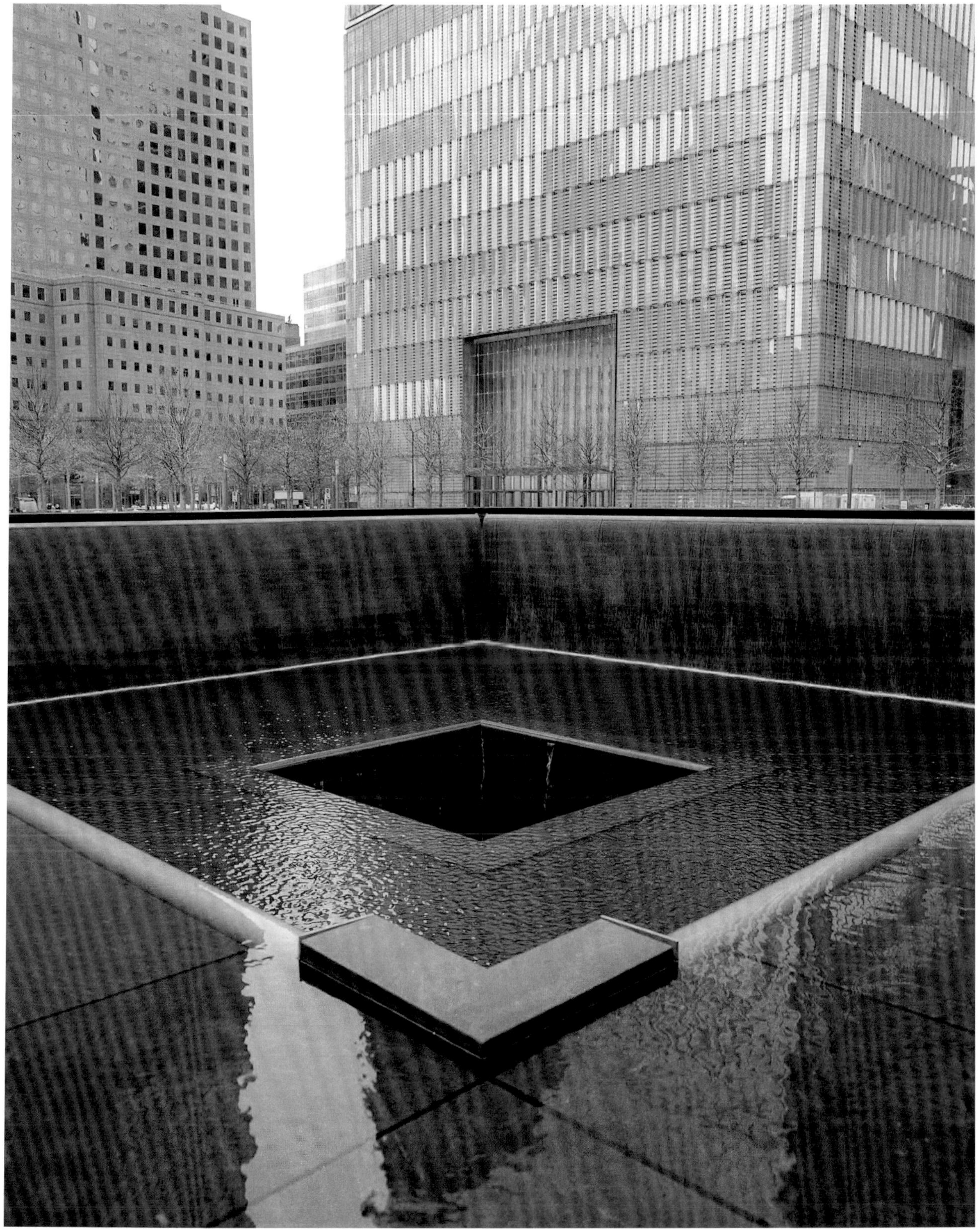

INDEX

ACKNOWLEDGMENTS

As with most of life, we would have had no chance at success without the help and grace of many people. A simple line of thanks seems inadequate in the face of their sometimes-Herculean efforts, but this book truly would not have been possible without them.

Mike Hall (CSM, RET) has been a mentor to all of us, first to Tom, going all the way back to the early-2000s. From vast wisdom, introductions to various veterans, and a deep commitment to transitioning veterans through The Three Rangers Foundation, he was always there for us.

Jason Van Camp and the entire Warrior Rising team put Dan through the start-up boot camp ringer. They believed in this project deeply and literally put their money where their mouths are in awarding us a small-business grant.

Merging Vets and Players, HunterSeven Foundation, Helicopters for Heroes, and Fairways for Warriors and their collective membership: many of the veterans in this book belong to these amazing organizations and many more veterans who are part of them came through for us as we ran into obstacles or issues.

Southeastern Camera and Shure Microphones gave their wholehearted company support to this project from day one.

The whole UnCanna crew was part of this journey, especially with their tincture products, which kept Tom's high-strung ass at least somewhat calm.

Ballast Books instantly saw our vision for this project and jumped on board.

Beau's dear friend, Emily Compagno, provided guidance as we navigated how to present this book to the public.

JC Glick gave friendship, wisdom, and a stream of amazing ideas, and a seemingly never ending Rolodex.

The steady stream of encouragement, inspiration, and mentorship from the National Veterans Memorial and Museum's senior staff has been invaluable.

Finally, Joseph Votel took the time to write an amazing foreword, and never ran out of words of encouragement and inspiration.

To all of them: Thank you, from the bottom of our hearts.

-Dan, Beau, and Tom

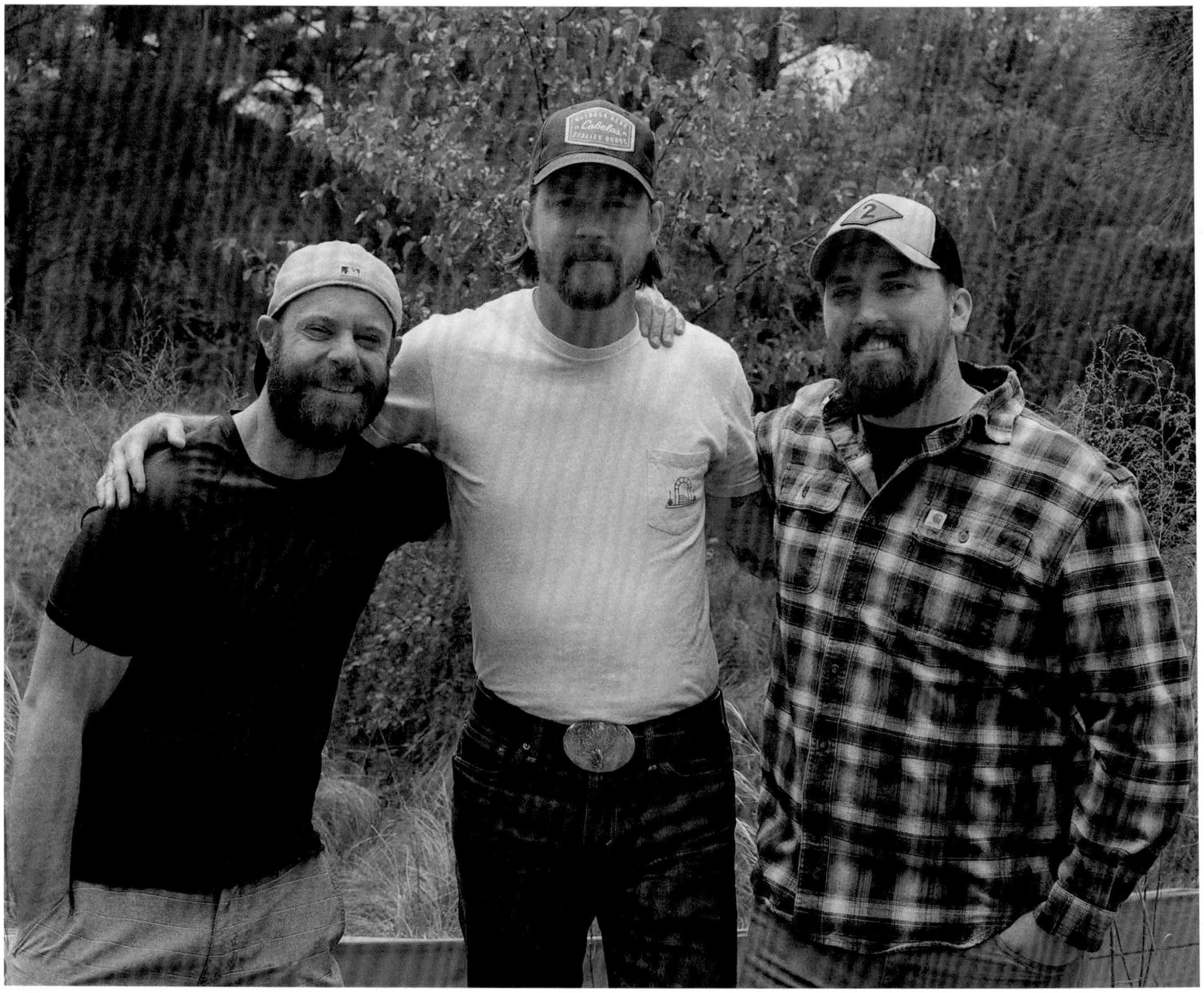

(*Left*) Tom Amenta, (*Center*) Beau Simmons, (*Right*) Dan Blakeley

THE TWENTY-YEAR WAR IS PRODUCED & DESIGNED BY:

UNITED VALOR

United Valor fosters authentic engagement, by helping the men and women who served this country. We continuously produce relevant content by sharing untold stories of veterans, accelerating the influence and reach of veteran-owned businesses, and promoting the image that paves the way for veteran success.

CONTACT US:

Website: www.weareunitedvalor.com
Email: info@weareunitedvalor.com

THE TWENTY YEAR WAR